Strategic Appraisal

T0146317

United States
Air and Space Power
in the 21st Century

Edited by Zalmay Khalilzad, Jeremy Shapiro

Prepared for the United States Air Force
Approved for public release; distribution unlimited

RAND
Project AIR FORCE

The research reported here was sponsored by the United States Air Force under Contract F49642-01-C-0003. Further information may be obtained from the Strategic Planning Division, Directorate of Plans, Hq USAF.

ISBN: 0-8330-2954-1

Cover design by Maritta Tapanainen

Published 2002 by RAND
1700 Main Street, P.O. Box 2138, Santa Monica, CA 90407-2138
1200 South Hayes Street, Arlington, VA 22202-5050
201 North Craig Street, Suite 102, Pittsburgh, PA 15213-1516
RAND URL: http://www.rand.org/
To order RAND documents or to obtain additional information, contact Distribution Services: Telephone: (310) 451-7002; Fax: (310) 451-6915; Email: order@rand.org

Aerospace power has become the archetypal expression of the U.S. ability to project force in the modern world. Throughout the world, U.S. aerospace power—and thus, the U.S. Air Force (USAF)—plays a critical, and often primary, role in securing U.S. interests, in promoting American values, and in protecting human rights. While the USAF has had significant success in employing aerospace power in the recent past, emerging trends in international relations, in technology, and in our own domestic society will create a wide variety of new challenges and new opportunities for U.S. aerospace power. Meeting these challenges and exploiting these opportunities will require careful planning, wise investments, and thoughtful training, as well as difficult cultural adaptations within the USAF. This book identifies many of these challenges and opportunities in a wide variety of issue areas and assesses the degree to which the USAF is prepared to meet them.

While the work was carried out under the auspices of the Strategy and Doctrine program of RAND's Project AIR FORCE, which is sponsored by the U.S. Air Force, this volume draws on the expertise of researchers from across RAND in a variety of related disciplines. The primary audience of this work consists of Air Force leaders and planners, but it should be of interest to others concerned about national security issues.

The *Strategic Appraisal* series is intended to review, for a broad audience, issues bearing on national security and defense planning. *Strategic Appraisal: The Changing Role of Information in Warfare* analyzed the effects of new information technologies on military operations. *Strategic Appraisal 1997: Strategy and Defense Planning for the 21st Century* dealt with the challenges the U.S. military faces in meeting the changing demands made upon it in a changing world. *Strategic Appraisal 1996* assessed challenges to U.S. interests around the world, focusing on key nations and regions.

The views expressed here are those of the authors. They do not necessarily reflect those of RAND or its clients. The research described here was conducted before the September 11, 2001, attacks on the United States and the subsequent U.S. campaign against Al Qaeda and other terrorist groups.

PROJECT AIR FORCE

Project AIR FORCE, a division of RAND, is the Air Force federally funded research and development center (FFRDC) for studies and analyses. It provides the Air Force with independent analyses of policy alternatives affecting the development, employment, combat readiness, and support of current and future aerospace forces. Research is performed in four programs: Aerospace Force Development; Manpower, Personnel, and Training; Resource Management; and Strategy and Doctrine.

CONTENTS

FIGURES

TABLES

ACKNOWLEDGMENTS

Fitting together so many independently minded researchers into a single volume is roughly akin to herding cats. The editors could never have accomplished this complex task without the diligent efforts of numerous colleagues. We would first like to thank the chapter authors and reviewers for their hard work and patience in a long process. We are also indebted to Phyllis Gilmore for her unwavering vision of properly written English prose and her unparalleled perseverance in shepherding the book to publication. Thanks are also due to C. Richard Neu and Edward R. Harshberger for their oversight of the review process and to Natalie Crawford for her support of the Strategic Appraisal series. We would also like to recognize Maj Gen John Barry, USAF, Maj Laura Olsen, USAF, and Todd Sample for their assistance and for sponsoring and organizing the Strategic Appraisal Seminar Series for the USAF staff. Finally, our greatest debt is owed to Luetta Pope, without whose administrative assistance we would not have found our offices much less have cleared the various obstacles to publication.

ABBREVIATIONS

AADC	area air defense commander
ABI	airborne interceptor
ABL	airborne laser
ABM	antiballistic missile
A/C	aircraft
ACS	Agile Combat Support
AEF	aerospace expeditionary force
AEW	Aerospace Expeditionary Wing
AFPAM	Air Force pamphlet
AF/XO	Air Force Deputy Chief of Staff, Air and Space Operations
AF/XOOA	Air Force Readiness Center
AF/XPM	Headquarters Air Force Directorate of Manpower and Organization
AIDS	acquired immune deficiency syndrome
AMC	Air Mobility Command
AMX-C	Air Mobility Express–Commercial
AMX-M	Air Mobility Express–Military
AOC	air operations center

AOR	area of responsibility
ASAT	antisatellite
ATACMS	Army Tactical Missile System
ATC	automatic target cueing
ATR	automatic target recognition
AWACS	Airborne Warning and Control System
BDA	battle damage assessment
BIDS	Biological Integrated Detection System
BM/C⁴ISR	battle management and command, control, communications, computers, intelligence, surveillance, and reconnaissance
BMD	ballistic missile defense
BMDO	Ballistic Missile Defense Organization
BWC	Biological Weapons Convention
C³	command, control, and communications
C⁴ISR	command, control, communications, computers, intelligence, surveillance, and reconnaissance
CAP	combat air patrol
CEC	cooperative engagement capability
CEM	combined effects munition
CINC	commander in chief
CNA	Center for Naval Analyses
CNN	Cable News Network
COMSTAC	Commercial Space Transportation Advisory Committee
CONOP	concept of operation
CONUS	continental United States

CRMAF	Commission on Roles and Missions of the Armed Forces
CSL	CONUS support location
DMZ	demilitarized zone
DOC	U.S. Department of Commerce
DoD	U.S. Department of Defense
DOT	U.S. Department of Transportation
DRID	Defense Reform Initiative Directive
DSP	Defense Support Program (satellites)
DSUP	Defense System Upgrade Program
EAF	Expeditionary Aerospace Force
EELV	Evolved Expendable Launch Vehicle
ELN	Army of National Liberation
ESA	electronically steerable antenna
EU	European Union
EW	electronic warfare
FAA	Federal Aviation Administration
FARC	Revolutionary Armed Forces of Colombia
FEBA	Forward Edge of the Battle Area
FIA	Future Imagery Architecture
FLIR	forward-looking infrared
FMSE	Fuel Mobility Support Equipment
FOL	forward operating location
FOR	follow-on operating requirement
FSL	forward support location
FW	fighter wing
FY	fiscal year

GBS	Global Broadcast System (satellites)
GLONASS	Russian system similar to the U.S. GPS
GMTI	ground moving target indication
GMTT	ground moving target track
GPS	Global Positioning System
HEMP	high-altitude electromagnetic pulses
HSI	hyperspectral imaging
IADS	integrated air defense system
IBS	Integrated Broadcast System
ICBM	intercontinental ballistic missile
ID	identification
IDF	Israeli Defence Forces
IMINT	imagery intelligence
INF	Intermediate-Range Nuclear Forces
IOR	initial operating requirement
IPB	intelligence preparation of the battlespace
JDN	Joint Data Network
JASSM	Joint Air-to-Surface Standoff Missile
JDAM	Joint Direct Attack Munition
JFACC	Joint Forces Air Component Command
JFLCC	Joint Force Land Component Commander
JFMCC	Joint Force Marine Component Commander
JIRD	Joint Interim Requirements Document
JMRR	Joint Monthly Readiness Review
JSF	Joint Strike Fighter
JSOW	Joint Standoff Weapon
JSTARS	Joint Strategic Tracking and Radar System

JTIDS	Joint Tactical Information Distribution System
LANTIRN	Low Altitude Navigation and Targeting Infrared for Night
LEO	low earth orbit
LIDAR	light detection and ranging
LOCASS	Low-Cost Autonomous Strike System
LODIS	Low-Cost Dispenser
LOG C^2	Logistics Command and Control
LPP	launch predition point
MAC	Military Airlift Command (predecessor to today's Air Mobility Command)
MAJCOM	major command
MASINT	measurement and signature intelligence
MEO	most efficient organization
MICAP	mission capable
MIRV	multiple independently targeted reentry vehicle
MOB	main operating base
MOE	measure of effectiveness
MOPP	mission-oriented protective posture
MTI	moving target indicator
MTID	moving target identification
MTW	major theater war
NAD	Navy Area Defense
NASA	National Aeronautics and Space Administration
NATO	North Atlantic Treaty Organization
NBC	nuclear, biological, or chemical
NIMA	National Imagery and Mapping Agency

NMD	national missile defense
NOAA	National Oceanic and Atmospheric Administration
NPOESS	National Polar-Orbiting Operational Environmental Satellite System
NPR	Nuclear Posture Review
NPT	Non-Proliferation Treaty
NRO	National Reconnaissance Office
NTM	national technical means
NTW	Navy Theater-Wide
O&M	operations and maintenance
O&S	operations and support
OJT	on-the-job training
OMB	Office of Management and Budget
ONA	Operation Noble Anvil (the air war over Serbia)
OPELINT	operational electronic intelligence
OPTEMPO	operational tempo
OSD	Office of the Secretary of Defense
PAC-3	Patriot (Advanced Capability)
PDD	Presidential Decision Directive
PGM	precision guided munition
Pk	probability of kill
PSYOP	psychological operation
Pre-Po	prepositioned
QDR	Quadrennial Defense Review
R&D	research and development
RAP	Ready Aircrew Program
RLV	reusable launch vehicle

S&T	science and technology
SAF/AQC	Assistant Secretary of the Air Force (Acquisition)
SALT	Strategic Arms Limitation Talks
SAM	surface-to-air missile
SAR	synthetic aperture radar
SBIRS-High	Space-Based Infrared System–High
SDB	small-diameter bomb
SDIO	Strategic Defense Initiative Office
SEAD	suppression of enemy air defenses
SFW	sensor-fused weapon
SIGINT	signals intelligence
SIOP	Single Integrated Operational Plan
SLBMs	submarine-launched ballistic missiles
SOF	special operations forces
SORTS	Status of Resources and Training System
SPECTRE	spectroscopic excitation and classification of trace effluents
SSBN	nuclear submarine
SSC	smaller-scale contingency or conflict
SSM	small smart munition
START	Simplified Tool for Assessment of Regional Threats (RAND theater-level campaign model)
START	Strategic Arms Reduction Treaty
STRATCOM	U.S. Strategic Command
SWA	Southwest Asia
TBM	theater ballistic missile
TDY	temporary duty

TEL	transporter-erector-launcher
THAAD	Theater High-Altitude Area Defense
TMD	theater missile defense
TPFDL	Time Phased Force Deployment List
UAE	United Arab Emirates
UAV	unmanned aerial vehicle
UCAV	unmanned combat aerial vehicle
UGS	unattended ground sensors
UN	United Nations
UNOSOM	United Nations Operation in Somalia
UNSCOM	United Nations Special Commission
USAF	U.S. Air Force
USAFE/LG	U.S. Air Forces in Europe Director of Logistics
U.S.C.	United States Code
UTA	unmanned tactical aircraft
VX	a nerve gas
WCMD	wind-corrected munitions dispenser
WMD	weapons of mass destruction
WRM	war readiness materiel
WWX	Worldwide Express

Chapter One

INTRODUCTION: THE PRICE OF SUCCESS

Jeremy Shapiro

Since the end of the Cold War, the air instrument has become America's weapon of first resort to handle nearly all varieties of contingencies from disaster relief to major theater wars. When called upon, the U.S. Air Force (USAF) has performed with such a startling degree of precision and power that many claim its new capabilities have revolutionized warfare itself. At the same time, as many of the chapters in this volume reveal, the Air Force faces tremendous challenges in adjusting to the new and expanded role it plays in U.S. national security strategy.

These observations present something of a paradox. At the precise moment that the USAF seems to have achieved its apogee of value and efficacy, the institution itself faces an array of challenges, from privatization to modernization to the integration of the space mission, that threaten to overshadow its recent achievements. In this way, the Air Force is very much a victim of its own success—the nation seems to demand ever more from those who have historically delivered the most. Whatever the reasons, it is clear from the scale of the challenges the Air Force faces that it cannot afford to rest on its laurels for even a moment. The world is clearly changing at a rapid pace, and the USAF must adjust as quickly possible to the new challenges and new tasks that such changes create.

The purpose of this volume—the fourth in RAND's *Strategic Appraisal* series—is to aid in that effort by helping readers understand better the capabilities that the USAF can bring to bear in support of U.S. interests and the ability of the USAF to meet the challenging demands of a changing technological and security environment. The contributors examine the geopolitical context in which U.S.

aerospace power must operate; the choices the Air Force faces in a variety of issue areas, from nuclear weapons to space; and the requirements for supporting future forces. This chapter sums up the issues explored by dividing them into problems that remain from previous eras, new problems that have emerged of late, emerging challenges, and emerging opportunities.

WHAT HAS STAYED THE SAME

Change is clearly the idiom of our age. The end of the Cold War, the emergence of new information technologies, and the ongoing integration of global markets have all engendered radical change in a relatively short time. The coincidental start of the new millennium even seems to have neatly delineated the new age on our calendars. Nonetheless, often unnoticed in this whirl of dynamism are the many less stirring but no less critical pockets of stability. For any military organization, and especially one like the Air Force that is culturally prone to look to high-technology innovative solutions, forgetting the past is at least as dangerous as failing to adjust to the future. A great deal was learned during the Cold War about how to create military power and how to apply it to political problems in the modern age. The end of that struggle has not automatically invalidated the lessons.

Two issues from this volume highlight the continued salience of Cold War experience to current dilemmas. The first is the issue of nuclear weapons. These weapons were the foundation of U.S. strategy during the Cold War and, according to some, the technological innovation that spared us the horrors of a third world war. Yet as Glenn Buchan documents (Chapter Seven), current U.S. nuclear policy is an unfortunate combination of stasis and neglect. Nuclear weapons are often seen as an irrelevant and even immoral anachronism in the post–Cold War world. Official U.S. policy toward nuclear weapons encourages this view by encompassing apparently contradictory undertakings: eliminating the U.S. nuclear arsenal in keeping with the Nuclear Non-Proliferation Treaty yet, at the same time, espousing continued adherence to a policy of strategic deterrence against Russia and maintaining U.S. nuclear forces on alert. Prudence requires maintaining the U.S. nuclear arsenal in a world in which nuclear knowledge cannot be unlearned, yet public relations require downplaying the existence and utility of nuclear weapons in the national military strategy. The result is a policy that, as Buchan demonstrates, is slowly being crushed under the weight of its own contradictions, as weapons deteriorate and nuclear knowledge is lost.

Buchan presents many conceivable solutions to this dilemma, from eliminating nuclear weapons altogether to incorporating them fully into U.S. warfighting doctrine. Each solution has its individual advantages and disadvantages, but all require the U.S. national military strategy to have the courage of its convictions: If the U.S. military wishes to retain such weapons over the long term, it must make a plausible case that these weapons serve a purpose that justifies the moral indignation they arouse and must ensure that the weapons are well maintained and deployed consistently with their purpose. The clearest role for U.S. nuclear forces is to continue to provide a deterrent force but against a wider variety of threats than during the Cold War. This implies maintaining survivable forces and command and control, a force of almost any reasonable size, and an adequate mix of forces to hedge against technical or operational failures. It also implies de-emphasizing rigid targeting plans aimed at specific adversaries and building flexibility into the force.

The second area of stability is access to bases. As David Shlapak's short history of access issues in Chapter Nine demonstrates, the ability to base assets abroad and to secure overflight rights has always been a critical element of U.S. power projection. Even in the Cold War, with a known adversary and reliable allies on its periphery, basing issues became a critical enabler of USAF actions. This reality reflected the fact that the Cold War was ultimately a global struggle that required the United States to exert influence in a variety of far-flung regions. In that struggle, the capacity to project and sustain military power over great distances formed the glue that bound the U.S. alliance structure and therefore became a critical element of U.S. influence in the world.

One lesson of that conflict was that a single adversary could capitalize on an increasingly small world to convert an argument over Europe into a global struggle with many fronts. In the future, as information and communication technologies render that world even smaller, a variety of adversaries will effectively perform the same task but in even less-predictable ways. Thus, the capacity to project and sustain military power, and therefore the issue of access, will become still more central to U.S. military power. Indeed, as

Shlapak demonstrates, the types of contingencies that are likely to crop up in the next decade or two will most likely occur in areas where the United States faces significant basing uncertainties, particularly the Middle East and East Asia. At the same time, the proliferation of missile and weapons of mass destruction (WMD) technologies has rendered many existing close-in USAF bases less secure. Despite the achievement of staging bombing raids on Yugoslavia from the continental United States during Operation Allied Force in 1999, current technology will not allow the United States to respond to this problem by relying exclusively, or even mainly, on extended-range operations from U.S. territory. Rather, the United States needs a diversified portfolio of strategies and relationships that mirrors and expands on its Cold War experience in worldwide struggle. This portfolio would include not only maintaining the current main operating bases overseas but also planning for uncertainty in access by means of flexible deployment and employment plans.

WHAT HAS CHANGED

Despite these important continuities, much has changed for the USAF in recent years. More specifically, the USAF has not yet caught up with several consequential changes in the threat environment. Two particular issues of this type come across in the contributions to this volume.

Smaller-Scale Contingencies

The first change involves the increased importance of smaller-scale contingencies (SSCs), such as conducting humanitarian operations and patrolling no-fly zones, in U.S. strategy and priorities. The U.S. military, particularly the Air Force and the Army, has tended to treat these operations as diversions from their principal mission, major theater war. In recent years, however, this position has become untenable because such contingencies have proven to be the principal occupation of these services, a situation that most observers expect to continue for some time. Again, this state of affairs reflects not so much a failure of planning as the U.S. military's continued success in preparing for and thus deterring major theater wars. This observation implies that the USAF must not allow its capacity for larger contingencies to erode as it prepares to deal with the smaller contingencies that are likely to continue to preoccupy U.S. policy-makers.

However, as Don Stevens, Jack Gibson, and David Ochmanek make clear in Chapter Four, SSCs, because of their frequency and longevity, can no longer be considered lesser-included cases of major theater wars. According to Carl Dahlman and David Thaler in Chapter Twelve, nearly continuous "peacetime" operations, such as the no-fly zones in Iraq (from 1991 to the present), have eroded the readiness of the Air Force in ways that have not even begun to show. Unlike the Navy and Marine Corps, the Air Force has not structured its personnel and maintenance policies for long, low-intensity and peacetime deployments. Thus, to undertake these operations, the Air Force has been sacrificing its future readiness, particularly by reducing its opportunities and capacity for training.

The development of Air Expeditionary Forces, intended to allow an orderly and predictable rotation of personnel through peacetime deployments, represents a start toward a solution, rather than a complete solution, for what must ultimately be a wholesale shift within the Air Force away from making neat distinctions between times of war and peace. To allow the Air Force to undertake frequent SSCs without impairing future readiness, these changes will need to reverberate throughout the service and go well beyond personnel policies and deployment schedules. Logistics provides a prime example of the type of wholesale shift required. The current system is designed for heavy deployments with fairly long lead times. To support Air Expeditionary Forces and the missions they are likely to undertake, the Air Force will need a logistics system that, like the one Robert Tripp and his coauthors lay out in Chapter Ten, allows quick deployments through fairly unprepared bases.

The Threat of Weapons of Mass Destruction

The second consequential change with which the USAF has not fully caught up is the potential for its opponents to use WMD. In Chapter Three, Daniel Byman and his coauthors note that we can expect adversaries to respond to U.S. conventional superiority by resorting to asymmetric strategies that will probably include use of WMD. Operationally, use of such weapons threatens the ability of the U.S. military to deploy to a theater and to operate out of close-in bases.

Politically, these weapons threaten to fracture U.S. coalitions and to undermine public support for U.S. intervention through the prospect of massive casualties. Indeed, as Richard Mesic argues in Chapter Eight, any use of WMD is likely to change the very nature and scope of conflict—and undoubtedly not in a way that favors the American way of war.

While awareness of the WMD threat is widespread, the response to this threat has so far been ineffective. Politicians and military organizations prefer to deal with vulnerability by attempting to eliminate it: If the enemy builds a bigger battering ram, the natural response is to build a thicker door. Unfortunately, though that response is appropriate, no door appears to be thick enough when it comes to WMD. Active defenses, particularly theater missile defenses, have great value but cannot be 100-percent effective against WMD, even when combined with preemptive counterforce attacks.

Effective responses will probably require an integrated approach that combines multilayered active measures with passive defenses (chemical suits, nuclear hardening, etc.) and political strategies that range from deterrence through threats of retaliation and denial and through active measures to enforce norms against proliferation. Such an integration would clearly require cooperation not just across the services, as is now familiar (albeit imperfect), but also with the government agencies responsible for civil defense and diplomacy. The approach might also include force mix adjustments to improve the Air Force's ability to operate effectively when based farther from target areas, as Shlapak suggests in Chapter Nine, and even to move some functions into space, specifically to avoid the WMD threat.

EMERGING CHALLENGES

Beyond the immediate problems, this Strategic Appraisal highlights several emerging challenges that the USAF will need to meet in the relatively near future. In Chapter Five, Robert Preston and John Baker highlight two emerging challenges in the area of space. First, the continued commercial viability of the U.S. space industry appears to be at risk because of competition from terrestrial alternatives and because export controls limit the industry's ability to compete internationally. Preston and Baker recommend that the government take active measures to employ its space industry to shape international capabilities and to maintain the U.S. advantage in space technology. Notwithstanding such measures, however, the Air Force should expect to increase its role as a supplier of space capabilities for areas in which the commercial sector lacks a viable business case, such as surveillance, warning, and protected communications.

The second emerging space challenge results from the possibility, even likelihood, of attacks on U.S. assets in space. As space systems increasingly integrate into military activity at the theater and even tactical levels, adversaries will begin to see space as contributing directly to military capability. They will therefore have the desire, and most likely the capability, to bring the fight to space and to attack U.S. space assets. Although Preston and Baker do not believe the USAF needs to take a position on the weaponization of space, they do recommend that the USAF prepare to defend critical U.S. space assets that will be an economic and military center of gravity for the United States.

Another challenge is the well-publicized but perhaps not-so-well-understood difficulties in maintaining military readiness. As Dahlmann and Thaler emphasize in Chapter Twelve, current readiness measures have understated USAF readiness problems. Current measures only reveal how ready units are for operational tasks, but military units have an additional critical requirement to maintain the human and physical capital of the force through training and maintenance. Because current metrics have failed to measure lost opportunities for on-the-job training, they have missed much of the effect on USAF units of declining recruitment and retention and frequent contingencies. The booming economy has meant that skilled workers are in high demand, while frequent contingencies have forced the USAF to sacrifice on-the-job training on the altar of operational demand. The result is that the USAF skill mix is deteriorating more than is widely known, especially in pilots and maintenance personnel. Much of the cause lies outside the Air Force, but, as Dahlmann and Thaler emphasize, the first step in meeting this emerging challenge will be to develop a readiness metric that can adequately communicate the problem to the political leadership.

Finally, the various technological developments surveyed in this volume together form the dim outlines of a more distant, yet no less important, challenge. According to many analysts, the increased importance of information and sensors implies that the first, most important battle in any conflict may be the fight for information superiority. In such a battle, the opening shot is likely to be a "sensor shot" that attempts to disable a nation's capacity to collect, process, or disseminate information. Unfortunately, some analysts fear that, concurrent with this trend, the United States may be building a national information architecture that is vulnerable to a first "information strike" that could disable or delay the U.S. ability to respond to an attack. This development is particularly frightening because the presence of such a first strike capability means that the offense-defense technological balance would appear to be shifting toward the offense. A critically poised and vulnerable information infrastructure might thus create the need to preempt potential attacks, creating the same type of hair-trigger, lose-or-use-it proposition that so concerned nuclear analysts during the Cold War.

This development, while not imminent, remains a possibility in the not-too-distant future. For the USAF, this type of warfare would put a premium on developing effective information defenses not just for its own systems but for other military and even civilian information infrastructures. Failing effective defenses, a hair trigger would also have an important effect on the USAF capacity to operate in coalition with allies. The rapid crisis dynamics implied above would require developing standing coalitions that have done enough training together to operate effectively from the outset of a crisis. An offense-dominant information war would render the current model of ad hoc coalition formation unworkable and would necessitate paying greater attention to integrating forces and capabilities with allies.

EMERGING OPPORTUNITIES

While the challenges detailed above are certainly daunting, many emerging opportunities, both technological and organizational, will help the U.S. military and the USAF in particular to meet the challenges ahead. This *Strategic Appraisal* highlights four such opportunities that have great potential and that often receive less attention than they deserve.

The first such opportunity comes from the promise of a new generation of munitions. Munitions receive far less attention than the more glamorous combat aircraft that deliver them. Nonetheless, effective, advanced munitions will be critical for realizing the potential of the next generation of aircraft and indeed should influence decisions about which aircraft to buy. New standoff weapons, particularly the Joint Air-to-Surface Standoff Missile (JASSM) and the Joint Direct Attack Munition (JDAM), and new smaller munitions, such as the Low-Cost Autonomous Strike System (LOCASS), offer dramatically improved performance at relatively low cost. Indeed, as Stevens, Gibson, and Ochmanek make clear in Chapter Four, the single most cost-effective action the USAF can take is to buy more standoff weapons, particularly JASSMs, so that its nonstealthy platforms can participate in the early part of an air campaign. Similarly, the availability of effective smaller weapons, such as LOCASS, will make aircraft with internal weapon bays (such as the F-22 and the Joint Strike Fighter) much more effective and favors the creation of an attack variant of the F-22 with a larger weapon bay, the F-22E. The authors emphasize, however, that, to take advantage of these dramatic improvements in munitions and realize the full capabilities of its next-generation combat aircraft, the Air Force will need to spend more than is currently programmed on munitions.

Another emerging opportunity highlighted in this volume is the possibility of improving Air Force outsourcing practices to achieve fairly dramatic savings in support services. Although outsourcing efforts were an integral part of every proposal to restructure Air Force support services during the 1990s, outsourcing has not to date produced the degree of savings that its apostles prophesied. In Chapter Eleven, Frank Camm asserts that this lackluster performance resulted from the inappropriate assumption that outsourcing could save money simply by transferring the provision of services to the private sector. However, outsourcing per se will not produce savings for the Air Force; rather, competition will. Whether the public or private bidder won (and public bidders have often won), competitions have created substantial savings and could potentially save even more.

Because of the focus on outsourcing as an inherent good, the USAF has often ignored how important the details of an outsourcing program are to ensuring competitive provision of services and therefore to the success of the program. Camm suggests a new process of "strategic sourcing" to determine what services should be outsourced, to ensure competition for providing services to the Air Force, and to align outsourcing with the strategic goals of the USAF.

Looked at through this lens, the Air Force appears to have been too conservative in determining what services are eligible for outsourcing and is missing important chances to learn from best commercial practice and to achieve substantial savings.

The third emerging opportunity comes in the area of ballistic missile defense. While missile defense has become one of the most visible national security issues, public debate to date has focused on national missile defense. However, the division between theater and national missile defense, while enshrined in the vocabulary of the political debate, is artificial. There are no technical differences in many areas, and the idea of a "national" missile defense system inappropriately signals U.S. allies that they will be excluded from the protective umbrella of missile defense. This implies that there needs to be much more emphasis on and understanding of the problems and opportunities of theater missile defense and its role in the U.S. national security strategy.

In fact, as Mesic points out in Chapter Eight, defense against WMD and ballistic missiles should be thought of as a system of systems that includes active missile defenses systems of all types (terminal, midcourse and boost phase), counterforce options, passive defenses and a battle management system to link the whole system together. While no one system can be 100-percent effective, Mesic emphasizes that modest capabilities can make a dramatic difference when combined into a system of systems. For this reason, the battle management system is perhaps the key element of a multilayered ballistic missile defense. The Air Force has important contributions to make to this system of systems in designing and operating the battle management system and in contributing promising theater missile defense systems, such as the airborne laser, to a layered missile defense architecture that can protect both the United States and its allies.

The final emerging opportunity highlighted in this volume concerns the contribution the USAF can make to securing the U.S. capacity to establish and retain information superiority over its opponents. In future conflicts, winning the contest for information superiority will allow the United States to secure the high ground in high-tech battles. In Chapter Six, Brian Nichiporuk demonstrates how the United States can use new information technologies to achieve information superiority and provides counters for some of the most appealing asymmetric strategies likely to be used against the United States. He presents four information warfare concepts of operation for how the United States might, with relatively little expenditure of blood or treasure, effectively diminish the utility of enemy WMD and preserve U.S. power-projection capability in the face of attempts to deny access.

GETTING PAST SUCCESS

Successful organizations rarely adapt to new challenges successfully. Their past record of unbroken triumph instills in them a confidence, some would say a hubris, in their current way of doing things that impedes their ability to recognize and to adapt to changes in their environment. It is for this reason that so few commercial companies have demonstrated an unblemished record of profitability over the long term. Eventually, nearly all large corporations have stumbled as they failed to recognize an emerging opportunity or challenge. While such corporations can usually recover from such a misstep, the USAF does not have the luxury of failing to respond to new realities. Avoiding that unhappy outcome will require getting beyond the successes of recent years and planning for an uncertain future in which past achievement does not guarantee future success.

Moreover, that planning must begin immediately; the decisions taken today—on force mix, on information technology, and on a host of other issues—will have ramifications far into the uncertain future. We cannot know the future, but to operate successfully in an environment with long-term planning horizons, we must have an opinion on it—one that should be informed by research and wisdom. While these features will not provide certainty, lack of certainty cannot justify inaction. The purpose of this volume has been to point out areas in which the accumulated research and wisdom of recent RAND work can point to actions that can be taken immediately to prepare for an uncertain future.

We recognize, however, that the Air Force faces trade-offs in implementing many of these recommendations. Resources are limited, and spending funds to improve readiness, for example, necessarily reduces funds available to modernize the force. Understanding the trade-offs and prioritizing the various demands on defense resources—those detailed here as well as many others—was beyond the scope of this study. This absence should not be taken as an implication that such a task is easy or unnecessary. In many ways, choosing among the challenges and opportunities detailed here, as well as several not discussed, is the most difficult and pressing task the USAF faces.

Part I

THE GEOPOLITICAL CONTEXT FOR
AEROSPACE POWER

Chapter Two

FORCES FOR WHAT? GEOPOLITICAL CONTEXT AND AIR FORCE CAPABILITIES

Zalmay Khalilzad,[1] David Ochmanek, and Jeremy Shapiro

The utility of any military force can only be judged in the light of potential needs for it. The United States is a global power, indeed the only global power, and as such has interests that span all regions of the world. Prudent U.S. defense planning therefore requires a view of likely geopolitical developments and trends throughout the world. Predictions in this realm are notoriously prone to error, but defense planning demands that we peer cautiously into the future. Toward that end, this chapter sets a geopolitical context for the rest of the volume and delineates U.S. goals and interests within that context. Finally, it provides an overview of the capabilities that the USAF can bring to bear in support of the goals, as well as some of the pressing challenges the USAF faces.

THE GEOPOLITICAL CONTEXT

Despite the inherent uncertainty in world events, two factors are reasonably likely to be particularly important to U.S. defense planning:

- the evolution of the international system—especially with regard to U.S. relations with other great powers and political developments in regions of interest to the United States

- U.S. goals—the decisions the United States makes about what it wants to accomplish in world affairs.

---

[1]Zalmay Khalilzad completed his contribution to this chapter in November 2000, prior to leaving RAND for a position in the U.S. government. The views expressed herein are solely those of the authors and do not necessarily reflect the position of the U.S. government.

Evolution of the International System

The international security environment is dynamic and will evolve in the coming years. At present, the global system consists of four categories of powers. The U.S. stands alone as the only power capable of projecting and sustaining large-scale military forces intercontinentally. Next, there is a group of great powers that play an important role in shaping regional and occasionally global politics. Examples include France, Russia, and Japan. Some of these powers are uneasy with U.S. preeminence and would like the world to evolve toward a multipolar system. Some might even consider joining a coalition to balance the power of the United States, at least on certain issues. The third category encompasses major regional powers—some allied with the United States and others hostile. Such states include Iran, North Korea, Turkey, and Australia. It is the hostile regional powers that have been the focus of U.S. military strategy and defense planning and the target of the most intense uses of U.S. military power over the past ten years. The fourth group is a larger number of states, including many states in Africa and South Asia, with limited relative power. Many of these states have been the scenes of humanitarian crises that ultimately involved U.S. forces.

As this categorization implies, a peer competitor similar in scope to the Soviet Union is very unlikely to appear in the near future. Such a competitor could conceivably emerge if there is a catastrophic and unpredictable turn of events resulting in significant and continued deterioration in the relative position of the United States. Such events could range from the fragmentation of U.S. alliances in Asia and Europe to the takeover of the world energy resources by a hostile power, an unsuccessful U.S. involvement in a major war, the rise of a coalition of great powers opposed to the United States, or the domination of East Asia by a hostile power. All such events are extremely unlikely.

In the absence of a peer competitor, relations between the United States and current and emerging great powers are likely grow in importance. These relations are in a state of slow, but steady flux—a marked contrast to the stasis of the Cold War. As a result, U.S. dominance is likely to be challenged on a number of fronts. One or more new great powers are likely to emerge in the course of the next 25 years. Both new and existing great-power competitors will become more ready and able to challenge the United States, especially on particular issues of high salience to them. Moreover, because of technological diffusion, even regional powers may acquire the ability to attack the United States directly with weapons of mass destruction delivered by ballistic or cruise missiles.[2]

Paradoxically, while the growing strength of great powers and regional states will present challenges to the United States, the relative importance of nonstate actors will also increase. Technological, economic, and social trends are weakening states relative to nonstate actors—although the pace at which this is happening is difficult to determine.[3] Nonstate actors—such as human rights organizations, religious and ethnic advocacy groups, and environmental or single-issue lobbies—will grow in importance, numbers, and influence on the international scene. The challenge of these organizations has already been felt, even by the most powerful states. Civil society movements that were instrumental in the creation of the international land-mine treaty and the international criminal court and in the protests against the World Trade Organization, among others, will continue to influence the public agenda in developed countries. These movements will often challenge even democratic governments to be more responsive in their foreign policies to various societal concerns, such as the environment, health, and human rights, including women's rights, rather than to the geopolitical concerns that have traditionally motivated statesmen. These influences will present a challenge to policymakers and military officers not used to operating in such an environment.

More nefariously, the same trends in communication and information technologies that favor groups in civil society will also empower transnational terrorist organizations and criminal gangs to challenge state power on a multitude of fronts ranging from piracy to drug trafficking. As the events of September 11, 2001, demonstrated, the sophistication and reach of such groups has grown enormously in recent years. As a result, many of these now specifically require tai-

---

[2]Countries that may soon have this capability include North Korea, Iran, and Iraq, as well as Russia and China. See National Intelligence Council (1999).

[3]Proponents of the view that trends are weakening states relative to nonstate actors include Guehenno (1995), Friedman (1999), and Arquilla and Ronfeldt (1996). Taking the opposite view is Hirst and Thompson (1996).

lored deterrence and defense strategies that go beyond ordinary law enforcement efforts.

The increasing influence of nonstate actors is symptomatic of the general trend toward globalization of economic and cultural life throughout the world. This development will proceed apace and may even deepen to include areas now thought of as almost exclusively national, such as the defense industry. The result of all these interconnections will be a blurring of national boundaries and an increasing salience of world events for the United States. By 2025, an internal or external security problem in almost any part of the world will, to a greater degree than today, be a security problem in all parts of the world.

As result of this trend, U.S. interests will become more global, often cutting across traditional regional and bureaucratic boundaries, although Europe, Asia, and the greater Middle East will remain foremost in U.S. strategic thinking. More subtly, the increasing pressure that globalization is putting on traditional societies—in terms of the destruction of traditional structures, widening income disparities within and between countries, and periodic financial crises—may cause a backlash against globalization. Such a backlash may well be directed at the United States, the country often seen as the principal architect and beneficiary of globalization.[4]

The potential for such backlashes highlights the degree to which U.S. security in the future will depend on various global economic dynamics that are well beyond the control of the U.S. military or the U.S. government. A world economy that generates sustained and reasonably equitable growth will soften many of the potentially hard problems detailed here. Conversely, global economic distress or even localized economic crises will exacerbate all types of international tensions and could dramatically undermine U.S. security.

Within this global environment, U.S. planning will continue to have an important regional dimension. Here, we highlight the most critical regions for U.S. defense planning and the USAF.

Asia: Drifting Toward Rivalry? Among the regions of the world, Asia is likely to experience the most dramatic and important changes in

---

[4]For a review of the tensions caused by globalization, see Gilpin (2000).

the coming 25 years and will require the most changes in U.S. policy and military posture. The current U.S. military posture in Asia stems from the needs to defend South Korea and to promote regional stability. Recently, however, the primary locus of international terrorism has shifted from the Middle East to Asia, particularly South and Southeast Asia; at the same time, the most likely site for interstate rivalry is moving south, away from Northeast Asia

It is possible that the North Korean threat will evaporate, either because of political unification of the peninsula or because of a political accommodation—short of near-term unification—that removes or dramatically reduces the threat to South Korea. North Korea, perhaps encouraged by China, might push indirectly for total U.S. withdrawal from South Korea as it negotiates for and reaches an accommodation with South Korea and seeks to improve relations with the United States. The evolution of the U.S. posture in South Korea will in turn affect the U.S. military posture in Japan. Tokyo might use the opportunity of a Korean accommodation to push for changes and perhaps reductions in the U.S. military presence in Japan.[5]

Such changes on the Korean Peninsula would take place in the context of broader changes in Asia—which point in the direction of potentially increased rivalry—perhaps similar to what Europe went through at the end of the 19th and beginning of the 20th century. Maintaining stability is likely to become more difficult because the region is producing two rising great powers: China and India. They are likely to be competitive with each other and with the other regional great power, Japan. The role of the region's other potential great power, Russia, remains uncertain and will depend on whether the country can solve its pressing internal economic and social problems.

At present, China appears to have a purposeful strategy for the long term. Beijing seeks regional primacy and a reduced U.S. role and presence in the region. It wants to be in a position to counter U.S. intervention in East Asian regional affairs. It is concerned about the U.S. national missile defense program and about theater missile defenses for East Asia. Beijing believes that such systems would

_____

[5]On these points, see Wolf et al. (2000), pp. 77–79; Bracken (1998); and Sokolski (2001).

strengthen the U.S. position in the region and undermine the People's Republic's coercive options against Taiwan.

At present, the Chinese government's priority is to foster economic growth as a means of developing a modern, technologically advanced society. However, that near-term priority appears to support a longer-term and purposeful Chinese strategy to achieve regional primacy and a reduced U.S. presence in the region. China wants to be in a position to counter U.S. intervention in Chinese and Asian regional affairs and to reduce U.S.-Japanese military cooperation. Beijing also remains steadfastly opposed to any Taiwanese moves toward independence (Roberts, Manning, and Montaperto, 2000).[6]

China's behavior, however, will also depend on its potentially problematic domestic political evolution. The current program of economic reform has replaced communist ideology with a more pragmatic (and successful) approach. In Deng Xiaoping's phrase, "It doesn't matter whether the cat is black or white, as long as it catches mice." Nonetheless, the absence of ideological justification for one-party communist rule exposes China's leaders to questions about the regime's legitimacy and its hold on power. This lack of legitimacy, as well as dramatic regional economic disparities, a rising nationalism among the population, and the increasing social dislocation that has resulted from economic reform, threaten China's long-term stability.

In the meantime, Chinese military power is growing, and there is a strong possibility that, as that power grows, Beijing will become ever more assertive on regional issues, particularly with regard to Taiwan. A mainland Chinese use of force against Taiwan might well lead to U.S. and even Japanese military involvement in defense of Taiwan— and the possibility of a protracted military confrontation that could have a determining role in the U.S. regional military posture.

How the other Asian great powers—Russia, India, and Japan—will adjust to China's rise and its possible push for primacy is the single biggest question affecting Asia's future security environment. Some of these powers might "bandwagon" with China, while others might well seek to balance it—by building up their own military power, by

---

[6]On Chinese military options toward Taiwan, see Shlapak, Orletsky, and Wilson (2000); O'Hanlon (2000); and Shambaugh (2000).

aligning themselves with the United States, or by a combination of the two.[7]

At present, most Asian governments are concerned about China's long-term intentions and about U.S. staying power. For the near term, they believe that China will continue to focus on building its "comprehensive power,"[8] a goal that requires Beijing to exercise restraint in throwing its weight around. Similarly, fear of abandonment by the United States is not widespread among Asian governments.

For its part, Russia is ambivalent. Moscow is assisting Chinese military modernization and shares Chinese opposition to the U.S. position as the preeminent power. But it does not trust China and fears that it might become a victim of future Chinese expansionism. However, at present, Russian concerns about the United States appear to play a larger role in Moscow's diplomatic calculations.

Japan is likely to follow a mixed approach—build up its own military power and remain allied with the United States. There is a heightened sense in Japan of a potential challenge from China. Should China push for regional hegemony and should the United States withdraw from the region, the result is likely to be a renationalization of Japanese security policy and competition between Tokyo and Beijing for regional influence. Japan would thus seek to become a more normal regional security actor and revise its constitution to allow the right of collective self-defense to effectively confront or balance against China.

India is also likely to seek to balance against China. The Indian nuclear test in 1998 was explicitly linked to Chinese military power. The Indian ballistic missile program is also is informed by New Delhi's concerns about China.[9] A dangerous triangle involving China,

---

[7]This discussion of alliances in Asia is based on Khalilzad et al. (2001), pp. 3–42. Also see Blackwill and Dibb (2000) and Tow (1999).

[8]When Chinese leaders refer to building China into a "comprehensive power," they generally mean that they seek to create a modern China that would rank among the leading nations in all dimensions of national power—political, economic, military, and technological.

[9]On the Chinese factor in the Indian nuclear program, see Chellaney (1998–1999) and "Aiming Missiles" (1998).

India, and Pakistan in South Asia has the potential to unleash major wars that could include the use of nuclear weapons. The tense situation in Kashmir could easily escalate to major conflicts and might involve China on Pakistan's side or might lead to the use of nuclear weapons or both. There is also the possibility of a destabilizing nuclear arms race between India and Pakistan—as each might think it is falling behind the other in acquiring more or better capabilities.

These various developments would pose many risks to regional stability and might even result in major conflicts. There could be competition among various great or potential great powers (Russia, India, China, and Japan) in Southeast, Northeast, and South Asia. In such an environment, some states, such as Thailand and Pakistan, might move closer to China; others, such as Vietnam and the Philippines, might join an anti-Chinese coalition. Increased instability in such places as Southeast Asia would present targets of opportunity for competing great powers. In fact, the region faces the risk of Balkanization as such regional states as Indonesia threaten to fragment.

At the moment, bilateral relationships, especially with Japan, Australia, and South Korea form the basis for U.S. engagement in Asia. The Asian region lacks strong regional security institutions that could moderate these drifts toward rivalry, preserve regional stability, and prevent the domination of the region by any one power.[10] One reason for this lack of multilateral security arrangements is that, unlike Europe, Asia is not yet seen as a geopolitical whole. Events in, for example, Northeast Asia are assumed to have little impact in South Asia. However, as this short excursion into Asian geopolitics has demonstrated, Asian regional issues from Pakistan to Korea are increasingly linked.

Europe: A New Partnership? Another key long-term factor affecting the future of the international system is the prospect for European unification and relations between a united Europe and the United States. With Europe, as with the democracies of Asia, the United States shares not only common ideals but also immense stakes in an increasingly integrated global economy. Since the end of the Cold War, the U.S. alliance with Europe has focused on integrating east-

---

[10]On current multilateral institutions in Asia, see Paal (1999).

central Europe into the Western alliance, on integrating Russia into the world community, and on stabilizing the Balkans. In recent years, the North Atlantic Treaty Organization (NATO) has also begun to focus on problems from the south—terrorism and the proliferation of weapons of mass destruction.

The U.S. partnerships in Europe are less in need of dramatic change. Significant adaptations have already been made since the end of the Cold War, and there is a commitment to make further changes, such as bringing more states into NATO and into the European Union (EU). However, there are some issues that might disrupt the alliance. The first potential problem is the U.S. complaint that the European NATO allies are not doing enough to contribute their fair share in combined missions. The disparity between U.S. and allied capabilities for air combat and large-scale power-projection operations was evident in operations Allied Force and Enduring Freedom. The second issue is the related European opposition to key U.S. policies, such as national missile defense. Third and more speculatively, it is possible that as European unification progresses, the EU nations might drift away from the alliance with the United States. There is a growing resolve among European leaders to add a defense component to the European project and to develop forces that are more modern and more deployable. Increasing trade friction and economic competition between the EU and the United States may exacerbate the tensions that an independent EU defense program would cause. The effects of these developments on transatlantic relations in general and on NATO in particular remain uncertain, but one can clearly detect a growing divergence between the United States and some European countries—particularly France but also such traditionally staunch allies as the United Kingdom and Germany.[11]

The final problem that is likely to trouble transatlantic relations is the continuing instability in the Balkans. Despite the fall of Slobodan Milosevic from power in October 2000, the Balkans remain the most troubled part of Europe west of the old Soviet frontier. Numerous possibilities for the reemergence of conflict persist in Bosnia, in Kosovo, and in Macedonia. The United States has yet to decide on the precise nature of its role in the Balkans. However, a large-scale

---

[11]On U.S.-European differences, see Gordon (2001), Haas (2000), and Gordon (2000).

U.S. withdrawal from the Balkans could easily exacerbate tensions in the NATO alliance nearly as much as a new war in the region would. Some level of U.S. military commitment to the region is thus likely to remain for the indefinite future.

The Russian Wildcard. The biggest wildcard affecting the U.S. posture in both Europe and Asia involves the uncertain evolution of reform and democracy within Russia and of relations between Russia and the West. Many Russians believe that the United States has taken advantage of Russian weakness. Relations could deteriorate over disagreements on a number of issues, including further NATO expansion; the deployment of a U.S. national missile defense systems; Russian arms sales to various states, particularly Iran, Iraq, and China; and access to energy resources in the Caspian basin. Russian cooperation in the initial stages of the war on terrorism has not fundamentally altered this equation. Given that Russia is the only country on earth that has the capacity to completely devastate the United States, these uncertainties will make managing the relationship with Russia one of the highest priorities of the next 25 years. Of course, developments within Russia remain dramatically unclear, even relative to the uncertainties discussed thus far.

Russia will also figure prominently in Asia. The emerging Russian-Chinese partnership is more a reflection of Russian weakness than of a genuine convergence of interests. Nonetheless, while Russia remains weak, it will continue to seek close relations with China, both to counter U.S. influence and to get much-needed funds from sales of military equipment. If Russia recovers somewhat, such considerations will diminish and more structural issues will assert themselves in Russian-Chinese relations. In particular, the growing Chinese migration to and influence in the sparsely populated Russian Far East will likely cause strains in the relationship.[12]

The Middle East—New Fault Lines? There is also the potential for significant change—both positive and negative—in the Middle East that would affect U.S. objectives, strategy, and military missions. The Persian Gulf remains a vital source of energy and therefore a critical

---

[12]For just a few of the widely varied views on the present and future of Russia and U.S.-Russian relations, see Blank (2000), Nunn and Stulberg (2000), McFaul (1999–2000), Carnegie Endowment for International Peace (2000), and Khalilzad et al. (2001), pp. 32–35.

region of U.S. interest. Moreover, the United States remains the region's ultimate security manager, playing a critical role in regional stability. The key factors affecting the future of the region include the following:

- how the power struggle in Iran is resolved and the future of the Iranian missile and nuclear programs
- Iraq's future direction and U.S. strategy toward Baghdad
- the future of the Arab-Israeli peace process
- the effects of globalization on the region
- the emergence of new leaders.

For the short term, Iran has become more focused on its internal power struggle. The desire on the part of the more pragmatic president, Mohammad Khatami, and the recently elected reformist parliament to improve Iran's regional and global ties and a greater desire to improve Iran's failing economy all point in a positive direction. Nonetheless, fundamentalist currents in Iran remain quite strong, and which side will prevail remains highly uncertain. Indeed, the current division of power is likely to continue and to produce increased instability. In any case, the United States is not in a strong position to influence the outcome of the current power struggle. If Iran is dominated by pragmatists, it is likely to distance itself from the use of terror, to show greater respect for human rights, and to seek reasonable state-to-state relations with the other nations. But even such an Iran is likely to continue to pursue nuclear weapons and long-range missiles and to seek regional primacy. U.S. allies in the Gulf are likely to remain suspicious of long-term Iranian intentions. Thus, even if Iran does become more pragmatic, Washington will need to retain at least some of the elements of its current containment policy.

Another important challenge continues to be the continuing inability of the United States, its allies, and the Iraqi opposition to get rid of the Iraqi strongman, Saddam Hussein. Through the end of the Clinton administration, a gap existed between the declared U.S. objective of deposing Saddam Hussein and the actual strategy, which could best be characterized as containment plus harassment. It is possible that, under the new administration, U.S. strategy might be brought up to the level of the objective or that the United States might reduce its objective, creating a more typical type of containment strategy—by ending the no-fly zones, for example. Of course, inconsistency remains a viable option and the United States might continue on the current path for some time to come.[13]

Another potential source of significant change springs from the ongoing Arab-Israeli conflict. The collapse of the peace negotiations in 2000 led to the "Al Aqsa" intifada, a bitter and violent series of demonstrations, shootings, mortar attacks, and terrorist incidents. Palestinian youths regularly engage in violent demonstrations against Israel. Radical Islamist groups, such as HAMAS and the Palestine Islamic Jihad, regularly conduct terrorism against targets in Israel. The Palestinian Authority often tolerates or even encourages this violence. The Israeli government, in turn, uses considerable force to put down unrest, often increasing tension.

The continued conflict poses several challenges to U.S. interests. Politically, the United States is often called upon to mediate or to pressure the parties to the conflict. In addition, the United States may be asked to provide peacekeeping troops along the Israeli-Syrian border or perhaps elsewhere, such as near Palestinian-Israeli flashpoints, as it already does in the Sinai. The conflict also affects U.S. security concerns outside the region. U.S. involvement in the Arab-Israeli conflict also complicates U.S. relations with other Arab allies, raising the political costs of their cooperation with the United States, and often spurs, or at least serves as a rhetorical justification for, terrorist attacks against U.S. targets.

Latin America: The Return of Instability and Authoritarianism? In recent years, the trends in Latin America have been mostly positive. Democracy and free-market economics have spread to an increasing number of countries in the region. Now, though, the trends are mixed. Drug trafficking networks and antidemocratic forces have made important gains of late and threaten to undo many of the positive developments of recent years.

The situation in Colombia is the most extreme manifestation of these trends. Drug traffickers and guerrillas—the leftist Revolutionary Armed Forces of Colombia (FARC) and the smaller Army of National

---

[13]On Iraq, see Byman (2000–2001). On Iran, see Chubin and Green (1998). On the military balance in the Persian Gulf, see Cordesman (1997).

Liberation (ELN) have come together and intensified their challenge to the Colombian state. The two groups control large pieces of territory, which the guerrillas use as safe havens for the drug trade and for attacking the rest of the country (Pardo, 2000). U.S. policy has focused on helping the Colombian government's antidrug efforts. It is uncertain whether this approach can succeed in eradicating the drug trade without the Colombian government defeating the guerrillas and gaining control of the territory in which the drugs are grown and processed.

Another policy that might enjoy more success in restraining the drug trade would involve promoting a reasonable settlement between the government and its opponents. This is the policy the Colombian government favors, although whether the FARC or the ELN wants the same thing remains unclear. It is possible that the three sides might not be able to make and keep an agreement that deals with U.S. concerns. The guerrillas, in particular, might believe that time is on their side. The war could therefore easily continue and might lead to greater FARC military successes, raising the real possibility that Colombia might emerge as a "narcostate." Alternatively, it might fragment into several de facto ministates controlled by different guerrilla groups and drug traffickers. Either outcome would probably lead to Colombian instability spilling over into neighboring states. Two FARC "fronts" already operate in Panama. Similar activities are taking place in Venezuela and Ecuador.

Another potential risk is that Venezuela—the second-largest supplier of crude oil to the United States—might turn hostile (U.S. Department of Energy, 2001). President Hugo Chavez appears to be heading toward establishing an authoritarian regime by threatening to destroy the checks and balances inherent in a democratic constitutional system. He has appointed dozens of military officers to government positions and has turned to the military to perform a wide range of public services and administrative tasks. Chavez also appears to have links with the FARC and has sought to distance his armed forces from those of the United States. He has pulled out of joint exercises with the U.S. military and expressed open admiration for Fidel Castro. It is possible that he may set up a populist Peronist-style regime with some Cuban characteristics. Should the Colombian government successfully prosecute a military strategy against the FARC, Chavez might well get involved in the Colombian conflict in support of the opposition, raising prospects for a regional conflict between a U.S.-backed democratic Colombia and a hostile, authoritarian Venezuela.[14]

Africa: More of the Same? There are a significant number of conflicts—both civil and regional—across Africa. The region is likely to continue to have pockets of instability and conflicts caused by poor leadership, failures of nation-building, a desire for regional dominance, and control over resources. These are old problems taking new forms and played by new actors in different subregions. But Africa will also be affected by new problems, such as AIDS. Africa's problems will produce humanitarian crises and call for outside intervention—to make or keep peace and to provide humanitarian assistance in the context of internal or regional conflicts.

Direct U.S. interests in Africa are few, but in an unusual way, Africa represents an important challenge to United States. U.S. internationalism is founded on the oft-expressed belief that U.S. interests and values require a stable and prosperous world order. Under these circumstances, U.S. governments cannot stand by while an entire region descends into chaos, poverty, and war without undermining the rationale that sustains public support for a leadership role in international affairs. Unfortunately, the U.S. capacity to address the myriad problems that plague Africa, from disease to civil war, is quite limited. Nonetheless, U.S. military efforts to conduct disaster relief, most recently in Mozambique, have saved countless lives and demonstrated a degree of commitment to the region. In the longer term, such initiatives as efforts to provide debt relief and to provide training programs to create an African peacekeeping or peacemaking force, will allow the United States to promote African stability actively at reasonable cost.

Summary. As this brief survey of regional trends implies, the absence of a global competitor will not free the United States from having to confront a host of difficult regional problems and competitors. Terrorism is the most pressing global problem, but regional problems will endure. Asia is likely to pose the most demanding challenges, and the U.S. posture in this region is most likely to need to undergo

---

[14]On the relationship between Venezuela and the Colombian conflict, see Rabasa and Chalk (2001), p. 87.

major changes. However, the threats from potential regional hege-
mons and rogues will not be limited to Asia and may even arise in the
western hemisphere. Moreover, specific regions of currently limited
national interest, particularly sub-Saharan Africa, may well occupy
U.S. policymakers and military forces because of their potential for
generating dramatic crises.

U.S. Goals

Beyond the geopolitical trends outlined above, the country's decisions about what role the United States seeks for itself on the world
stage and how it decides to prioritize its various goals will also play a
key role in determining the challenges that the U.S. military will to
have to face. Since the end of the Cold War, the United States has
been engaged in a continuous although often muted debate on what
its national security strategy should be. Despite a host of bitter
arguments that characterized the domestic politics of foreign policy
during the Cold War, the United States did have reasonable domestic
consensus on what it wanted to achieve—avoidance of nuclear war
and containment of the Soviet Union—although there was great
disagreement on how to achieve the goals. Today, such consensus,
beyond the need to prosecute the war on terrorism, is harder to
maintain.

Indeed, several visions of the U.S. role in the world now compete for
the mantle of "U.S. grand strategy" (Posen and Ross, 1996–1997). The
United States could pursue a strategy of primacy—seeking to preclude the rise of a global competitor or a system of multiple, roughly
equal, great powers—in which it would assert unilateral leadership of
the world community, with all the burdens and benefits that such
leadership implies.[15] Alternatively, the United States could seek to
share leadership with allies and other like-minded powers to achieve
stability at lower costs but also with less control. A variation on this
shared-leadership model could be a kind of geographic division of
labor. For example, Europe could take the responsibility for some or
all security issues in Europe; the United States and Europe and their
Asian and Middle Eastern friends and allies could share responsibility for the greater Middle East; while the United States and its Asian

---

[15]On primacy (often termed *preeminence*), see Joffe (1995).

friends and allies would focus on Asia. There could, of course, be alternative permutations. More radically, the United States could self-consciously limit its engagements to selective areas of vital interest, perhaps even retreating into an isolated fortress America.[16]

Theoretically, any of these strategic choices is possible. Indeed, one can identify U.S. policy actions in the post–Cold War period that have been consistent with the full range of possible strategies sketched above. In this environment, defense planning has become even more difficult than usual. The "two major theater war" concept, even in its most recent incarnation, is less a strategy than a convenient post–Cold War metric for determining the size of the military.[17]

Realistically, the United States has two options in the near future: unilateral or shared global leadership. Because the United States has economic, cultural, and political ties to almost every region on earth, withdrawal and isolationism are unlikely and would certainly carry greater risks. This policy has few adherents in the United States, except on the radical fringes of the political spectrum. Post–Cold War trends only increase and broaden U.S. global involvement as technology, trade, and capital flows continue to increase the interconnections between nations. As former Secretary of State Henry Stimson commented over 50 years ago, the United States can never again

> be an island to herself. No private program and no public policy, in any sector of our national life, can now escape from the compelling fact that if it is not framed with reference to the world, it is framed with perfect futility. (Quoted in Leuchtenburg, 1995, Ch. 1.)

At the other end of the range of choices, few in the United States have shown much interest in achieving primacy. Indeed, recent experience has made quite clear that U.S. domestic considerations will prevent the United States from incurring substantial costs, in terms of either blood or treasure, during operations not considered absolutely vital to U.S. national security (Larson, 1996; Burk, 1999). Simi-

---

[16]One advocate of "selective engagement" is Art (1998–1999). Advocates of retreating to America are rare, but one example is Gholz, Press, and Sapolsky (1997).

[17]Agreeing with this statement is U.S. Commission on National Security/21st Century (2000), pp. 14–15.

larly, such considerations will dictate adherence to evolving standards of morality and international law regarding limitations on damage to nonmilitary targets and on suffering imposed on noncombatants. Thus, the opinions of the U.S. public and elite are likely to impose substantial constraints not only on the use of military force but also on how it is used. These constraints will become looser as the level of national interest in an operation rises, but they will never disappear entirely.

As a result, the most likely U.S. course is to strive to remain the world's principal military power, arriving eventually at some compromise between the strategic visions of "unilateralism" and "shared global leadership." Such an equilibrium must involve four principal features: maintenance of clearly superior military capabilities, continued U.S. military involvement in selected humanitarian and regional crises, rebalanced partnerships that give more responsibility to allies within and across existing alliances, and improvements in the military capacities of the allies.

The NATO alliance in Europe appears headed in this direction. In Asia, this type of alliance is a more distant possibility but is no less necessary for achieving U.S. goals. In the Middle East, there is even less of a basis for creating broad-based alliances; the United States will thus need to rely on multiple, bilateral relationships to achieve its goals. On a political level, the desire of the United States to promote regional stability and to avoid regional hegemony will dictate increasing attention to relations, including military relations, with potential great powers, particularly China, Russia, and India.

This vision unfortunately provides somewhat less-than-precise guidance for determining which of the wide variety of possible crises around the world will generate a U.S. military response. For example, it was not evident a priori that the goal of maintaining a reasonably stable and just global order required U.S. intervention in Somalia, Haiti, or Bosnia. U.S. involvement in these countries could hardly have been predicted well in advance by any generic reference to U.S. strategy and, indeed, came as a surprise to many observers. In retrospect, it seems clear that the decision to intervene in each case was based not only on the intrinsic importance of the mission itself for U.S. goals but also on the highly idiosyncratic international and domestic political context of the moment. This observation highlights the fact that simple, generalized rules for predicting or limiting

U.S. uses of force will almost inevitably founder on the shoals of the domestic and international political imperatives of the day.

U.S. REQUIREMENTS FOR MILITARY FORCES

The unique U.S. role in international affairs generates requirements for unique military capabilities. To ensure stability in the various volatile regions of the world and to protect its citizens at home from terrorists of global reach, the United States must be able to influence events far from its shores. More to the point, as the de facto guarantor of stability in a variety of regions around the world, the United States must participate in the defense of the independence and territorial integrity of a large number of allies in Europe, East Asia, and the greater Middle East. These roles require the armed forces of the United States to be able to project military power effectively over great distances and to sustain a high tempo of operations in theaters far from home.

While power projection is the centerpiece of U.S. defense planning, the geopolitical context and U.S. goals also dictate a wide array of other missions, including

- promoting alliance relations and demonstrating U.S. commitment through routine engagement operations, or "overseas presence"

- deterring and defeating the use of weapons of mass destruction, both those threatening the U.S. homeland and those threatening U.S. forces and allies abroad

- countering international terrorism

- intervening to resolve inter- or intrastate conflicts that affect important U.S. interests broadly defined

- monitoring compliance with peace agreements

- providing humanitarian relief to victims of disasters.

This is a long list of missions, many of which call for specialized equipment and training. Being prepared to undertake all of them places a premium on forces that are characterized by flexibility: the ability to undertake a wide variety of missions under a wide range of circumstances.

Besides having to be able to conduct a large number of qualitatively different types of missions, U.S. armed forces face the additional challenge of having to accomplish their missions under demanding conditions. Deploying and sustaining forces far from their home bases is enormously challenging. Modern military forces conducting high-tempo operations require large amounts of manpower, spare parts, support equipment, munitions, and other support assets in addition to the weapon platforms. To conduct a campaign, all this must often be moved rapidly to a theater from elsewhere. Of course, U.S. defense planners can, to some extent, anticipate where possible aggression might call for a large-scale U.S. military response and can position supplies, equipment, and other assets accordingly. But given the wide variety of global challenges and the inherent uncertainty of geopolitical events, U.S. forces cannot be everywhere all the time. An enduring aspect of U.S. military operations is that the enemy typically enjoys the initiative at the outset of a conflict, choosing the time, place, and nature of the opening engagements. Even in the presence of the unparalleled global and theater surveillance capabilities the United States enjoys, surprise remains distinctly plausible. So rapidity of response—deployability—is a characteristic that U.S. political leaders and military commanders prize highly.

Other factors that are equally endemic in the post–Cold War strategic landscape also shape the "demand function" for U.S. military capabilities. One is what might be called asymmetry of stakes. Confrontations in Iraq, Serbia, and elsewhere have shown that enemy leaders are prepared to sacrifice much—their nation's economic base, the welfare of their people, human lives—to prevail against the United States. This is in part because the survival of their regimes (as well as their own personal fates) could well hinge on the outcome of the conflict. The United States, which typically intervenes in defense of interests that are less than truly vital, is generally not prepared to make commensurate sacrifices to achieve its objectives.

This is not to say that Americans are inherently "casualty shy." Research has shown repeatedly that Americans, by and large, support their government's decisions to go to war when they understand that important interests or values are at stake and when they perceive a plausible prospect of success. But the connection is inexorable: The human costs and risks of a military operation must be broadly commensurate with the interests at stake. Hence, if we expect to employ forces in the defense of rather amorphous interests (such as halting genocide or bringing stability to far-flung regions, such as the Balkans), or even for more important interests (such as control over the flow of oil from the Persian Gulf), U.S. forces must offer the nation's leaders options that allow them to minimize and to control the risk of friendly casualties.

Another characteristic of U.S. military operations is concern over what is sometimes euphemistically termed "collateral damage," that is, damage to nonmilitary targets and suffering imposed on non-combatants. In today's pervasive media environment, when Western television correspondents can broadcast pictures of bomb damage around the world within minutes of impact, U.S. policy leaders will rightly insist that U.S. military operations avoid causing substantial civilian suffering. Aware of this consideration, adversaries will work tirelessly to portray military operations directed against them as inhumane and unconnected to stated Western objectives. The best and most moral way to overcome such efforts is to develop and maintain highly accurate information about the military and civil infrastructures of potential enemy nations and to conduct strikes on targets with great precision and discrimination.[18]

In short, the forces that can expect to be called upon most frequently to support U.S. national security strategy will be those that are

1. rapidly deployable over great distances

2. able to accomplish their missions while minimizing the risk of casualties

3. able to neutralize an enemy's forces and exert coercive leverage on enemy leaders while minimizing harm to noncombatants

4. able to undertake a wide range of missions.

---

[18]Chapter Three offers a more detailed examination of the importance of minimizing the suffering of innocents as part of "the American Way of coercion."

THE MATURATION OF U.S. AEROSPACE POWER: CAPABILITIES OF TODAY'S FORCES

Developments in such areas as sensors, computing, miniaturization, and precision manufacturing techniques have enabled aerospace forces to accomplish these tasks far more effectively than even in the fairly recent past. It is instructive to review just how much things have changed since World War II.

Defeating Enemy Air Attacks

Perhaps the first responsibility of any air force to the joint forces commander is to prevent the enemy's air forces from attacking one's own surface forces, populations, and support facilities in rear areas. U.S. air forces have historically been quite successful in this regard, such that U.S. forces and allies now largely take the ability to operate in sanctuary from enemy air attacks for granted.[19] Since the demise of the Soviet Union, U.S. capabilities to defend against air attacks have simply overmatched the air forces of any potential adversary. Nevertheless, such regional powers as China are modernizing their air attack capabilities, and new systems, such as cruise missiles and beyond-visual-range air-to-air missiles, are available that, if employed competently, could challenge the capabilities of present U.S. forces to defend vital assets.

Destroying Fixed Targets

The destructive power of an explosive weapon is a linear function of its explosive power and a squared function of its accuracy. Recent major changes in the accuracy of air-delivered weapons have enabled quite dramatic changes in their effectiveness. During World War II, the typical long-range bomber could deliver its load of bombs with a circular error probable of around 1,000 m.[20] This meant that it

---

[19]In an instance of the exception proving the rule, one can gauge the success of U.S. air defense by the extent to which its three major failures—Pearl Harbor, the Kasserene Pass, and Japan's kamikaze attacks on the U.S. fleet—stand out in the history of U.S. military operations of World War II.

[20]That is, one-half of its bombs, on average, would be expected to fall within a circle whose center point was the aimpoint and whose radius was around 1,000 m.

took upwards of 240 bombs to be confident of destroying a single, "hard" target, such as a bridge or a command post (see Figure 2.1, which shows the change over time in the number of bombs required to drop a bridge). It also meant that aerial attacks on military targets would, perforce, result in heavy damage to people and facilities in the general vicinity of the targets. The situation was little changed ten years later in Korea or, indeed, ten years after that in the early stages of the war in Vietnam. Only the invention of the laser-guided bomb, first used in Vietnam in 1970, brought a marked improvement in the effectiveness and efficiency of aerial attacks on fixed targets.

Today, laser guidance for air-delivered ordnance has become both more accurate and more robust. Circular error probables are now down to about 3 m for the most modern platforms, and automation

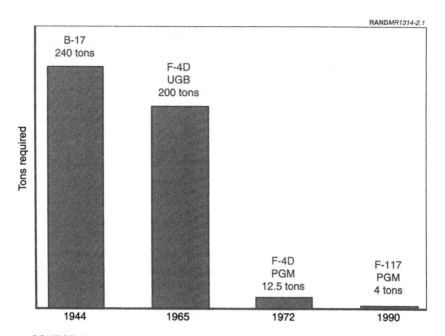

RANDMR1314-2.1

SOURCE: Mark (1994), pp. 236, 387.
NOTE: Data represent tons of bombs required to drop a bridge span with 90-percent confidence.

Figure 2.1—Attacking Fixed Targets: Improved Accuracy Has
Yielded Major Increases in Effectiveness

allows the pilots of single-seat aircraft to deliver laser-guided bombs. Perhaps more importantly, the advent of weapons guided by signals from the Global Positioning System means that modern aircraft can attack fixed targets with accuracies approaching those of laser-guided bombs in virtually all weather conditions. And the attack can be undertaken with little risk to people and facilities in the surrounding area.

Destroying Mechanized Ground Forces

As guarantor of the territorial integrity and sovereignty of allied nations, the armed forces of the United States must be prepared to halt and to defeat invasions by mechanized ground forces. Such armored thrusts were the dominant form of warfare in most of the 20th century. Defeating this type of warfare was also the crux of the problem assigned to U.S. forces during the Korean and Gulf Wars, and this requirement could arise again in these or other theaters. As with attacks on fixed targets, new systems and the associated concepts of operation have led to great improvements in the capabilities of aerospace forces to aid in achieving this objective.

Figure 2.2 illustrates the trend in U.S. sortie effectiveness against moving armored columns in the six decades since the beginning of World War II. From the 1940s, when general-purpose bombs would be dropped (with accuracies similar to those achieved against fixed targets), through the 1970s, when first-generation guided weapons were fielded, some progress in sortie lethality was made. Still, joint force commanders throughout this period regarded airpower as useful primarily as a means of delaying or disrupting an enemy's armored forces. As such, it was seen as an adjunct to one's own heavy ground forces, which alone had the ability to destroy an opponent's armored units in large numbers. The limitations on the lethality of air interdiction in this era were magnified by the fact that effective attacks could not be prosecuted at night or in poor weather, giving enemy ground forces ample opportunities to avoid attacking aircraft in many situations.

Much has changed since the 1970s. Most important, the Air Force has fielded a new generation of "smart" anti-armor weapons that allow aircraft to engage and attack several armored vehicles in a sin-

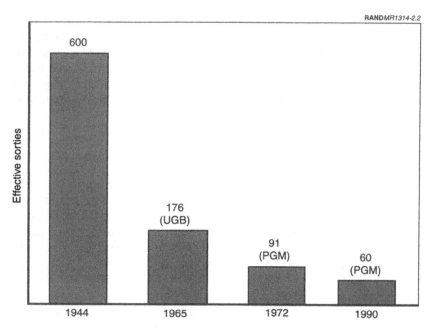

SOURCES: Mark (1994), pp. 236, 387; Scales (1998).
NOTE: Data represent attacks required to kill 30 armored vehicles.

**Figure 2.2—Comparative Lethality of Air-Delivered Weapons Against
Moving Armored Columns (1940s–1990s)**

gle pass, day or night, and in almost all types of weather. The effectiveness of these weapons, which are currently available only in fairly modest numbers, still suffers from important limitations imposed by terrain, deficiencies in accurate targeting information, and such enemy countermeasures as dispersal and mobile air defense systems. Nonetheless, much progress has already been made and more is promised in the future as improved munitions and information systems come online.

Information and Its Uses

The value of these enhancements in the lethality and precision of air attacks is magnified greatly by concomitant improvements in the ability of U.S. forces to acquire and to use information. Most notably,

U.S. commanders and forces have a more accurate and comprehensive picture of enemy forces and activities today than at any time in the history of modern warfare. Sensors onboard satellites and aircraft can detect movements of sizable formations of enemy forces well inside enemy territory and can relay data about them instantaneously to assessment centers. With information derived from the raw data, commanders can direct the attack assets available to them to engage the most lucrative targets, in some cases providing aircrews their targets only after they have taken off. The ability to manipulate and transmit large amounts of information quickly and securely means that commanders can orchestrate the employment of larger numbers of assets more nimbly than ever before.

The United States is, however, far from achieving perfect awareness of the battlefield and almost certainly never will achieve it. The mistaken bombing of the Chinese embassy and the accidental destruction of civilian convoys during Operation Allied Force starkly illustrate the current limitations of the U.S. information complex. Nonetheless, the significance of the progress already achieved can be appreciated by contrasting today's situation with the state of affairs common in prior eras. Historically, at any one time during an extended conflict, only a very small portion of each side's fielded forces was likely to be engaged in combat with the enemy. Most assets were idle most of the time. Of those engaged in operations, most lacked targets. And of those that had (or thought they had) targets, most were expending rounds that missed them. If this is an accurate depiction of the norm for past conflicts, one can see that even small changes in the portion of the force actually engaged can result in major changes in outcome. By dominating the contest to acquire and use information, a combatant force can operate at a high tempo and a high level of efficiency, while denying the same to his opponent. Moreover, accurate information about the location and disposition of enemy forces and targets is essential in reducing the potential for damage to nonmilitary assets.

Survivability

Flying aircraft, in peace or war, has always been a risky enterprise. From the Wright brothers through the early years of the jet age, accident rates alone were fairly daunting. Aircrews flying into combat faced the serious prospect that they would not return. Figure 2.3

shows how dramatically things have changed, at least for U.S. airmen. The chart shows average rates of USAF aircraft losses experienced in four conflicts.[21] Loss rates have been reduced by more than two orders of magnitude (a factor of more than 100) since World War II. New generations of surface-to-air missiles (SAMs), as well as improved enemy tactics, threaten to reverse, or least halt, this trend. For the time being, however, the combination of timely information about enemy air defenses, dominant air-to-air fighters, stealth aircraft, electronic jamming assets, weapons for SAM suppression, standoff attacks, and superb training have created a situation in

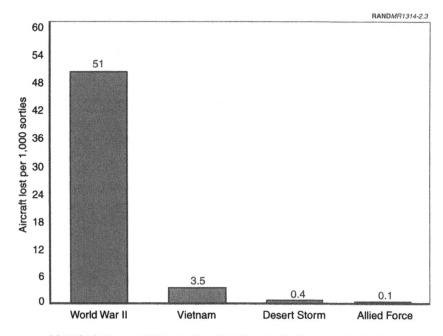

SOURCES: Murray (1983), McCrea (1976), DoD (1992), and HQ USAF (2000).

Figure 2.3—Comparison of Average Loss Rates of USAF Aircraft

[21]These are averages over the course of each conflict. In every case, loss rates were substantially higher than average in the opening phase of the operation and decreased as the enemy drew down its forces and as the United States and its allies developed new equipment and tactics to counter air defenses.

which U.S. fixed-wing combat aircraft have operated with something approaching impunity over large parts of the enemy's territory.

The significance of this trend for U.S. military operations is profound. Before Operation Desert Storm, it was expected that control of operations in the air would be contested throughout the duration of any large-scale conflict and that the enemy could continue to extract some price for efforts to attack his assets. In Desert Storm, however, following an intense period in which U.S. forces concentrated on suppressing and dismantling the opponent's air defense system, U.S. air forces operated with virtual impunity. This meant that these forces could be far more effective in locating and attacking targets than they had been in the past and that many aspects of the conflict suddenly became quite one-sided. The United States and its allies could employ airpower to pursue their objectives without giving the enemy substantial leverage in terms of lost aircraft or of killed or captured aircrews. This situation has proven to be quite demoralizing to enemy forces exposed to air attacks.[22] It has also given U.S. policymakers new coercive options to support diplomacy.

Implications for U.S. Joint Operations

The developments outlined above have mutually reinforcing effects. Accurate attacks on fixed targets, such as bridges and roads, can slow the rate of advance of enemy ground forces, giving U.S. reinforcements more time to arrive and creating more lucrative targets for interdiction. Access to quality information about the location of targets and threats can allow aircrews to plan missions so as to minimize their exposure to enemy air defenses. And a more comprehensive and accurate picture of the battlefield allows commanders to employ more of the forces available to best effect. U.S. defense planners are only now beginning to grasp the potential implications of

---

[22]For example, Iraqi prisoners of war reported that the inability of their air defenses to protect them or to down coalition aircraft induced feelings of helplessness, betrayal, and resignation both in the rank and file and in the officer corps. This accounts, in large measure, for the facts that between 20 and 40 percent of the Iraqi troops in the Kuwait theater of operations (between approximately 100,000 and 200,000 troops) deserted their units before the coalition's ground offensive began and that another approximately 85,000 surrendered to advancing troops rather than fight. See U.S. Department of Defense (DoD) (1992), pp. 34, 662–663. For a thorough assessment of the psychological effects of air operations, see Hosmer (1996).

these developments even as much of the progress is threatened by developments both within and beyond the USAF.

CHALLENGES FOR THE USAF

This chapter has thus far focused on the enormous progress the USAF has made in recent years. Despite the accomplishments, however, substantial challenges remain, and new, more imposing challenges can be discerned on the horizon. It would be foolhardy to assume that, absent substantial continued efforts to innovate and modernize, U.S. aerospace forces can retain the advantages they currently enjoy over potential enemies. Adversaries around the world have carefully studied U.S. military operations and are working to challenge capabilities that they recognize are central to U.S. strategy for power projection. Too, the USAF faces institutional challenges that will have to be addressed if the force is to remain healthy over the coming years. The bulk of this volume is devoted to providing a more-detailed understanding of the problems some of these challenges and their possible solutions pose. Here, we highlight some of the most pressing challenges that confront the Air Force.

On the most general level, if U.S. aerospace forces are to retain the extensive advantages they now enjoy over the forces of regional adversaries, they will need to continue to develop new operational concepts and supporting capabilities in a number of areas. Perhaps first among the challenges all types of U.S. expeditionary forces face is to ensure that they can gain access to facilities and operating areas in theaters of conflict.[23] Because U.S. military forces have such overwhelming superiority once they are employed in sizable numbers, adversaries will place a high priority on slowing U.S. deployment in the hope of being able to seize their primary objectives prior to the arrival of the bulk of the U.S. forces.

Besides gaining access to the theater, air forces will have to work harder to sustain their freedom to operate in enemy airspace. The successful performances of U.S. and allied aircraft against the enemy's air defenses in operations Desert Storm and Allied Force were achieved against SAM systems that were designed and fielded in the

---

[23]Chapter Nine provides a more detailed exploration of access issues.

1960s and 1970s and fighter-interceptors from the 1970s and 1980s. As adversaries acquire more-modern SAMs, such as the Russian-built SA-10 and -12, and fourth-generation fighters, such as the Su-27, U.S. forces will need entirely new systems and concepts for defeating these threats.

Adversaries are developing new tactics, as well as new systems, in the effort to counter U.S. power-projection operations. In Operation Allied Force, Serbian army and police units were largely able to elude observation and attack from the air by dispersing and hiding under trees or in towns. And when the units moved, they often complicated NATO's efforts to attack them by using civilians as human shields, mixing the automobiles, trucks, and other vehicles of Albanian civilians into convoys of military vehicles. Such tactics may not always be feasible in the context of larger-scale operations, but U.S. air forces must expect to be called upon again to attack fielded forces under difficult operational circumstances.

Modernization and Recapitalization

In addition to changing enemy capabilities, another impetus to modernization is the increasing cost of operating fleets of aircraft that are growing ever more long in the tooth. Today, the average fighter aircraft in the U.S. Air Force is 20 years old. Similarly, the workhorse of the airlift fleet is still the C-141, which entered service in the mid-1960s, though many are now being by replaced by the C-17. Many of these aircraft have been flown for thousands of hours more than their original design lives, and, given currently projected budgets, it will be years before many are replaced by newer designs. Maintaining aircraft of this vintage is imposing unanticipated costs—both economic and operational—on the Air Force. Not only does it require more money and manpower to maintain older aircraft, but even with these investments, fewer sorties are available from each aircraft for training or for operations.[24]

Last in line for modernization is the Air Force's fleet of more than 500 aerial refueling aircraft. These aircraft play vital roles in joint opera-

---

[24]Chapter Four gives a more detailed evaluation of the options for modernizing the Air Force's fleet of combat aircraft—fighters and bombers—as well as key types of munitions they deliver.

tions, enabling the rapid deployment of combat and support aircraft to distant theaters and extending the range, payload, and endurance of the aircraft in operations.[25] Remarkably, 90 percent of these are variants of the Boeing 707—the world's first commercial jet transport, which made its debut in the 1950s. As with the fighters and airlifters, the costs associated with sustaining these venerable airframes is growing and is increasingly difficult to predict, given that no one has much experience with maintaining and operating fleets of aircraft this old.

Human Capital

Much of this chapter has focused on hardware. While the aircraft, missiles, satellites, munitions, and other items of equipment the Air Force operates are the most visible manifestations of capability, they are not the most important. Rather, the most significant factors determining the capabilities of the force are the qualities of the people who operate and maintain the equipment and who command the units the force comprises. Without doubt, the people serving in the Air Force today are highly capable. Nevertheless, the service has lost considerable talent over the past several years, and sustained, deliberate efforts will be needed to regain it.[26]

This problem is being felt most acutely among experienced aircrews and aircraft maintenance personnel. Many units in the USAF are short in these critical areas by 20 percent or more. There are several reasons for this, including personnel policies during the downsizing that has taken place over the past decade and competition for talent from a strong economy. But perhaps the most significant factor is the increased stress placed on people by the pace and nature of work they have recently been called upon to do.

---

[25]Operation Allied Force—NATO's effort to coerce Yugoslavia into accepting its terms for settling the conflict in Kosovo—provides one illustration of the importance of aerial refueling in modern operations. At its peak, NATO employed more than 200 tanker aircraft (the bulk of which were from the USAF) to support the operations of approximately 700 combat aircraft (about 500 were U.S. aircraft). This constituted nearly 70 percent of the tanker aircraft in active-duty USAF units, despite the fact that the overall level of forces employed in Allied Force was substantially less than that called for by a major theater conflict.

[26]Chapter Twelve presents a detailed assessment of some of the causes and effects of shortfalls in key personnel in the Air Force.

Since the end of the Gulf War, the Air Force has experienced a form of culture shock. Traditionally, Air Force personnel in peacetime have engaged primarily in training to prepare for possible future conflicts. Even Air Force units stationed abroad during the Cold War spent the bulk of their time training at home station. The chaotic geopolitical circumstances of the 1990s and the expanded role of the United States in various regions around the world have caused an explosion in the demand for USAF operations in the form of largely ad hoc deployments. For example, the Air Force has been called upon to deter Saddam Hussein from repressing Kurds in Northern Iraq and from once again threatening Kuwait and Saudi Arabia, to convince the Bosnian Serbs to cease their atrocities against their Muslim and Croat neighbors, and to compel Slobodan Milosevic to accept NATO's terms in Kosovo. To these "shooting contingencies" has been added a host of other operations, ranging from combined exercises with allies in Europe, East Asia, and the Gulf, to patrolling drug smuggling routes between North and South America, to the transport of troops and humanitarian relief supplies to such places as Somalia, Rwanda, and Haiti.

U.S. defense strategy calls upon the armed forces to perform such operations as these to shape the international security environment in desirable ways, and the Air Force has shown that its capabilities are highly relevant to these purposes. But such operations have done more to the Air Force than simply add to its workload; they have also required new modes of operation, demanding that people spend months at a time away from home and that units find ways to accomplish their training missions even though they must spend time in locations lacking suitable training facilities and opportunities. The leadership of the Air Force has taken steps to alleviate some of the most serious problems associated with new modes of operation. It has increased reenlistment bonuses and incentive pay for people in the most critical career fields and, in a major initiative, designated ten Aerospace Expeditionary Forces that conduct the full range of USAF operations on a rotation basis, thus permitting personnel in operational units to anticipate their likely times of deployment. Nonetheless, much remains to be done if the Air Force is to maintain its traditionally highly capable workforce.

CONCLUSION: CREATING OPTIONS

In sum, the capabilities of U.S. aerospace forces have matured rapidly over the past several decades, yielding improvements that, in some key areas, are measured in orders of magnitude. Because of these capabilities, U.S. leaders and allies have become accustomed to lopsided victories. But many challenges loom ahead. As this brief tour of the world and the Air Force implies, the United States faces an array of difficult challenges far greater in scope, if not in difficulty, than it faced during the Cold War.

To support U.S. goals, the nation must have a force able to deter or decisively defeat a range of potential adversaries. U.S. forces must be lethal and flexible and must have strategic reach, yet must also protect the United States itself against potential attacks, large and small. Given this requirement, the most basic job of the U.S. armed forces is to provide political decisionmakers with options in an uncertain world. When an unexpected crisis arrives in an unfamiliar region, the President and the Secretary of Defense will demand options from U.S. military forces. Lack of warning will not excuse an inability to meet the demands.

In recent years, the USAF has proved quite adept at providing such options in a variety of difficult circumstances. However, it is becoming increasingly apparent that the USAF is living on borrowed time. To support the wide range of U.S. interests and to give policymakers options, the Air Force must, among other things, continue to develop new concepts and systems, modernize and recapitalize its aging fleet of combat and support aircraft, and continue to reinvent the way it does business. It is our hope that this volume, in some small way, will help point the way toward achieving these goals.

REFERENCES

"Aiming Missiles: Missile Race on Asia's Subcontinent Has Plenty to Do with China, *The Economist,* May 9, 1998, p. 42.

Arquilla, John, and David F. Ronfeldt, *The Advent of Netwar,* Santa Monica, Calif.: RAND, MR-789-OSD, 1996.

Art, Robert J., "Geopolitics Updated: The Strategy of Selective Engagement," *International Security*, Vol. 23, No. 3, Winter 1998–1999, pp. 79–113.

Blackwill, Robert D., and Paul Dibb, eds., *America's Asian Alliances*, Cambridge, Mass.: MIT Press, 2000.

Blank, Stephen, "Russia as Rogue Proliferator," *Orbis*, Vol. 44, No. 1, Winter 2000, pp. 91–107.

Bracken, Paul, "How to Think About Korean Unification," *Orbis*, Vol. 42, No. 3, Summer 1998, pp. 409–422.

Burk, James, "Public Support for Peacekeeping in Lebanon and Somalia: Assessing the Casualties Hypothesis," *Political Science Quarterly*, Vol. 114, No. 1, Spring 1999, pp. 53–78.

Byman, Daniel L., "After the Storm: U.S. Policy Toward Iraq Since 1991," *Political Science Quarterly*, Vol. 115, No. 4, Winter 2000–2001, pp. 493–516.

Carnegie Endowment for International Peace, "An Agenda for Renewal: U.S.-Russian Relations," Washington, D.C., 2000.

Chellaney, Brahma, "After the Tests: India's Options," *Survival*, Vol. 40, No. 4, Winter 1998–1999, pp. 93–111.

Chubin, Shahram, and Jerrold Green, "Engaging Iran: A U.S. Strategy," *Survival*, Vol. 40, No. 3, Autumn 1998, pp. 153–169.

Cordesman, Anthony H., *U.S. Forces in the Middle East: Resources and Capabilities*, Washington, D.C.: Center for Strategic and International Studies, 1997.

Friedman, Thomas, *The Lexus and the Olive Tree*, New York: Farrar, Straus, and Giroux, 1999.

Gholz, Eugene, Daryl G. Press, and Harvey M. Sapolsky, "Come Home, America: The Strategy of Restraint in the Face of Temptation," *International Security*, Vol. 21, No. 4, Spring 1997, pp. 5–48.

Gilpin, Robert, *The Challenge of Global Capitalism: The World Economy in the 21st Century*, Princeton, N.J.: Princeton University Press, 2000.

Gordon, Philip H, "Their Own Army? Making European Defense Work," *Foreign Affairs*, Vol. 79, No. 4, July–August 2000, pp. 12–17.

_____, "Bush, Missile Defence and the Atlantic Alliance," *Survival*, Vol. 43, No. 1, Spring 2001, pp. 17–36.

Guehenno, Jean-Marie, *The End of the Nation-State*, Minneapolis, Minn.: University of Minnesota Press, 1995.

Haas, Richard N., ed., *Transatlantic Tensions: The United States, Europe, and Problem Countries*, Washington, D.C.: The Brookings Institution, 2000.

Headquarters, U.S. Air Force, *The Air War Over Serbia: Initial Report*, Washington, D.C., 2000.

Hirst, Paul, and Grahame Thompson, *Globalization in Question: The International Economy and the Possibilities of Governance*, Oxford: Blackwell Publishers, 1996.

Hosmer, Stephen, *Psychological Effects of U.S. Air Operations in Four Wars 1941–1991: Lessons for U.S. Commanders*, Santa Monica, Calif.: RAND, MR-576-AF, 1996.

Joffe, Josef, "'Bismarck' or 'Britain'? Toward an American Grand Strategy After Bipolarity," *International Security*, Vol. 19, No. 4, Spring 1995, pp. 94–117.

Khalilzad, Zalmay, et al., *The United States and Asia: Toward a New Strategy and Force Posture*, Santa Monica, Calif.: RAND, MR-1315-AF, 2001.

Larson, Eric C., *Casualties and Consensus: The Historical Role of Casualties in Domestic Support for U.S. Military Operations*, Santa Monica, Calif.: RAND, MR-726-RC, 1996.

Leuchtenburg, William E., *The FDR Years: On Roosevelt and His Legacy*, New York: Columbia University Press, 1995.

Mark, Eduard, Center for Air Force History, *Aerial Interdiction in Three Wars*, Washington, D.C.: Government Printing Office, 1994.

McCrea, Michael M., *US Navy, Marine Corps, and Air Force Fixed-Wing Aircraft Losses and Damage in Southeast Asia (1962–1973)*, Arlington, Va.: Center for Naval Analyses, August 1976.

McFaul, Michael, "Getting Russia Right," *Foreign Policy*, No. 117, Winter 1999–2000, pp. 58–73.

Murray, Williamson, *Strategy for Defeat: The Luftwaffe 1933–1945*, Maxwell Air Force Base, Ala.: Air University Press, 1983.

National Intelligence Council, *Foreign Missile Developments and the Ballistic Missile Threat to the United States Through 2015*, September 1999. Online at http://www.fas.org/irp/threat/missile/nie99msl.htm (as of December 12, 2001).

Nunn, Sam, and Adam N. Stulberg, "The Many Faces of Modern Russia," *Foreign Affairs*, Vol. 79, No. 2, March–April 2000, pp. 45–62.

O'Hanlon, Michael, "Why China Cannot Conquer Taiwan," *International Security*, Vol. 25, No. 2, Fall 2000, pp. 51–86.

Paal, Douglas, *Nesting the Alliances in the Emerging Context of Asia-Pacific Multilateral Processes*, Stanford, Calif.: Stanford University, Asia Pacific Research Center, July 1999.

Pardo, Rafael, "Colombia's Two-Front War," *Foreign Affairs*, Vol. 79, No. 4, July–August 2000, pp. 64–74.

Posen, Barry R., and Andrew L. Ross, "Competing Vision for U.S. Grand Strategy," *International Security*, Vol. 21, No. 3, Winter 1996–1997, pp. 5–53.

Rabasa, Angel, and Peter Chalk, *Colombian Labyrinth: The Synergy of Drugs and Insurgency and Its Implications for Regional Security*, Santa Monica, Calif.: RAND, MR-1339-AF, 2001.

Roberts, Brad, Robert A. Manning, and Ronald N. Montaperto, "China: The Forgotten Nuclear Power," *Foreign Affairs*, Vol. 79, No. 4, July–August 2000, p. 57.

Scales, MG Robert USA, presentation for the Future Operating Environment session of the Professional Military Education for the 21st Century Warrior Conference, Naval Postgraduate School, Monterey, January 15–16, 1998, slide 5. Online at http://web.nps.navy.mil/FutureWarrior/Presentations/Scales/sld005.htm (as of December 12, 2001).

Shambaugh, David, "A Matter of Time: Taiwan's Eroding Military Advantage," *Washington Quarterly*, Vol. 23, No. 2, Spring 2000, pp. 119–133.

Shlapak, David A., David T. Orletsky, and Barry A. Wilson, *Dire Strait? Military Aspects of the China-Taiwan Confrontation and Options for U.S. Policy*, Santa Monica, Calif.: RAND, MR-1217-SRF, 2000.

Sokolski, Henry D., ed., *Planning for a Peaceful Korea*, Carlisle, Pa.: Strategic Studies Institute, 2001.

Tow, William T., *Assessing U.S. Bilateral Security Alliances in the Asia Pacific's "Southern Rim": Why the San Francisco System Endures*, Stanford, Calif.: Stanford University, Asia/Pacific Research Center, October 1999.

U.S. Commission on National Security/21st Century, *Seeking a National Strategy: A Concert for Preserving Security and Promoting Freedom*, April 15, 2000. Online at http://www.nssg.gov/PhaseII.pdf (as of December 12, 2001).

U.S. Department of Defense, *Conduct of the Persian Gulf War: Final Report to Congress*, April 1992.

U.S. Department of Energy, Energy Information Administration, *Annual Energy Review 2000*, Table 5.4. Online at http://www.eia.doe.gov/aer (as of May 15, 2001).

Wolf, Jr., Charles, et al., *Asian Economic Trends and Their Security Implications*, Santa Monica, Calif.: RAND, MR-1143-OSD/A, 2000.

Chapter Three

THE FUTURE OF U.S. COERCIVE AIRPOWER

Daniel L. Byman, Matthew C. Waxman, and Jeremy Shapiro

The capitulation of Serbia after 78 days of NATO bombing in spring 1999 heralded for many the arrival of a new era in the use of coercive airpower. Lt Gen Michael Short, who ran the bombing campaign, has argued that "NATO got every one of the terms it had stipulated in Rambouillet and beyond Rambouillet, and I credit this as a victory for air power."[1] For airpower advocates like General Short, the Kosovo experience means that, in the future, the United States and its allies will have in airpower a coercive instrument capable of effectively modifying rogue state behavior.[2]

Even before the Kosovo crisis, many American political leaders appeared to view airpower almost as a first resort to enhance the credibility of their threats or to demonstrate their resolve. Economic sanctions are now widely viewed as ineffective and cruel; weapons of mass destruction (WMD) are imprecise and disproportionate; and ground forces are too risky and too slow to use for such nuanced purposes. In contrast, attacks from the air—be they isolated cruise missiles attacks against Sudan and Afghanistan, prolonged intimidation of Iraq, or a concerted bombing campaign against Serbia—offer U.S. political leaders the possibility of a coercive option that is precise, scalable, rapid, and relatively risk-free. As a result, the United

---

[1]As quoted in Whitney (1999).

[2]The research in this chapter was conducted before the September 11, 2001, attacks on the United States and the subsequent U.S. campaign against Al Qaeda and other terrorist groups. We contend that many of the propositions advanced in this chapter still hold true. However, the focus of most of this chapter is on more-limited conflicts that affect U.S. interests less directly than did the September 11th attacks.

States has frequently deployed airpower for coercive purposes in recent years. This trend will very likely continue.

Any assessment of coercive U.S. airpower should consider that, whatever success airpower may have achieved in the Kosovo crisis, it did so against a recent U.S. record in coercive crises that is mixed at best. While the military lesson of the Kosovo campaign was that airpower is the most critical tool in modern warfare, the political lesson of that and other recent coercive episodes was that the capacity of U.S. airpower to fulfill a coercive role is very much in doubt. Despite the overwhelming power the United States and its allies can and have deployed, U.S. threats are often not perceived as credible, forcing the United States to make good on its threats or further damage its credibility.[3] Against Somali militants, Serb nationalists, the Iraqi dictator, and other adversaries, U.S. coercive efforts have at times failed or even backfired. Despite the lopsided U.S. edge in raw power, regional foes often refuse to blink, preferring to defy U.S. threats and ultimatums. Moreover, even when the United States carries out its threats, adversary resistance often increases rather than decreases. Against both Iraq in 1991 and Serbia in 1999, the United States ended up conducting full-blown air campaigns that involved large amounts of force. These failures seem puzzling in the light of most understandings of U.S. coercive capacity that emphasize U.S. military superiority, especially in the air.[4]

An analysis of this puzzle must begin with a better understanding of the nature of coercion and a realistic appreciation of how U.S. airpower can be used as a part of coercive strategies. *Coercion*, in this context, is defined as the threat of force, as well as the limited use of actual force to back up the threat, to induce an adversary to behave differently than it otherwise would. Coercion is not destruction. Although partially destroying an adversary's means of resistance may often be necessary to increase the effect and credibility of coercive threats, coercion succeeds when the adversary gives in while it still

---

[3]At times, this credibility problem stems from the statements of U.S. officials. In Kosovo, for example, several senior U.S. officials noted that the air campaign would be limited and that ground troops would not be used and otherwise made it clear that an adversary willing to withstand the initial strikes might outlast the United States.

[4]For 20th-century instances of coercion, see Byman and Waxman (2002).

has the power to resist. Coercion can be understood in opposition to what Thomas Schelling (1966, p. 3) termed "brute force":

> [B]rute force succeeds when it is used, whereas the power to hurt is most successful when held in reserve. It is the threat of damage, or of more damage to come, that can make someone yield or comply. It is latent violence that can influence someone's choice . . . .

Coercion may be thought of, then, as getting the adversary to act a certain way via anything short of brute force.[5]

Coercion generally requires the credible threat of "pain" beyond the benefits an adversary sees in resisting.[6] The authors who first explored the nature of coercion and articulated this formulation, however, did not anticipate a world in which the United States would impose constraints on its own use of force, and therefore on the level of pain that it can deliver, because overall U.S. interests in a particular crisis might be exceptionally limited. Adversaries can capitalize on these constraints and win a coercive contest despite being militarily, politically, and economically inferior.

These self-imposed constraints stem to some degree from the new strategic reality the United States and its allies face today. Most obviously, the threat of nuclear annihilation no longer colors coercive crises as it did during the Cold War. Decisionmakers today do not view local conflicts in Angola, Somalia, or other developing nations through the lens of nuclear superpower competition. The stakes also differ considerably from the Cold War era, both for the United States and for its adversaries. The Soviets often challenged the United States in the developing world to expand their influence or to probe U.S. credibility. For adversaries today, however, the coercive contest is often a question of their very survival: Saddam Hussein, Slobodan

---

[5]Most properly, coercion and brute force should be seen as a spectrum. At the low end are limited threats or uses of force (e.g., the 1998 U.S. cruise missile strike on terrorist training camps in Afghanistan); at the higher end are coercive campaigns that are part of an overall war effort (e.g., the "Linebacker" bombings during the Vietnam War). U.S. operations in Bosnia in 1995 (Operation Deliberate Force) were largely coercive; the 1999 operations (Operation Allied Force) over Kosovo began as a coercive campaign and moved toward brute forces as it intensified. For more on this point, see Byman and Waxman (2002).

[6]This description draws heavily on broader theoretical works on coercion, including Schelling (1966) and George and Simons (1994).

Milosevic, Mohammad Farah Aideed, and other foes might lose power, and even their lives, if they make the wrong move in response to U.S. pressure. The United States and its allies, in contrast, are often involved for humanitarian reasons or to preserve the credibility of a specific alliance. Such concerns, while important, pose no immediate threat to the national security of the United States or its allies.

This chapter aims to explain the recent mixed U.S. record on coercion by explaining how adversaries can influence and exploit self-imposed constraints that can weaken, and ultimately defeat, U.S. coercive airpower. This task requires exploring the phenomenon from an adversary's point of view. Whether a reactive, adaptive foe will yield in the face of threats depends not simply on the level of threat or motivation present at the time the United States issues its demands but also on the self-imposed limits the United States places, or is perceived to be placing, on the use of force. More generally, this requires understanding that coercion is a dynamic process: Adversaries shape their strategy to exploit U.S. weaknesses. Thus, understanding the success or failure of U.S. coercive diplomacy means understanding how adversaries can manipulate key features of U.S.-style coercion to reduce the costs inflicted or to convince the United States to abandon its effort.

The first section will define an "American way of coercion." It will argue that the U.S. theory and practice of coercion place severe limits on the U.S. ability to coerce and, more precisely, opens up numerous strategies for adaptive, reactive foes to exploit specific U.S. weaknesses. The second section will detail six such strategies and provide examples of how such strategies have been used in the past. The final section will assess the implications of these weaknesses and make some recommendations about how the USAF can counter some of the strategies outlined. Many of these limits on coercion stem from deep-seated U.S. cultural and societal norms that are not readily addressable. The limits will therefore need to be recognized and managed rather than overcome.

THE AMERICAN WAY OF COERCION

In recent years, the United States has evolved a certain style of coercion that reflects its societal preferences about how and under what circumstances the nation is willing to threaten or use force in crises.

This American way of coercion has five key characteristics that are relevant to understanding how enemies might attempt to counter it. This section describes each characteristic with reference to particular crises. All five factors, of course, represent generalizations, and the degree to which they are present in any given episode varies, as do their strengths relative to each other. However, the strength of these characteristics, or the perception among adversaries that they are strong, has been particularly high in recent encounters. Together, these factors provide means by which adversaries can undermine U.S. coercive strategies.

A Preference for Multilateralism

The United States generally prefers to operate within a broad coalition and to isolate its opponents from international support, both moral and material. With only a few exceptions, all U.S. coercive military operations since the end of the Cold War have been prosecuted under the auspices of the United Nations (UN), NATO, or ad hoc collections of allies or partners.[7] The preference for multilateralism stems from many factors. Sometimes coalition partners provide the bulk of ground troops. Allies also contribute basing and overflight privileges, on which effective military operations depend. In such cases, coalition unity becomes central to planners and decisionmakers. In many instances, however, the value of multilateralism is largely political: International cooperation infuses military action with legitimacy in the eyes of domestic and international audiences. In such cases, the United States, while preferring to act multilaterally, may be willing to forego foreign support rather than adapt its strategy to placate international dissent, especially if vital U.S. interests are at stake. While the United States often asserts its right and willingness to act unilaterally when threats demand it, lone coercive operations stand out as the exception far more than the rule.

An Intolerance for Casualties

The application of military force is also dictated by a sensitivity to casualties, particularly with regard to U.S. servicemen. Contrary to

---

[7]Some of the difficulties of conducting coercive operations through multinational coalitions are analyzed in Waxman (1997).

the predictions of those who saw Desert Storm as putting the Vietnam experience to rest, the relatively low American death total seems to have fed perceptions among policymakers and military planners that public expectations of a "bloodless" foreign policy had risen. The U.S. withdrawal from Somalia following the October 1993 deaths of 18 U.S. servicemen in Mogadishu evinces the strong pull that even low casualty levels can exert on U.S. policy. Even the extended deployment of U.S. ground forces to enforce the Dayton peace accords in the former Yugoslavia only confirms this tendency. Unlike those of most other NATO partners, U.S. troops patrol in convoys and have avoided actions likely to provoke hostile responses from local factions. Although several empirical studies have shown that U.S. casualty sensitivity depends heavily on a number of other variables,[8] this sensitivity affects policy decisions both prior to conflicts (e.g., whether to intervene) and during them (e.g., what forces to use and how to use them).

Aversion to Civilian Suffering

Sensitivity to casualties has often manifested itself as an aversion to inflicting suffering on those deemed innocents, and this consideration shapes the U.S. application of force. Despite the vast devastation that resulted from the U.S. bombing in Vietnam, for example, concerns that harm to civilians would damage the U.S. image at home and abroad severely constrained U.S. targeting choices (*Pentagon Papers*, Vol. 4, pp. 44–45; Hosmer, 1987, p. 59). More recently, coalition air campaign planning during the Gulf War and subsequent strikes on Iraq placed a premium on avoiding collateral damage in Baghdad and other population centers, especially as U.S. regional partners voiced concerns about U.S. policy causing possible civilian suffering (DoD, 1992, pp. 611–616). The increasing pervasiveness of the global media has magnified policymakers' fears of the domestic reaction to civilian tragedies resulting from the use of force.

As with sensitivity to U.S. casualties, this sensitivity varies with context and has not recently been tested in a high-stakes crisis. But pub-

---

[8]The variables most frequently mentioned are interests at stake, likelihood of victory, and elite consensus. For such conclusions and evidence drawn from public opinion data in a broad range of case studies, see Larson (1996).

lic statements by administration officials, the devotion of resources to develop "nonlethal" weapon technologies, and self-imposed restraints including restrictive rules of engagement during military operations all testify to an increasingly heightened sensitivity among decisionmakers to collateral damage and civilian suffering resulting from U.S. military operations.

A Preference for and a Belief in Technological Solutions

A long-standing tenet of the American way of war has been a reliance on materiel over manpower, high technology over low-technology mass (Weigley, 1977; Sapolsky and Shapiro, 1996). The U.S. military, and indeed U.S. society in general, tends to believe in the possibility of clean, technological solutions to even seemingly intractable social and political problems. What Eliot Cohen (1994) has dubbed "the mystique of U.S. air power" stems from the U.S. penchant for applying a high-technology force of latest-generation fighters, fantastically precise bombs, and intelligent cruise missiles to what has historically been a very grimy affair.

Not only do such high-technology instruments provide the necessary target discrimination to satisfy the public's demand for minimizing civilian suffering, but they also allow U.S. forces to bring effective firepower to bear without placing significant numbers of U.S. personnel in danger. Cruise missile attacks, which promise pinpoint accuracy without requiring any pilot, have increasingly become the option of first resort when coercive force is deemed necessary. The use of cruise missiles to attack suspected terrorist targets in Afghanistan and Sudan (August 1998) and their threatened use against Iraqi forces (November 1998) reflected this tendency, even at the expense of predictably degraded strategic effectiveness (Mann, 1998).

A Commitment to International Norms

Related to the aversion to foreign civilian casualties is a traditional commitment to international legal norms, particularly the principles of proportionality and combatant-noncombatant discrimination. Again, this partly stems from the demands of ensuring perceived legitimacy among audiences at home and abroad. Because most U.S.

military adventures are justified, at least rhetorically, in terms of the sanctity of international principles, policymakers generally desire to conform U.S. actions as much as possible to international legal obligations. These concerns often manifest themselves in carefully tailored and often highly restrictive rules of engagement. To be sure, Washington bends interpretations of legal obligations to serve its agenda when the stakes are high. But when possible, operations are tailored to comply with internationally recognized constraints. Opponents may argue that it is easy for the United States to follow rules that it wrote and that ultimately support its hegemony. Nonetheless, the recent controversies over land mines and the international criminal court show that the formation of international norms is not entirely in U.S. hands and that norms may bind the United States in ways it did not anticipate and does not enjoy.

Summary

The strengths of all five factors—coalition maintenance, casualty intolerance, aversion to civilian suffering, a preference for technological solutions, and a commitment to norms—vary from crisis to crisis as stakes and broader strategic priorities change. While U.S. policymakers in the post–Cold War period have always invoked an interest-based (as well as a morally based) argument for the U.S. use of coercion, the arguments have usually been hotly disputed among domestic audiences. In such an environment, the characteristics outlined above have regularly shaped the U.S. use of force. As this strategic environment is likely to continue for some time, the United States will often find itself unable to threaten or inflict pain anywhere nearly commensurate with its military power. Equally important, these characteristics will be ripe for an adversary to exploit, allowing it, in the extreme, to preempt or even force an end to a U.S. coercive campaign.

ADVERSARY COUNTERCOERCIVE STRATEGIES: A TAXONOMY

That the factors enumerated in the previous section govern U.S. coercive use of force is no secret. Adversaries have had ample opportunity to study patterns of U.S. behavior, and the lessons of previous crises are not lost on adversary leaderships. But what do they do with this information?

The most important use of this information is deciding whether to challenge the United States and its allies in the first place. A rational adversary measures the likely U.S. response before it acts. If the anticipated price is too high, it will not cause an affront at all. Saddam Hussein may have wanted to invade Kuwait when he built up troops along the border in 1994, but, in the face of rapid U.S. deployment to the region, he recognized that the U.S. response would be devastating, and he desisted.

Casualty sensitivity, concern over allied relations, and the other factors discussed earlier create the perception in adversaries' eyes that the United States lacks the will to carry out threats. Adversary leaders from Slobodan Milosevic to Saddam Hussein have evoked the image of Vietnam to suggest that they can outlast the United States. Similarly, resistance to U.S. naval forces attempting to transport troops to Haiti in late 1993 was widely attributed to the rapid pullout from Somalia following the deaths of U.S. servicemen earlier that year (Conversino, 1997).

In general, U.S. adversaries have tended to overlearn this lesson and have thus underestimated U.S. will when responding to coercive challenges. Both the Iraqi and Serbian regimes, for example, seemed to have believed that the U.S. threats with regards to Kuwait (1991) and Kosovo (1999) would not be carried out. In this assumption, they were mistaken, as they learned at great cost. At the same time, the inability of the United States to communicate its true resolve represents a failure for American coercion and costs the United States both blood and treasure. Adversary optimism is a result of the American way of coercion and of the open and contentious nature of the American political process. These features make it easy for adversaries to doubt American will and to hope that countercoercion strategies that play on self-imposed political constraints can deter or defeat American intervention.

In this way, the key tenets of American-style coercion give rise to a broad set of countermoves that adversaries routinely employ to neutralize U.S. coercive strategies. Six types of adversary countercoercion strategies stem from the American way of coercion, most of which have been seen in recent crises.[9] This organization is not meant to imply the existence of clear lines that distinguish among these strategies; many countercoercive strategies bridge these types, because they play simultaneously on several features of American-style coercion. When these countermoves are effective, coercion is likely to fail because the United States will not threaten to impose sufficient costs on an adversary or because adversary counterpressure will force the United States to end its campaign prematurely, or both. Note, however, that a countercoercive strategy need not "succeed" for U.S. coercion to nonetheless fail—if an adversary expects, correctly or incorrectly, U.S. will to erode in the face of a counterstrategy, the adversary will likely hold out.

Create Innocent Suffering

An increasingly common countercoercive strategy is to exploit U.S. sensitivity to civilian casualties and suffering. As noted earlier, this sensitivity often generates severe restraints on the U.S. use of force. Once force is employed, adversaries may capitalize on the inevitable tragic consequences befalling their civilian populations to erode support, among the American and international public, for continued or escalating coercive operations. Such strategies are a natural device for adversaries who cannot hope to compete militarily with U.S. strength—indeed, the potential effectiveness of these strategies is probably enhanced by the fact that the adversary is far weaker and can claim to be a victim.

During the Rolling Thunder campaign, the North Vietnamese government frequently used the damage U.S. air strikes caused as propaganda, often with great success in undercutting U.S. coercive pressure. North Vietnamese authorities repeatedly asserted that U.S. air attacks were directed at Red River Valley dikes, with devastating effects on the local civilian population, thus playing on U.S. decisionmakers' fears that such allegations would destroy public and foreign support. Concern over this issue caused the Johnson administration to emphasize that these dikes were off limits for campaign planners, and these targeting restrictions gave the North Vietnamese

---

[9]A range of other responses might be included, which we have not explored. They include "salami-slicing" tactics and false compliance. We thank Eric Larson for pointing out these additional tactics.

freedom to place antiaircraft and other military assets at the sites, countering the air campaign operationally (Parks, 1982).

U.S. concern for civilian casualties and a strong desire to uphold humanitarian norms, including the principle of combatant-noncombatant discrimination, provide tremendous incentives for adversaries to utilize "human shields" to deter U.S. strikes. During the Gulf War, the Iraqi government intentionally placed military assets in civilian populated areas in an effort to immunize them from attack. Somali leader Aideed deliberately mingled his forces with civilians, making it almost impossible for UN forces to target them. Although Aideed's methods to shield his factional forces differed from those of Saddam Hussein, both relied on the same premise—that civilian casualty sensitivity would deter escalation by UN forces and, failing that, would undermine support for continued military operations. At the very least, such tactics make U.S. planners more reluctant to conduct coercive strikes.

Civilian suffering–based strategies are effective not only because of the sensitivity among U.S. decisionmakers but also because several features common to adversary regimes facilitate such strategies. Some dictatorial regimes may not depend on popular support and have less to fear from allowing civilians to suffer than might a democracy. Saddam Hussein for many years rejected the UN "oil for food" deal, even though it would have reduced privations among the Iraqi people, because he saw popular suffering as a tool for ending Iraq's international isolation (Baram, 1998, pp. 65–66). Moreover, dictatorial regimes' tight control over media and well-oiled internal propaganda machines allow adversary leaderships to control broadcast accounts of incidents (though this is not to suggest that the United States and its allies may not also strive to "sanitize" information about conflicts). While manipulating the content of information flowing to its own population, these regimes can influence the timing and, indirectly, the substance of information disseminated abroad by selectively permitting journalistic inspection. Sudan, for years having virtually blacked itself out in the international media, welcomed television crews with open arms when the August 1998 cruise missile strike destroyed a pharmaceutical facility in Khartoum. Saddam Hussein similarly permits interviews with suffering Iraqi people, but he is careful to ensure that they place the blame entirely on the international community.

Adversaries' willingness and ability to exploit the extreme U.S. sensitivity to these sorts of accusations often has debilitating, reinforcing effects on coercive strategies. U.S. sensitivity *invites* adversary practices designed to put at risk the very civilians the United States seeks to protect. Typical U.S. responses to these practices then *reward* them, by placing further constraints on its own threats of force. For instance, the United States often responds to charges that it has attacked civilian targets, even if largely as a result of the adversary's own efforts to collocate civilian and military assets of persons, by restricting its own rules of engagement or placing additional limits on targeting. As noted earlier, North Vietnamese allegations of U.S. attacks on Red River Valley dikes caused the Johnson administration to denounce publicly any intention to conduct such strikes. This response ratified the immunity North Vietnamese officials desired for military assets placed near the dikes and vindicated the target commingling-plus-propaganda strategy. Similarly, after the North Vietnamese accused the United States of blatantly attacking civilian areas, causing massive suffering, during December 1966 air strikes against railway targets near Hanoi, Washington responded by prohibiting attacks on all targets within 10 nmi of Hanoi without specific presidential approval (Hosmer, 1987, p. 61; *Pentagon Papers*, Vol. 4, p. 135).

An identical pattern of encouraging, then rewarding, civilian suffering–based strategies occurred during the Gulf War. Iraqi leadership placed civilians in the Al Firdos bunker, which U.S. intelligence believed to be a command and control facility. On the night of February 13, 1991, U.S. F-117 strikes destroyed the bunker, killing dozens of civilians, a tragedy that the Iraqis attempted to exploit in the media. The incident ultimately turned out to have little impact on U.S. public support (Mueller, 1994, p. 79). However, the U.S. political leadership took no further chances: Attacks on Baghdad were suspended for several days. Thereafter, the George H. W. Bush administration (which had previously been committed to avoiding the micromanagement of target selection that had plagued the Johnson administration's efforts in Vietnam) required that Baghdad targets be cleared beforehand with the Chairman of the Joint Chiefs of Staff (Gordon and Trainor, 1995, pp. 326–327).

These incidents reveal that U.S. decisionmakers respond to adversary counterstrategies almost as predictably as they create the conditions for such strategies in the first place. Suffering of innocents, whether directly attributable to U.S. actions or not, will usually weaken U.S. resolve and allied cohesion. Even when domestic and international support for further U.S. coercive strikes does not erode, the United States imposes a range of operational restrictions on its activities for fear that further similar episodes will. Washington thus limits the coercive pressure it can bring to bear.

Shatter Alliances

Because the United States exhibits a strong desire to conduct coercive operations multilaterally rather than unilaterally, coalition unity itself becomes a vulnerable center of gravity that adversaries frequently attempt to exploit. The adversary capacity to shatter the U.S. alliance will not only sap U.S. political will but will also weaken U.S. operational capacity by, in many cases, denying the U.S. forward bases, transit rights, and allied combat capacity. Because each coalition member brings its own set of interests to the table, the need to maintain cohesion often places downward pressure on the level of practicable force available for coercive threats. The greatest impact of coalition dynamics on coercive operations therefore often goes unseen—it affects what the United States does *not* do militarily.

Once coercive operations are under way, adversaries often take a number of active steps to exploit the U.S. priority on coalition maintenance. Saddam Hussein attempted to widen coalition splits at several key junctures in the Gulf War and subsequent crises. Prior to the coalition ground assault, his attempted negotiations with the Soviet Union not only nearly averted the war but also caused some allies to question the need for military action. Iraq similarly tried to dislodge Arab support for coalition operations by linking resolution of the Kuwaiti crisis to the Arab-Israeli dispute, thereby driving a wedge between the Arab states and the U.S.-Israeli partnership.

Aside from such techniques as linkage and negotiations designed to fracture coalition support, adversaries often direct their military operations so as to undermine support among key coalition members. Even adversary military actions that, from a purely military standpoint, have little effect can have potentially enormous strategic consequences. Iraqi Scud attacks against Israel probably reflected Saddam's calculations that drawing Israel into the conflict would have destroyed coalition unity. While unsuccessful in that respect, they did nevertheless cause the United States to divert air sorties from more militarily effective efforts to hunting Scuds. Low-level violence by Aideed's faction following air assaults on his compounds caused a major crisis within the UN Operation in Somalia (UNOSOM) II; small-scale attacks on UN personnel fed intense Italian opposition to what Rome viewed as an escalating conflict, creating dissension within the UN coalition.

As with strategies designed to cause civilian suffering, the United States often reacts to attempts to shatter alliances in ways that create damaging, reinforcing effects. Coalition dissent often leads the United States to restrict the further use of force to repair diplomatic rifts. As a result, these restrictions reward the adversary's coalition-splitting efforts and encourage further such ploys.

In January 1993, the United States shot down several Iraqi aircraft and launched air strikes against military targets in response to Iraqi incursions and deployment of antiaircraft missiles in protected zones. The resulting widespread opposition to U.S. military action among coalition partners gave rise to speculation that Saddam had deliberately incited U.S. reprisals to win Arab support for the lifting of sanctions (Fineman, 1993).[10] Turkey, which provided key air bases supporting no-fly zone enforcement, worried that an extended conflict could contribute to its own crisis involving separatist Kurds. Russia, under pressure from nationalist hard-liners eager to reestablish economic ties with Iraq, criticized U.S. air strikes as inconsistent with international law and unauthorized by the UN Security Council. Arab states, fearing public backlash in response to U.S. military action against a regional power, urged Washington to call off further strikes (Brown, 1993; Robinson, 1993; Wright, 1993). Vocal criticism from Gulf War partners may have emboldened Saddam Hussein to test the coalition's resolve in the future and convinced him that provocation might be an effective strategy for breaking international efforts to isolate it. Meanwhile, U.S. planners were forced to acknowledge that coalition dissent necessitated limits on potential

---

[10]As one example of coalition resistance to U.S. strikes, a former Egyptian ambassador to the United States urged "a pause from the policy of military escalation against Iraq in order to stop the rapid erosion of favorable Arab public opinion which was the base of support for allied action against Saddam Hussein in the Gulf War" (El Reedy, 1993).

escalatory options. Even though, for the most part, the coalition held (relations have, however, continued to grow more and more strained), the U.S. response undermined its own coercive leverage by acknowledging its intention to trade potency of follow-on strikes for coalition support.[11]

Coalition-busting efforts have not as yet proven effective at inducing a total collapse of allied support. But they have proven effective in reducing the level or type of force a coalition employs or threatens. Subsequent responses to Iraqi provocations, until the 1998 Operation Desert Fox, were cautious with regard to the use of force. Even Operation Desert Fox demonstrated the effectiveness of this strategy. Because the anti-Iraq alliance had been reduced to only the United States and the United Kingdom, its operational effectiveness was increased, but its legitimacy and therefore its political effectiveness had decreased. The Iraqi regime apparently did not believe that the United States could sustain its attacks in the face of so much international opposition, and Desert Fox thus ultimately failed to force Iraq to accept arms monitors.

Create Counteralliances

Like shattering alliances, creating counteralliances will reduce the adversary's economic and military isolation and undermine the U.S. contention that the adversary in question is a rogue nation. Countries with great-power allies or even simply neutral outlets through which they can get supplies and weapons are much less vulnerable to U.S. economic and military pressure and can impose diplomatic costs on the coercing power.

Although Serbia ultimately failed to gain Russian support for its cause in Kosovo, the attempt demonstrates the potential of this strategy for resisting coercion. Had Serbia won strong Russian support, it would have gained a means of resistance and diplomatic escalation. The price to NATO of continued war in Kosovo would have meant alienating a major power whose goodwill most Western states eagerly sought. Initially, Russia pressed NATO to end the bombing as a prelude to a diplomatic settlement, and, as late as the

---

[11]For an analysis of attempts to coerce Iraq, see Byman, Pollack, and Waxman (1998).

end of May 1999, Russia publicly touted its opposition to NATO.[12] Although evidence is not available, Milosevic probably looked at Russia's rhetorical support and condemnation of the NATO campaign as an indication that Moscow would champion Belgrade's cause on the international arena. Russia's switch from Serbia's friend to NATO's de facto partner prevented Milosevic from imposing diplomatic costs on NATO, eroding Serbian leaders' hopes of a potentially favorable settlement to the conflict.[13]

There is little that airpower or any other military instrument can do directly to neutralize such diplomatic efforts. Russia's unwillingness (or inability) to help Belgrade was most likely a product of Moscow's own limits and Serbia's unattractiveness as a diplomatic ally. At the same time, diplomacy is in no way isolated from battlefield events. Without the constant battering of the air campaign, Russia's pressure on Belgrade probably would have accomplished little. On the other hand, the most significant diplomatic windfall for Serbia occurred when a U.S. warplane hit—very precisely—the Chinese embassy. These events only highlight that all U.S. coercive efforts are likely to take place in a diplomatic crucible, where seemingly small events can easily affect both U.S. and adversary alliances, and thus the ultimate success of the coercion attempt.

Create Actual or Prospective U.S. or Allied Casualties

Ho Chi Minh is often quoted as claiming that his Vietnamese forces would suffer ten times as many losses as those of the United States but that he would still win that exchange in the long run. The underlying assumption was that the United States lacked the motivation to sustain American casualties—particularly when the stakes were much higher for North Vietnam than for the United States. Saddam Hussein similarly postulated that the United States would be unable

---

[12]Russian envoy Vicktor Chernomyrdin even authored an opinion piece for *The Washington Post* that criticized the NATO campaign (Chernomyrdin, 1999).

[13]Sir Michael Jackson, NATO's commander in Kosovo, concluded that Russia's decision to back NATO's position on June 3 "was the single event that appeared to me to have the greatest significance in ending the war . . ." (quoted in Gilligan, 1999). General Clark also refers to Serbia's "isolation" as a major factor in Milosevic's ultimate decisionmaking (see "Interview: General Wesley Clark," 1999).

to stomach the losses he could inflict in a Gulf ground war. According to one account,

> Saddam strongly believed that the United States' Achilles' heel was its extreme sensitivity to casualties, and he was determined to exploit this weakness to the full. As he told the American Ambassador to Baghdad, April Glaspie, shortly before the invasion of Kuwait: "Yours is a society which cannot accept 10,000 dead in one battle." (Freedman and Karsh, 1993, p. 276.)

In retrospect, Saddam may have underestimated U.S. resolve—American planners prepared for upward of 10,000 casualties in the lead-up to Desert Storm—though the coalition rout made the point moot. But clearly, the U.S. concern for American casualties remains, especially in the eyes of adversaries, perhaps the *weakest point* in its otherwise overpowering military capability.

When the United States is protecting peripheral interests against a highly motivated adversary, direct attacks on U.S. personnel have historically eroded policymakers' willingness to continue the operation, especially when a clear and quick U.S. victory seemed unattainable. The 1983 Beirut Marine barracks bombing, resulting in 241 U.S. deaths, catalyzed the U.S. decision to euphemistically "redeploy" its forces to the sea. A decade later, the deaths of 18 U.S. soldiers in Somalia, this time covered gruesomely on television, solidified the administration's decision to pull out U.S. forces.

Like strategies that cause civilian suffering, measures that put American or allied personnel at risk often complement efforts to fracture coalitions by fueling resentment among specific members of the alliance. There is often a severely uneven distribution of risk in multilateral operations between coalition partners. The United States often pushed for air strikes against Bosnian Serb forces, for example, when it was the British and French, with forces on the ground, that would have borne the brunt of reprisals.

The U.S. penchant for utilizing high-technology, low-personnel-danger instruments often limits adversary options for attacking U.S. forces directly. In such cases, adversaries are likely to exploit the extreme concerns of the United States and its allies about actions that place their own noncombat personnel at risk. The Bosnian Serbs frequently used lightly armed UN peacekeepers or aid workers as hostages to stave off NATO air strikes during the Yugoslav crisis. The

Serbs' ability to countercoerce the Western powers became readily apparent in April 1993 when NATO began enforcing the no-fly zone. Although the Serbs issued no specific threats, the UN suspended aid flights the day before the first NATO air patrols for fear of reprisals (Tanner, 1993). On a number of occasions, the Serbs responded to NATO air strikes against military installations by detaining peace-keeping personnel on the ground. In all these cases, the Serbs threat-ened the weakest points of the allied effort—the vulnerability of humanitarian assistance and ground personnel—successfully to up the ante and deter immediate follow-up strikes. The use of hostages further indicates that many adversaries, while exploiting U.S. adher-ence to legal norms, are willing to violate them flagrantly. Even with-out matching the Western powers militarily, the Serbs were able to manipulate the allies' cost-benefit equation with relative ease.

Although hostage-taking has proven an effective deterrent to U.S. and allied strikes, adversaries may hesitate to execute, as opposed to injure or kill in combat, U.S. or allied personnel for fear of a backlash, thereby negating potentially effective countercoercive strategies. Adversaries considering this strategy thus face a dilemma: Threaten-ing U.S. or allied personnel may cause coalition planners to rethink military action, but carrying out such threats may prompt the reverse. Images of UN peacekeepers chained to Serb military vehicles in the face of NATO threats in summer 1995, while effective as a deterrent to strikes, also further demonized the Serbs in Western eyes. Any deaths that might have resulted would likely have gener-ated intense calls for aggressive NATO retribution, just as the deaths of U.S. servicemen in Somalia and Lebanon provoked demands for revenge. Indeed, in both cases the United States swiftly overcame its aversion to inflicting casualties on adversary civilian populations when its servicemen were threatened. The firefight that killed 18 U.S. soldiers in Mogadishu also killed perhaps 1,000 Somalis, and the United States responded to the attacks in Lebanon by shelling Shi'a villages.

American and NATO officials widely hailed the use of air strikes dur-ing the 1999 Kosovo crisis as a success, as the Serbian president accepted allied demands to pull back his military forces and negoti-ated an international contingent on the ground to monitor compli-ance. These results, however, obscure what the debate behind the potential use of force, particularly the reluctance of U.S. officials to advocate operations that would put American troops at risk, portends for future coercive strategies and how adversaries may counter them. As NATO threatened air strikes, the Clinton administration failed to reach consensus on what types of military forces the United States would commit, with Secretary of Defense William Cohen openly recommending that the United States not insert ground forces. This reluctance, at the same time as the United States appeared enthusiastic about air attacks, corroborates recent evidence from the Iraqi weapon inspection crises and the U.S. attacks on suspected terrorist sites that the United States will employ high-risk instruments only in the rare circumstances in which threats are important enough to justify more muscular means.

Play Up Nationalism at Home

Adversaries can take measures not only to blunt the effect of U.S. attacks but also to increase their own military capacities and their political positions at home. The American way of coercion offers adversary regimes several ways to consolidate their positions and to extract greater military resources from their societies. In particular, the U.S. desire to minimize casualties and to fight in large, multinational coalitions means that the United States usually seeks to deploy oversized forces to achieve rapid, overwhelming victories and to attack civilian infrastructure. These tendencies give credence to the beliefs, already latent in many countries, that their homeland is a victim of hegemonic bullying and that their regime is under attack by a powerful aggressor intent on violating national sovereignty and independence. Clever adversaries will capitalize on these sentiments to play up nationalism at home, generating increased will to sustain sacrifices among the populace and an increased capacity to extract military and financial resources from the society.

While strategic attacks aimed at undermining regime support can contribute to coercion, popular or elite unrest in response to coercion often does not occur or takes time to develop. A recurring historical lesson has been that attempts to force an adversary's hand by targeting its populace's will to resist often backfire.[14] Coercion

---

[14]For various works on the psychological impact of bombing, see Hosmer (1996), Clodfelter (1989), Janis (1951), Futrell (1983), and Schaffer (1985). See also U.S. Strategic Bombing Survey (1946).

commonly stiffens an adversary's determination as both the leadership and the country as a whole unite against the coercer. A coercive threat itself may raise the cost of compliance for an adversary's leadership by provoking a nationalist backlash. Russian attempts to bomb the Chechens into submission during the 1994–1996 fighting simply led to unified defiance; even residents who formerly favored peaceful solutions—or favored fighting each other—united to expel the invader. U.S. operations in Somalia, though humanitarian, faced a similar problem when the United States attacked and killed several leaders of Aideed's clan. Although many clan leaders had been critical of Aideed's confrontational stance toward the United States, they united behind him when faced with a direct outside threat.[15] Defiance against a coercive threat can even enhance a leader's stature despite his or her inability to produce military success. Egyptian President Nasser lost the Suez War in 1956, but his unbending stance toward Israel, France, and Britain made him more popular than ever (Neff, 1981, p. 393). In Kosovo, spontaneous pro-Milosevic rallies occurred in response to the initial bombing. Over time, support fell, but only after a sustained and lengthy campaign.

Striking the proper balance between provoking unrest and inflaming nationalism is difficult, and the historical record does not itself offer a clear set of predictive conditions. History does, however, point to the need to anticipate obvious countercoercive strategies designed to strengthen an adversary regime vis-à-vis its opponents. In general, when coercive operations threaten to foster instability, whether wittingly or unwittingly, target regimes are well prepared to respond. As the United States regularly finds itself pitted against authoritarian regimes, it will find itself confronting governments skilled at maintaining order. Although these regimes at times have little domestic support, their police and intelligence services can prevent instability from toppling the leadership.[16] If widespread domestic unrest appears likely, regimes will preserve power by increasing police pres-

---

[15]For an excellent account of the air campaign in Chechnya and the Chechen response, see Lambeth (1996), pp. 365–388. On Somalia, see Drysdale (1997).

[16]The resilient ability of police states to maintain stability in the face of wartime hardships was a key finding of the U.S. Strategic Bombing Survey of World War II air operations against Germany. See Clodfelter (1989), p. 9.

ence, using mass arrests, and even slaughtering potential opposition members.

Slobodan Milosevic, for example, constructed an extensive police state to resist both internal and external pressure (Woodward, 1995, p. 293). Like many authoritarian regimes, the Serb leadership had also developed an extensive capacity to control information. In the months leading up to the Kosovo crisis, Milosevic displayed a pattern of cracking down on independent media each time tensions flared with the international community (Bird, 1998; Perlez, 1998). During Operation Allied Force, Milosevic shut down independent newspapers and radio stations inside Serbia, used state-run television to stoke nationalist reactions, electronically jammed some U.S. and NATO broadcasts intended for the Serbian populace, and prohibited Western press from much of Kosovo (while granting it permission to film bombed sites).

While air strikes may be extremely precise in a technological sense, finely tuning their political effects on an adversary population remains largely beyond the capability of military planners and political leaders. In different political and cultural contexts, very similar air campaigns can have very different, indeed opposite, effects. While the loss of electrical power and other modern amenities made the Serbian population restive and raised the specter of internal dissent, a similar campaign in Iraq has strengthened the regime by increasing the dependence of the population on government support. Thus, even destroying a particular target may not lead to the change in behavior sought—the true object of coercion. Understanding this relationship between a target's destruction and the desired outcome is difficult and requires insights into culture, psychology, and organizational behavior. It may also require a capability and willingness to engage in a propaganda war in the adversary's homeland.

Threaten Use of WMD

One strategy with which the United States has little direct experience but that seems likely to appear in the future is the threat to use WMD against U.S. troops and civilian targets or against its allies. Although the United States has only faced this problem once in recent times— when it confronted an opponent armed with chemical and biological weapons during the 1990–1991 Gulf War—many regional adversaries are acquiring or may soon acquire a range of nuclear, chemical, and biological weapons.[17] WMD may give regional adversaries a means of countering the vast U.S. conventional superiority and offsetting U.S. regional influence. The relative invulnerability of U.S. troops and even the U.S. homeland would dissolve. WMD would thus allow an adversary facing conventional threats from the United States to threaten massive costs, completely changing the nature of military and diplomatic crises.

Regional states are attracted to WMD for a range of reasons. WMD can serve as an effective deterrent against regional adversaries, as well as an efficient augmentation of conventional forces. Perhaps more importantly, these weapons are also status symbols. They demonstrate to the world, and to a regime's power base, that the leader commands respect. Finally, regional powers may seek WMD to guarantee a regime's hold on power. In such cases, WMD become weapons of last resort: things to be used when the regime is most threatened or the most vital interests are at stake (Wilkening and Watman, 1995, pp. 35–38).[18] Woven into all these motivations is the idea that WMD can serve as a direct counter to U.S. conventional military superiority. As General K. Sandurji, a former chief of staff of the Indian Army, noted after the U.S. military victory over Iraq: "the lesson of Desert Storm is don't mess with the United States without nuclear weapons."[19]

The American way of coercion may have fed General Sandurji's thesis in several respects. Most obviously, large WMD arsenals have the potential to kill or injure considerable numbers of American and allied personnel or civilians, capitalizing on U.S. casualty sensitivity. A determined and ruthless enemy possessing such weapons would

---

[17]The United States, Russia, Britain, France, China, India, Pakistan, and Israel all have nuclear weapons; Iraq is known to possess an extensive biological program; and China, Egypt, Ethiopia, India, Israel, Iran, Libya, Burma, North Korea, Pakistan, Russia, Serbia, South Korea, Syria, and Taiwan all are believed to possess chemical weapon programs. Regional states that possess or are seeking additional nuclear, chemical, or biological systems include Algeria, Cuba, Egypt, Ethiopia, India, Indonesia, Iran, Iraq, Israel, Laos, Libya, Myanmar, North Korea, Pakistan, South Korea, Sudan, Syria, Taiwan, Thailand, Vietnam, and Yugoslavia. Sources are Monterey Institute (2001) and Chandler (1996), pp. 8–9.

[18]For a treatment of state motivations, see Sagan (1996–1997), pp. 54–86.

[19]As quoted in Chandler (1996), p. 149.

inspire fear within a populace and, to a lesser extent, among the troops, making political leaders hesitate to challenge such an adversary. Even when these weapons fail to deter the United States, their presence will reduce U.S. military effectiveness as troops are forced to take cumbersome preventative countermeasures. The possibility of delivery via terrorists or special forces means that no one is safe from this threat, even in the U.S. homeland. In regional contests in which U.S. interests are only tangentially implicated, even the remote possibility that New York or Los Angeles might be under threat, could be enough to tip the balance against U.S. intervention.[20]

Moreover, an adversary's capacity to inflict costs on allied forces and populations grows immensely with the possession of WMD. Even the low probability that an adversary would resort to using WMD risks dividing a U.S.-led coalition. Given current limits on ballistic missile ranges of likely WMD-armed foes (e.g., Iran, Iraq, Libya, and North Korea), such regional allies as Israel, the Gulf states, and Japan might fear that they will be attacked if the United States squares off against a WMD-armed foe. An Iranian threat to use biological weapons against Riyadh might, for instance, dissuade the Saudis from allowing the United States to launch an offensive from their soil. Indeed, an adversary can also escalate by dragging in neutrals or noncoalition members, thus raising the political costs to the coercing countries. If the risks to allies are significantly higher than the risks to the United States, U.S. brinkmanship during a crisis might create allied resentment and lead to a loss of allied support.

Despite all these advantages, however, WMD nevertheless represents something of a quandary for U.S. adversaries. From an adversary's perspective, threatening and especially using WMD against U.S. or allied targets may be a disastrous countercoercive strategy. While many U.S. policymakers are quite risk averse, a core reason the United States is vulnerable to countercoercion is that most regional crises do not implicate stakes that are vital to U.S. interests. By threatening WMD, however, an adversary will dramatically raise the stakes of any given crisis, perhaps thereby increasing U.S. resolve

_____

[20]Richard Betts (1995) makes the point that the threat to destroy parts of one or two U.S. cities is puny compared to the standard used during the Cold War but that this may outweigh what are typically modest U.S. interests in a regional conflict.

and willingness to cause suffering. The introduction of WMD into a crisis affects our usual assumptions about how the United States will view coercion and may propel the leadership toward taking actions it would otherwise have avoided. Even more dramatically, the actual use of WMD would significantly free U.S. decisionmakers from the constraints outlined above and, indeed, would create political pressures that would force them to respond decisively.

In short, WMD is a powerful, but extraordinarily dangerous, counter-coercion tool for regional adversaries. This dilemma suggests that the most effective strategy for an adversary is to have an implicit WMD threat but to avoid any explicit threats and especially any verifiable use of WMD. In this way, the adversary might hope to avoid any stiffening in U.S. resolve, while preserving the capacity to frighten the U.S. public and U.S. allies.

THE FUTURE OF U.S. COERCIVE AIRPOWER

Against a range of potential adversaries, the United States will continue to have the capacity to bring massive airpower to bear and to do so without the realistic threat of retaliation in kind.[21] U.S. political leaders can therefore be expected to continue to call frequently on airpower for coercive purposes. As the previous section illustrates, how ever, the potential magnitude of force brought to bear is just one small component of the coercion process. To take full advantage of its capabilities, the United States must recognize that its self-imposed limits often prevent the success of coercion and provide opportunities for adversaries.

As noted, there is little that can or even should be done to remove the self-imposed constraints outlined in this chapter. The U.S. government and the USAF, in particular, have only very limited influence over such deeply rooted features of the American polity. Some of these constraints result from short- or long-term policy choices, but others reflect deeply entrenched cultural norms and values that have long guided American foreign policy. While recent political and social developments, such as the end of the Cold War and real-time media coverage, have exacerbated these tendencies, they did not

_____

[21]The use of terrorism by adversaries remains an exception.

create them. U.S. leaders could more precisely spell out interests, operate more unilaterally, restrict media coverage, and otherwise try to minimize these limits, but eliminating them—or even reducing them substantially—will prove impossible. A more constructive approach is to recognize the viability of these countercoercive strategies, take them into account when deciding to use coercive threats, and design coercive campaigns to mitigate the threats of these counterstrategies.

This approach means recognizing that the United States will often find itself caught in increasingly problematic feedback cycles. For example, adversary responses that cause rifts in coalitions may prompt the United States to alter its approach to repair the rupture, in turn emboldening the adversary to direct further efforts at coalition-splitting. Likewise, adversary efforts to exploit collateral damage (both real and fabricated) resulting from U.S. attacks may prompt the United States to restrict its own future efforts, both undermining the potency of its follow-on threats and encouraging further exploitation of suffering.

These feedback cycles mean that success in coercive air campaigns depends not only on the ability to threaten and inflict costs but also on the ability to design "counter-countercoercion" strategies that can mitigate the destructive cyclical effects that the adversary's counterstrategies outlined above can create. Toward this end, both political and operational coercive strategies should aim at building stronger domestic and international support. Because of the dynamic nature of coercive contests, public support, adversary resolve, and levels of available force will vary as crises unfold because of the moves and countermoves of each side. As a result, successful propagation of coercive strategies will require building sufficiently high and robust domestic and international support so that sudden drops do not induce U.S. policymakers to respond with the very restrictions on coercive strikes that the adversaries seek. It also requires projecting an image of resolve to adversaries. Adversary optimism that its countercoercive strategy will succeed can cause U.S. coercion to fail, or at least raise substantially the costs of success.

Where possible, the USAF should consider developing capabilities that can help U.S. policymakers avoid or overcome constraints on operations. As such, the goal should be to preserve escalation dominance by increasing the range of actions available to the United States while limiting those of adversaries. Capabilities that will help the United States work better with coalition members, avoid casualties, continue operations in a WMD environment, and otherwise minimize adversary countercoercion can help sustain an operation and improve its overall credibility.

Yet relying on improved capabilities is an imperfect solution. Historically, the United States has responded to similar strategic dilemmas with technological solutions, both to minimize U.S. casualties and to rebut accusations that the United States did not care about adversary civilian suffering. One answer to North Vietnam's attempt to exploit collateral damage was to introduce more-advanced precision-guided munitions against targets likely to draw propaganda efforts. When striking terrorist camps in Afghanistan and Sudan in 1998, Washington used cruise missiles, in part because they posed no threat to U.S. personnel, even though a manned-flight bombing mission could have inflicted greater damage.

Technological advances, particularly in precision targeting and intelligence capabilities, are indeed immensely useful for improving the potency of military threats and narrowing an adversary's counteroptions. However, while military assets that can strike valuable targets may be extremely precise in a technological sense, finely tuning their political effects—at home, among allies, and in the adversary's own populace—remains largely beyond the capability of planners and political leaders. The assessment of this relationship between targeting and desired political effects requires more than simple intelligence or bomb damage assessment. Attacks on a country's water and electrical supplies can cause fears of internal dissent and therefore surrender or can stoke nationalist passions and therefore add to an adversary's capacity. Technological prowess cannot detract from this fundamental dilemma.

At home and among allies, viewing technological progress as the solution to the countercoercive strategies described above can magnify the very problems it promises to alleviate. Overemphasis on technological solutions can unreasonably raise expectations about the tragic but inevitable destructive impact of military force. Future missteps, even if unrelated to technological capacity, become harder to explain and thus become more-effective propaganda for the adversary. The reluctance of so many commentators both at home and among U.S. allies to believe that the attack on the Chinese embassy in Belgrade could have been an accident reflects this dilemma.[22] Like the current policy emphasis on air strikes or cruise missile attacks as the preferred coercive instrument, exclusive emphasis on technological solutions diverts attention from difficult but necessary choices.

Improved planning can often help where technology cannot. It is vital for the USAF and other services to have a nuanced understanding of individual adversary's strategic center of gravity. Strikes on an electrical grid, for example, may intimidate some adversaries but embolden others: No simple rules apply for the strategic effects of targeting. In addition, planners should prepare more-robust alternative uses of force in case limited force fails. Such plans make policymakers' threats to escalate more effective or, if that is not credible, allow them to avoid unproductive blustering.

In general, U.S. interests are often better served by refraining from military action than by utilizing it under unworkable restrictions. Policymakers and the public often view airpower as a low-risk, low-commitment measure, satisfying demands for military action while avoiding the need to put U.S. personnel in harm's way. This type of thinking may make sense politically, but demonstrative attacks, without a credible threat of escalation, will lead only to humiliating retreat or inadvertent escalation. From a strategic point of view, such an outcome both undermines coercive strategies in a given crisis and decreases U.S. credibility in the long term, thereby inviting future challenges.

REFERENCES

Baram, Amatzia, *Building Toward Crisis: Saddam Husayn's Strategy for Survival*, Washington, D.C.: The Washington Institute for Near East Policy, 1998.

Betts, Richard K., "What Will It Take to Deter the United States?" *Parameters*, Winter 1995, pp. 70–79.

---

[22]For an example, see Sweeney et al. (1999).

Bird, Chris, "Kosovo Crisis: Yugoslav Media Fear Crackdown Amid War Fever," *Guardian*, October 8, 1998, p. 15.

Brown, John Murray, "Russia, Turkey Voice Concern About New Attacks Against Iraq," *Washington Post*, January 19, 1993, p. A17.

Byman, Daniel L., and Matthew C. Waxman, *The Dynamics of Coercion: American Foreign Policy and the Limits of Military Might*, New York: Cambridge University Press, 2002.

Byman, Daniel, Kenneth Pollack, and Matthew Waxman, "Coercing Saddam: Lessons from the Past," *Survival*, Spring 1998.

Chandler, Robert W., *Tomorrow's War, Today's Decisions*, McLean, Va.: AMCODA Press, 1996.

Chernomyrdin, Vicktor, "Impossible to Talk Peace with Bombs Falling," *Washington Post*, May 27, 1999, p. A39.

Clodfelter, Mark, *The Limits of Air Power: The American Bombing of North Vietnam*, New York: The Free Press, 1989.

Cohen, Eliot A., "The Mystique of U.S. Air Power," *Foreign Affairs*, Vol. 73, January–February 1994, p. 109.

Conversino, Mark J., "Sawdust Superpower: Perceptions of U.S. Casualty Tolerance," *Strategic Review*, Vol. 15, No. 1, Winter 1997, pp. 19–20.

Drysdale, John, "Foreign Military Intervention in Somalia," in Walter Clarke and Jeffrey Herbst, eds., *Learning from Somalia*, Boulder, Colo.: Westview Press, 1997.

DoD—*See* U.S. Department of Defense.

El Reedy, Abdel Raouf, "Striking the Right Balance," *Guardian Weekly*, January 31, 1993, p. 6.

Fineman, Mark, "Hussein's Moves Seen as Steps in Calculated Plan," *Los Angeles Times*, January 17, 1993, p. A1.

Freedman, Lawrence, and Efraim Karsh, *The Gulf Conflict 1990–1991*, Princeton, N.J.: Princeton University Press, 1993.

Futrell, Robert Frank, *The United States Air Force in Korea: 1950–1953*, Washington, D.C.: Office of Air Force History, United States Air Force, 1983.

George, Alexander, and William Simons, eds., *The Limits of Coercive Diplomacy*, 2nd ed., Boulder, Colo.: Westview Press, 1994.

Gilligan, Andrew, "Russia, Not Bombs, Brought End to War in Kosovo Says Jackson," *London Sunday Telegraph*, August 1, 1999 (electronic version).

Gordon, Michael R., and Bernard E. Trainor, *The Generals' War: The Inside Story of the Conflict in the Gulf*, Boston: Little, Brown and Company, 1995.

Gravel, Mike [Maurice Robert], ed., *The Pentagon Papers: The Defense Department History of United States Decisionmaking on Vietnam*, Vol. 4, Boston: Beacon Press, 1971–1972.

Hosmer, Stephen T., *Constraints on U.S. Strategy in Third World Conflicts*, New York: Crane Russak & Co., 1987.

_____, *Psychological Effects of U.S. Air Operations in Four Wars 1941–1991: Lessons for U.S. Commanders*, Santa Monica, Calif.: RAND, MR-576-AF, 1996.

"Interview: General Wesley Clark," *Jane's Defence Weekly* (electronic version), July 7, 1999, p. 40.

Janis, Irving Lester, *Air War and Emotional Stress: Psychological Studies of Bombing and Civilian Defense*, Westport, Conn.: Greenwood Press, 1951.

Lambeth, Benjamin S., "Russia's Air War in Chechnya," *Studies in Conflict and Terrorism*, Vol. 19, No. 4, 1996, pp. 365–388.

Larson, Eric V., *Casualties and Consensus: The Historical Role of Casualties in Domestic Support for U.S. Military Operations*, Santa Monica, Calif.: RAND, MR-726-RC, 1996.

Mann, Paul, "Strategists Question U.S. Steadfastness," *Aviation Week & Space Technology*, August 31, 1998, p. 32.

Monterey Institute of International Studies, Center for Nonproliferation Studies, "Chemical and Biological Weapons Resource Page," Monterey, Calif., 2001. Online at http://cns.miis.edu/research/cbw/possess.htm (as of December 12, 2001).

Mueller, John, *Policy and Opinion in the Gulf War*, Chicago: University of Chicago Press, 1994.

Neff, Donald, *Warriors at Suez*, New York: Simon & Schuster, 1981.

Parks, W. Hays, "Rolling Thunder and the Law of War," *Air University Review*, Vol. 33, No. 2, January–February 1982, pp. 11–13.

Pentagon Papers—See Gravel.

Perlez, Jane, "Serbia Shuts 2 More Papers, Saying They Created Panic," *New York Times*, October 15, 1998, p. A6.

Robinson, Eugene, "Criticism from Gulf War Allies Strains U.S.-Led Coalition," *Washington Post*, January 20, 1993, p. A25.

Sagan, Scott D., "Why Do States Build Nuclear Weapons?" *International Security*, Vol. 21, No. 3, Winter 1996–1997, pp. 54–86.

Sapolsky, Harvey M., and Jeremy Shapiro, "Casualties, Technology, and America's Future Wars," *Parameters*, Summer 1996, pp. 119–120.

Schaffer, Ronald, *Wings of Judgment: American Bombing in World War II*, New York: Oxford University Press, 1985.

Schelling, Thomas, *Arms and Influence*, New Haven, Conn.: Yale University Press, 1966.

Sweeney, John, et al., "The Raid on Belgrade: Why America Bombed the Chinese Embassy," *The Observer*, November 28, 1999, p. 18.

Tanner, Marcus "Aid Flights Halt on Eve of No-Fly Patrol," *The Independent* (London), April 12, 1993, p. 8.

U.S. Department of Defense, *Conduct of the Persian Gulf War: Final Report to Congress*, 1992.

U.S. Strategic Bombing Survey, "The Effects of Strategic Bombing on German Morale," Washington, D.C., December, 1946, in David MacIsaac, ed., *The United States Strategic Bombing Survey*, Vol. IV, New York: Garland Publishing, 1976.

Waxman, Matthew C., "Coalitions and Limits on Coercive Diplomacy," *Strategic Review*, Vol. 25, No. 1, Winter 1997, pp. 38–47.

Weigley, Russell F., *The American Way of War*, Bloomington: Indiana University Press, 1977.

Whitney, Craig R., "Air Wars Won't Stay Risk-Free, General Says," *New York Times*, June 18, 1999, p. A22.

Wilkening, Dean, and Kenneth Watman, *Nuclear Deterrence in a Regional Context*, Santa Monica, Calif.: RAND, MR-500-A/AF, 1995.

Woodward, Susan L., *Balkan Tragedy*, Washington, D.C.: The Brookings Institution, 1995.

Wright, Robin, "Diplomacy: U.S. Officials Concede That Discord Within 29-Nation Alliance Served to Limit Actions Against Iraq," *Los Angeles Times*, January 19, 1993, p. A10.

Part II

WHERE DOES THE USAF NEED TO GO?

Chapter Four

MODERNIZING THE COMBAT FORCES: NEAR-TERM OPTIONS

Donald Stevens, John Gibson, and David Ochmanek

This chapter offers insights about decisions that will be considered in the early years of this decade regarding modernization of the combat platforms the U.S. Air Force fields. The focus is on systems that either are currently fielded (the B-1 bomber), are in advanced development (the F-22), or are close to decisions on whether they are to become major programs (the Joint Strike Fighter [JSF]). Other efforts are under way that might provide new combat platforms in the more distant future, including designs for uninhabited combat air vehicles, and craft that transit space en route to their targets. But it is not possible to assess these systems with the same level of fidelity as the more mature systems treated here because the operational concepts for these and other concepts have not yet been sufficiently developed, and their performance parameters have not yet been sufficiently defined. We do assess near-term platform modernization options in light of the air-delivered weapons and munitions that are planned for the coming two decades or could be fielded within that time frame. Recognizing that the Air Force, like its sister services, plays important roles in advancing U.S. interests in peacetime, as well as in war, the chapter examines the demands of both conflict and routine, peacetime "presence" operations.

MISSIONS

Any consideration of force size and mix must begin with an understanding of what such forces are likely to be called on to do. Specifically, planners approaching the task of evaluating options for future forces must have some insights into the following questions:

- For what missions will U.S. forces need to prepare in the future?
- Under what conditions might these missions be carried out?
- What operational strategies will commanders likely employ in accomplishing their missions? Specifically, for planners of air and space forces, what roles can and should the forces play in these strategies?

U.S. national security strategy is ambitious and calls for active U.S. engagement in regions where the nation has important interests at stake. Such a strategy depends on U.S. military forces being prepared to conduct a wide range of missions. These missions, in turn, encompass activities U.S. forces undertake in peacetime, crisis, and wartime. The following are the most important activities:

- **Project stability and influence abroad in peacetime.** By stationing and deploying military forces overseas, conducting training with allied and friendly forces, and providing security assistance, U.S. forces demonstrate, to friends and foes alike, their ability to defend important national interests. Such activities underwrite deterrence and enhance interoperability among friendly forces. They also give personnel opportunities to become familiar with the environments in which they might operate in times of crisis or war.

- **Deter and defeat large-scale aggression.** U.S. forces are called on to be able to defeat major aggression in two distant theaters in overlapping time frames. Included in this mission is the ability "to rapidly defeat initial enemy advances [invasions] short of enemy objectives in two theaters, in close succession" (Cohen, 1997, p. 13).

- **Protect U.S. interests through smaller-scale operations,** including providing humanitarian assistance, peacekeeping, and disaster relief; patrolling no-fly zones; reinforcing allies; and conducting limited strikes and interventions.

- **Deter and defeat the use of weapons of mass destruction (WMD)** against the U.S. homeland, against U.S. forces abroad, and against the territory and assets of allies.

- **Deter and defeat terrorist attacks** on U.S. citizens and facilities at home and abroad and on deployed U.S. forces.

Obviously, these missions are not mutually exclusive: U.S. forces are regularly called on to perform many of them at once. Indeed, individual contingencies, such as a major theater conflict, may include the need to defeat WMD, to defeat terrorism, and to provide humanitarian relief to the victims of aggression. Also noteworthy is what might be termed the joint and multidisciplinary nature of modern military missions. None of the missions listed above falls exclusively on any single service. And combat forces are not the only ones affected. On the contrary, systems that permit the rapid movement of materiel, that provide information about the enemy, and that allow U.S. commanders to act swiftly to direct forces and operations play important roles in all military missions.

There has been considerable debate in the U.S. defense community over the past ten years about the relative weight DoD should assign to each of these missions while considering the allocation of resources. On the one hand, projecting stability and conducting smaller-scale operations are clearly the missions that U.S. forces are called on most frequently to perform. On the other hand, preparing to fight major theater conflicts (along with the need to defeat WMD) should arguably be the centerpiece of force planning, even though such conflicts are rare, because, aside from using WMD, adversaries can threaten truly important U.S. national interests only through large-scale conventional attacks. To some extent, of course, this is a false dichotomy: The United States only rarely has to engage in major wars because our forces are well prepared to win them, which usually deters potential adversaries from trying to attack. In any case, many of the capabilities needed to prevail in theater warfare are also essential to the conduct of smaller-scale operations.

These caveats notwithstanding, the U.S. defense program would look substantially different along a number of dimensions if forces as a whole were to be optimized for a high tempo of peacetime engagement in key regions and for policing trouble spots, as opposed to large-scale power projection. Under such a construct, the modernization of forces might receive less emphasis, which could free funds to sustain a larger force structure and to ensure robust operations and maintenance (O&M) accounts. This debate will continue: The growing emphasis on the use of military power in the pursuit of objectives that are largely humanitarian has the look of a long-term trend and should not be seen simply as an artifact of the Clinton administration. To a growing degree, national interests in the United States and other Western countries are being defined to encompass humanitarian dimensions. As our conception of who constitutes our "neighbor" expands and as global communications become ever more encompassing and ubiquitous, the line between national interests, as classically defined, and purely humanitarian concerns will become more blurred. Hence, forces may be asked to deter or stop actions that cause undue suffering to civilian populations, even of populations far from the locus of economic or strategic importance and even if it is necessary to operate within the country perpetrating the atrocities.

All that said, however, large-scale power projection—the ability to deploy sizable numbers of forces over long distances and to sustain them in high-tempo operations against a capable foe—is and should remain the cynosure of U.S. defense planning. Only the United States has the capability to project large-scale military power today, and it is this capability that sustains power balances in key regions of the world. As such, it is also the essential ingredient in the viability of U.S. strategic alliances—alliances that form the heart of this nation's national security strategy. Whatever else U.S. forces may be called on to do, force planners dare not compromise the ability of the forces to dominate combat operations in the "backyards" of potential adversaries around the world. That is a demanding task in light of changing threats and political challenges.

CONDITIONS AND CONSTRAINTS

Equally important as an enumeration of missions is an understanding of the conditions under which the missions are likely to be carried out and the constraints that may be placed on forces during operations. For conflicts involving all but the most important of national interests, U.S. military operations will certainly be constrained by domestic demands to minimize friendly casualties, to avoid inflicting suffering on innocent civilians, and to act in concert with allies.[1]

---

[1] These constraints are covered in more detail in Chapter Three.

However, these constraints do not derive from purely domestic considerations. U.S. threats to employ military power—be they implicit or explicit—are only credible to the extent that potential adversaries believe they will be carried out. Adversaries understand the constraints on U.S. military actions and are more likely to view U.S military threats as credible if the United States fields forces that can achieve national objectives despite any constraints. For defense planners, these considerations mean that they must continue to offer the nation's leaders military options that can be exercised with confidence that the risk of friendly and civilian casualties can be managed and can be held to a level commensurate with the interests the nation has at stake.

Planners should also anticipate that future U.S. military operations will most often be coalition affairs, rather than unilateral campaigns. Coalition operations mean that our forces can have some confidence of having constant access to airspace, ports, air bases, and other assets near the region of conflict. By the end of Operation Allied Force, for example, U.S. aircraft were able to operate from bases in more than eight countries, effectively surrounding Serbia.[2] Regrettably, however, coalition operations are often less than the sum of their parts. Bringing people and assets from multiple nations together unavoidably adds friction and inefficiencies to the planning and execution of an operation. And sharing decisionmaking authority with leaders from multiple nations can necessitate compromises in strategy that reduce operational effectiveness. These difficulties notwithstanding, the strategic and political benefits of coalitions will remain substantial enough that political leaders will almost always prefer to have partners when they go to war. That preference implies a need for U.S. forces and operational concepts to incorporate features that enhance interoperability across national lines.

Other conditions are especially pertinent to large-scale power-projection operations. If an enemy is going to challenge U.S. interests through overt aggression (such as Iraq's invasion of Kuwait in 1990, or another potential North Korean invasion of South Korea), it must be assumed that the attack will be undertaken so as to maximize the attacker's inherent advantages. This means, among other things, that

---

[2]For more on the importance of base access to U.S. operations, see Chapter Nine.

U.S. defense planners must expect to be surprised. Our opponents are not eager for a fight with the U.S. military. They would prefer to achieve their objectives without having to resort to force or, failing that, in a coup de main that succeeds before large-scale U.S. forces can be brought to the theater. Advanced surveillance systems, including satellites and airborne platforms, make it harder for enemy forces to prepare for an attack without being noticed, but these systems do not, by themselves, guarantee that U.S. forces will be deployed promptly. Some adversaries, such as the North Koreans, can routinely posture their forces such that little overt preparation is needed prior to an attack. And information about a possible attack is a necessary but not sufficient condition for reinforcement. Decisions to act must be made in Washington and in other capitals before forces can move. All this takes time.

So U.S. forces must be postured to respond rapidly to short-warning aggression. They do this in two ways: first, by having some of the most critical components of a defensive force (the forces themselves, as well as their munitions and other supplies) stationed or routinely deployed abroad close to potential regions of conflict and, second, by being able to deploy rapidly over long distances.

Related to this is the need for what might be called *high leverage* early in a conflict. Normally, U.S. forces arriving in a theater in the opening days of a major conflict will be badly outnumbered. Yet, if they are to prevent the enemy from achieving his objectives, they must be able to wrest the initiative away from the enemy and defeat his attack quickly. This means that the early arriving forces must have great qualitative superiority over the opposition if they are to succeed in their mission.[3]

Finally, planners must consider prospective changes in the capabilities of forces fielded by potential adversaries. The threat they pose does not stand still. A composite portrait of the military forces of a

---

[3]The relationship between the quality and quantity of forces is well known to defense analysts. Lanchester's famous equations show that the capability of a force can be expressed by B^2b, where B is the number of weapons or units available and b is an expression of their quality relative to an opponent's forces. Because the variable for quantity is squared, a force that is outnumbered two-to-one must have four times the quality of its opponent to be equal in capability. A force outnumbered four-to-one must have 16 times the quality. This is the basis for the observation that "quantity has a quality all its own."

representative "high end" regional adversary for 2005 to 2010 might look something like the following (DoD, 1993):

- a military centered on an army of perhaps 20 to 40 divisions, fielding between 5,000 and 10,000 armored vehicles, with upwards of 500,000 troops

- a modest-sized air force of perhaps 500 combat aircraft, a growing portion of which will be of fourth-generation design (e.g., Mig-29, Su-27, or Su-35); in addition, regional adversaries will field high-performance surface-to-air missile (SAM) systems, such as the SA-10 and SA-12, in substantial numbers

- a small navy, whose missions are focused on sea denial via mines, submarines, patrol craft, and antiship missiles delivered from surface ships and land-based launchers and aircraft

- between 100 and 1,000 surface-to-surface ballistic and cruise missiles[4]

- WMD, principally chemical and biological agents but also, in some cases, nuclear.

"Overlaying" these enemy capabilities onto the operational situation described above—a short-warning offensive that is launched far from the United States—provides an appreciation of the need to regain the initiative quickly and of the requirement to prevail while suffering, at most, modest numbers of casualties. When seen in this light, it becomes clearer why the United States today spends more on military forces, by an order of magnitude, than any other nation: U.S. forces are called on to perform unique, demanding, and important missions.

ROLES OF AIR AND SPACE FORCES

The missions and challenges outlined above apply to all elements of the U.S. military establishment. What, specifically, do they imply for

---

[4]Most of these will be Scud-type missiles and modifications and will have ranges from 300 to 900 km, although some will have ranges upwards of 2,500 km. These weapons will be more accurate than the missiles Iraq hurled at the coalition and Israel in 1991. This improved accuracy and the deployment of specialized conventional payloads (e.g., cluster bombs) will make these missiles effective against targets parked in the open.

air and space forces? Again, the answer turns on an understanding of what the forces are likely to be asked to do within a joint campaign.

Since the earliest days of military aviation, commanders have relied on airpower to play two crucial roles: to provide information about the location and disposition of enemy forces and to gain dominance in the air so as to deny the enemy the advantages conferred by effective air operations. The two roles have not fundamentally changed, except that the scope and timeliness of information that can be gathered by airborne and spaceborne platforms have grown dramatically over the past two or three decades. Today, with the sensors carried onboard such aircraft as the Joint Strategic Tracking and Radar System, it is all but impossible for a large (division-sized or larger), mechanized force to move undetected, assuming that U.S. reconnaissance assets are deployed to the region and that they enjoy freedom to operate. As sensors and platforms improve and proliferate, U.S. forces will be able to detect and identify smaller formations of surface forces, even in foliated or urbanized terrain.

Military air forces have also long been called on to attack enemy assets on the surface, including land and naval forces, in support of joint campaigns. In addition, airpower has been used to attack other elements of national power, such as military industries, lines of communication, national infrastructures, and the means of political control, both to reduce the enemy's ability to conduct military operations and in an attempt to coerce the enemy leadership into surrendering. In these areas, new capabilities have led to improvements that portend, in some circumstances, a revolution in the conduct of military operations. U.S. air forces demonstrated in Operation Desert Storm that, in favorable terrain, they could dominate operations not only in the air but also on the surface. While airpower was not, itself, able to compel a withdrawal of Iraqi ground forces from Kuwait, over 38 days of nearly incessant air operations shattered the fighting abilities of a large, combat-tested mechanized army.[5] Eight years later, in Operation Allied Force, airpower was

---

[5]For example, an estimated 40 percent of the Iraqi soldiers in the Kuwait Theater of Operations deserted prior to the coalition's ground attack in late February 1991. Many of those who remained offered only token resistance once the ground invasion began, as evinced by the surrender of more than 85,000 Iraqi officers and enlisted men during the 100-hour ground operation. And less than 20 percent of the Iraqis' tanks and 10

unable to impede the operations of the Serbian forces in Kosovo, but it was able to coerce the Serbian leader to accede to NATO's demands that his forces evacuate Kosovo and allow a NATO-led force to secure the province.

This transformation in the capabilities of modern airpower springs from several related developments:

- As noted previously, it is now possible to detect, locate, and identify large formations of enemy ground forces and some other types of targets quickly and over large areas.

- Improvements in the accuracy and lethality of conventional munitions have led to order-of-magnitude increases in the effectiveness of air-delivered weapons. Once, it required multiple sorties to be confident of destroying a single enemy installation or armored vehicle by air; now, it is possible, given the right munitions, to destroy multiple targets with a single sortie.

- New sensors onboard attack aircraft, combined with miniaturized guidance sets for munitions, permit effective attacks by aircraft at night and, increasingly, in poor weather, denying surface forces what for decades was a sanctuary from air attack.

- The combination of jamming, deception, standoff weapons, and limited numbers of stealthy aircraft has allowed U.S. air forces to limit the effectiveness of the enemy's surface-based air defenses, allowing the rest of the "air package" to do its work relatively unmolested, as long as low altitudes and certain highly defended areas can be avoided.

The net result of these developments has been a gradual but steady shift in the "division of labor" among forces for joint operations. These new capabilities, in conjunction with the rapid worldwide deployability air forces afford, have meant that airpower has become the instrument chosen in many situations to be the "leading edge" force at the commencement of an operation. For example, with only minimal ground forces in place on a day-to-day basis, the U.S. defense posture in Southwest Asia recognizes implicitly that air-

---

percent of their armored personnel carriers showed evidence of having attempted to resist during the ground attack. See Hosmer (1996), pp. 152–170.

power will be responsible for destroying the vast majority of armor in an Iraqi offensive should Baghdad again invade Kuwait or Saudi Arabia. And because of the risks of casualties inherent in a high-intensity land battle, airpower was called on in Operation Allied Force to compel Serbia to stop terrorizing Kosovar Albanians, despite its known limitations against small units in close terrain.

Airpower, in short, is playing greater roles than ever before in U.S. military operations. That said, we should not generally expect airpower to "win wars" on its own. Other things being equal, it will always be preferable to confront an enemy with a multidimensional problem. But, for a host of strategic, operational, and technical reasons, the air forces of all four services are providing qualitatively new capabilities at a time when the capabilities are highly applicable to a wide range of needs. The challenge USAF planners face is to ensure that air forces retain and, to the extent possible, expand the margins of superiority that they enjoy today as adversaries field an array of new capabilities aimed at defeating them. The remainder of this chapter explores in greater detail the key factors determining the capabilities the U.S. Air Force needs to support U.S. defense strategy and options for providing the capabilities.

MODERNIZATION—KEY CONSIDERATIONS

An Aging Fleet

With the exception of the Air Force's two "stealth" aircraft—the F-117 and the B-2—all the combat aircraft in its inventory today were designed in the late 1960s or early 1970s. At the beginning of 2000, the average USAF fighter aircraft was 13 years old. Thanks to the slow place of modernization, the fighter force, comprising nearly 1,600 aircraft, will have an average age of 20 years by 2010. Historically, this is the age at which the Air Force has retired fighter aircraft. In short, the Air Force is faced with a daunting task as it seeks to modernize its force of combat aircraft.

The F-16C, which constitutes more than one-half of the Air Force's fighter force structure, will begin to reach the end of its programmed service life in the next 10 to 15 years. The current plan is for the JSF to perform most missions the F-16 currently performs. The JSF is planned to be a family of aircraft consisting of three versions:

- conventional takeoff and landing

- short takeoff and vertical landing

- operating from the U.S. Navy's aircraft carriers.

The Air Force has committed to buying the first version and is also considering the second. The JSF is due to begin entering the force in 2010.

The F-22 Raptor will be taking over the F-15C's current air superiority mission. One result of the 1997 Quadrennial Defense Review was that DoD agreed that the Air Force should procure three wings of F-22s optimized for air superiority and that it should examine the value of procuring an additional two wings of a variant of the F-22 that could assume some of the interdiction missions the F-15E and the F-117 currently perform. The air superiority variant of the F-22 is scheduled to begin entering the force in 2004.

The heavy bomber force can carry large weapon loads over very long distances. Unfortunately, all three types need upgrades and, in the cases of the B-1 and B-2, extensive modernization to reach their full potential for conventional military operations.

For any of the fighters and bombers to be effective, the Air Force must buy adequate quantities of advanced munitions. Over the next ten years, the Air Force plans to spend $5.4 billion on 2,400 medium-range standoff missiles (Joint Air-to-Surface Standoff Missiles [JASSMs]), 6,000 shorter-range dispensing weapons (Joint Standoff Weapons [JSOWs]), 5,000 advanced antiarmor weapons (sensor-fused weapons [SFWs]), and 62,000 Global Positioning System (GPS)–guided bombs (Joint Direct Attack Munitions [JDAMs]). Several recent studies suggest that these amounts may be insufficient to support operations that can quickly and decisively achieve their objectives in future large-scale conflicts.

This chapter offers insights about how to invest the nation's resources in these assets—new fighter aircraft, upgrades to the heavy bomber fleet, and advanced air-to-ground munitions—so as to maximize military capabilities in a variety of circumstances. Decisions made regarding these programs will affect the Air Force's ability to support U.S. national security over the next three decades and beyond.

Analytical Approach

Setting a military modernization program involves integrating the demands of strategy, the challenges adversaries pose, and the opportunities emerging technologies present. One way to capture the most important aspects of each of these dimensions is to analyze the dynamics and outcomes of hypothetical conflicts (scenarios) that represent the kinds of challenges one's future forces might one day confront. Once defined, the scenarios can serve as yardsticks against which to measure the capabilities of alternative forces. In the context of an analysis of friendly and enemy campaigns, scenarios can also help provide insights into areas of weakness in the capabilities and strategies of both sides.

To evaluate and compare the capabilities of the primary force mix options available to the U.S. Air Force, we used four basic scenarios depicting large-scale theater conflicts and three scenarios of distinct types of smaller-scale conflicts (SSCs) that can generate demands for the ongoing deployment of USAF forces. Each of these is set in the year 2020.

Figure 4.1 summarizes the four scenarios of theater-level conflict, each of which depicts U.S. efforts to defeat a large-scale, combined-arms invasion (with varying degrees of assistance from allies). The four aggressors are North Korea, Iran, China, and Russia. We chose these scenarios not because they are likely, in the common definition of the term. (Major war is almost never likely in that sense.) Rather, these scenarios depict conflicts that could plausibly arise from an enduring clash of interests involving the United States and each of these states, and the operational challenges these scenarios pose are broadly representative of those U.S. forces might confront over the next two or three decades. In each scenario, the most important factors determining the success or failure of a particular mix of air forces were the size and composition of the enemy's air defense forces (interceptors and SAMs). Accordingly, we have arranged the four scenarios along two axes, showing the approximate number of modern interceptors and SAMs fielded by the enemy force in each.

FORCE MIX ALTERNATIVES

Having defined the scenarios, we then assessed, quantitatively and qualitatively, the ability of different mixes of combat aircraft to achieve U.S. and allied military objectives. In defining alternative mixes of USAF aircraft, we assumed that both the F-22 and JSF will reach fruition, but we examined some variations in the number of fighters of each type purchased and, for the F-22, the operational utility of two different variants of the aircraft, both of which enhance its air-to-ground attack capabilities.

The Air Force today fields a fighter force of approximately 20 fighter wing equivalents.[6] The 1993 Bottom-Up Review and the 1997 Quadrennial Defense Review deemed this force necessary in light of what was then seen as the most stressing requirement on U.S. general purpose forces: the need to fight and win two nearly simultaneous major regional conflicts. Experience in the intervening years has

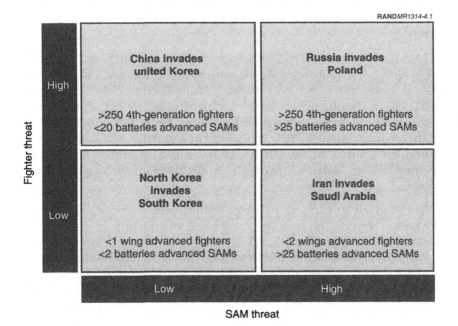

Figure 4.1—Theater Conflict Scenarios Examined

---

[6]Each fighter wing consists of 72 primary authorized aircraft along with associated personnel and support equipment.

shown that routine demands for USAF forces stationed and deployed abroad necessitate a force of at least this size to avoid exceeding a sustainable operational tempo (see Thaler and Norton, 1998). For these reasons, our baseline force structure for fighters assumes a continuing need for a fighter force of 20 fighter wing equivalents.

The baseline heavy bomber force consists of 16 B-2s, 70 B-1Bs, and 44 B-52Hs. Plans call for this force to be held constant over the period studied, with important upgrades to the avionics and communications programmed for each type. Since the Air Force has no plans to develop a new heavy bomber prior to 2030 and since no bomber production lines are open, the only options we examine regarding the bomber forces other than the baseline option are (1) to enhance its capabilities by buying more standoff weapons than planned for the B-1B and (2) to reduce the planned bomber force to free resources with which to modernize the fighter force at a faster rate.

This study refers to variations from the baseline force as "force mix options." These options are structured such that each would incur equal costs (procurement plus O&M) from 2000 to 2020. We selected 2020 as the far time horizon for the investment options developed in this study because periods beyond that are laden with too much uncertainty regarding the availability of resources, the rates of inflation, the nature of the threat, and so forth to permit meaningful comparisons. Table 4.1 provides an overview of the options examined.

Our analysis of the mix between fighters and bombers focused on whether the B-1 should be retired and the savings in O&M funds used to buy more F-22s or JSFs each year.[7] We examined the relationship between the effectiveness of the fighter and bomber forces and the weapons they deliver. This led us to explore cases in which the inventory of standoff weapons increased. This increase was paid for by slowing the rate of modernization of the fighter force. We

---

[7]We did not examine the option of retiring the B-52 because of its important role in the nation's nuclear deterrent force (in keeping with the START treaty, B-1Bs have been rendered incapable of delivering nuclear weapons) and because the B-52 is uniquely capable of delivering the conventional air-launched cruise missile, Harpoon antiship missiles, and some other conventional weapons.

Table 4.1

Variations in Fighter and Bomber Force Mixes

Fighter-Bomber Mix	
Retire B-1	Buy JSFs or F-22s at a faster rate
Keep B-1, buy additional standoff weapons	Reduce rate of fighter procurement
Fighter Mix	
Buy 2 additional wings of F-22Es or F-22Xs	Reduce size and rate of JSF buy

assumed that all the bombers retained in the force will be upgraded with the system improvements necessary to allow each bomber to receive and employ information about targets and to target the weapons they carry.

The primary issue facing decisionmakers regarding the mix among fighter aircraft concerns the value of developing and fielding a variant of the F-22 optimized for ground attack. The 1997 Quadrennial Defense Review recommended that the Air Force consider fielding two wings of a ground attack variant of the F-22A to replace the F-117 and F-15E (Cohen, 1997, p. 45). Two possible versions of such a variant have been defined: the F-22X and the more capable F-22E. We considered the addition of two wings of either of these variants, using a reduction in the JSF force to pay for them. The addition of two wings of air-to-ground F-22s would result in a change in the Air Force's fighter force structure of 2028 to five F-22 wings (as opposed to three) and 15 JSF wings (as opposed to 17). Of course, DoD decisionmakers might also consider forgoing production of the F-22 altogether. We examined this option but, as will be seen, found it impossible for U.S. forces to achieve their objectives in key scenarios without the F-22. For this reason and because of the manifest need to begin replacing older models of the F-15C, we did not examine options to cancel the F-22A.

APPROACH

Table 4.2 provides a full listing of the force structure options defined by each of the choices outlined above. The analysis that follows compares the effectiveness of each of these options with that of the baseline force in the year 2020.

Table 4.2
2020 Force Mix Composition

	Baseline	Retire B-1s, Buy JSF	Retire B-1s, Buy F-22X	Retire B-1s, Buy F-22E	Reduce JSF, Buy 2 FW F-22X	Reduce JSF, Buy 2 FW F-22E
F-22A	216	216	216	216	216	216
F-15C	30	30	30	30	30	30
F-15E	0	0	0	0	0	0
F-117	0	0	0	0	0	0
F-22X	0	0	139	0	144	0
F-22E	0	0	0	132	0	144
A-10	141	141	141	141	141	141
F-16 Block-30	171	0	32	39	260	278
F-16 Block-40	141	0	141	141	252	252
F-16 Block-50	162	54	162	162	174	174
JSF	591	1,014	591	591	238	220
B-1B	70	0	0	0	70	70
B-52H	44	44	44	44	44	44
B-2A	16	16	16	16	16	16

Comparisons of the force mix options were based on campaign analysis using the START model, which has been used extensively since the mid-1990s in analyses of theater-level conflicts.[8] Inputs include data and assumptions regarding the number and location of allied and enemy air and ground forces in theater, the operational strategies employed by both sides, and the rate at which U.S. forces enter the theater. START is capable of reporting on a number of measures of effectiveness (MOEs), but we found a great deal of correlation among them and so selected the movement of the enemy forward line of troops (FLOT) as our primary MOE. This campaign analysis is not intended to be predictive. That is, we do not attempt to assess the overall capabilities of the United States or its possible opponents 15 or more years into the future. Rather, we use the scenarios and the campaign analysis as yardsticks for comparing the relative effectiveness of different USAF force structures in the face of a range of possible challenges.

Our analysis of peacetime engagement operations focuses on determining the force structure required to support a plausible set of ongoing operations and commitments within the currently acceptable level of annual temporary duty (TDY) rates.

Table 4.3 shows our assumptions about key characteristics of the three variants of the F-22 examined in the study. The F-22A, now under development, is optimized for air-to-air combat. However, it can carry two 1,000-pound JDAM bombs (GBU-32) or eight small-diameter bombs (SDBs).[9] These weapons allow the F-22A to attack targets on the ground effectively if their location is provided to the platform from offboard sources. This dependence springs from the fact that the F-22A has relatively poor air-to-ground avionics and cannot itself find moving targets on the ground, though improvements are possible through upgrades to the software associated with the F-22's radar and other enhancements.

---

[8]Simplified Tool for Assessment of Regional Threats (START) is a theater-level campaign model developed by RAND.

[9]The SDB program aims to develop air-delivered bombs that, through greater accuracy and explosive yield, equal the destructive capacity of weapons two or more times larger. If the program is successful, an SDB weighing 250 pounds might be as effective against many types of targets as a currently available 500-pound bomb.

Table 4.3

Key Characteristics of Three F-22 Derivatives

	F-22A	F-22X	F-22E
Flyaway cost (1998 M$)	79.4	87.3	95
Annual O&S cost per wing (1998 M$)	228	242	242
Program	Data from 12/97 SAR	Proposal	Proposal
Missions	Primarily air-to-air	Multimission	Multimission
Performance	Can attack prelocated fixed targets	Same size weapons bay as F-22A	40-inch plug added to aircraft (4,000–8,000 lb increase in empty weight)
	Very low observable	Weapons bay altered to carry WCMD	10–30% increase in range
	30% F-15 deployment footprint	Same observables and deployment footprint as F-22A	Same observables and deployment footprint as F-22A
		Carries external stores	Carries external stores
Avionics	Poor air-to-ground avionics	Added SAR, MTI radar modes	Improved radar IFTS
Weapon carriage	Two 1,000-lb JDAMs or Eight SDBs internal	Two 1,000-lb JDAMs or Two WCMDs or Eight SDBs internal	Four 2,000-lb JDAMs or Four WCMDs or 16 SDBs internal

The F-22X has been proposed as an air-to-ground derivative of the F-22A with a relatively low incremental cost increase over the F-22A (a little over 10 percent). The primary difference between the F-22X and F-22A is the addition of synthetic aperture radar (SAR) and moving target indicator (MTI) radar modes, which allow the F-22X to find targets on the ground, including moving vehicles. The F-22X weapon bay is the same size as that of the F-22A, so the overall payload is the same. However, we assume that some minor modifications will be made to the F-22X weapon bay so it will be able to carry two CBU-103/5 wind-corrected munitions dispensers (WCMDs) delivering either SFW or combined effects munitions (CEMs). These weapons are effective against groups of vehicles. The cost assumptions for the F-22 derivatives are rough estimates; therefore, the analysis includes an examination of the possible effects of changes in the cost of the F-22 on the preferred mix of fighters. We assume that the F-22X would be built concurrently with the F-22A. If the F-22X were built after the completion of the F-22A buy, it could be cheaper than the F-22A, because of learning-curve effects and depending on the degree of commonality between the two variants.

The F-22E has been proposed as a more capable derivative of the F-22A and would be about 20-percent more expensive than the basic version. The F-22E has a 40-inch fuselage plug, which adds 4,000 to 8,000 pounds of empty weight to the aircraft. This additional size allows the F-22E to carry double the payload of the F-22A—either four 2,000 JDAMs (GBU-31), four WCMDs, or 16 SDBs internally. The fuselage plug also increases the F-22's range by 10 to 30 percent. In addition to the SAR and MTI radar modes, it has an internal forward-looking infrared (FLIR) and targeting system. Again, the cost numbers are rough estimates, and we will examine sensitivities to the cost estimates later.

Table 4.4 presents our assumptions regarding the JSF and its closest competitor, the F-16C. We based our JSF assumptions on the third update of the Joint Interim Requirements Document (JIRD-III).[10] We assumed that the JSF will have 30-percent more range than a simi-

---

[10]Published in 1998, JIRD-III defines and articulates at a fine level of detail the warfighting needs of the JSF.

Table 4.4

JSF and F-16 Block-60 Characteristics

	JSF	F-16 Block-60
Flyaway cost (1998 $M)	30.4	32.0
Annual O&S cost per wing (1998 $M)	138.0	198.0
Program data	From December 1997 SAR	Based on UAE sale
Missions	Multimission	Multimission
Performance	Based on JIRD-III	Two conformal tanks
	30% more range than F-16 Block-50	30% more range than F-16C Block-50
	Low observability configuration	Same observables and deployment footprint as F-16 Block-50
	50% of F-16 deployment footprint	
Avionics	Internal FLIR tracking system	Internal FLIR tracking system
	Active ESA radar with SAR and MTI modes	Active ESA radar with SAR and MTI modes
	Advanced countermeasures	Advanced countermeasures
Weapon carriage	Two 2,000-lb JDAMs internal	Same as F-16C Block-50

larly configured F-16C Block-50 (that is, not carrying external fuel). The JSF incorporates "stealth" features into its design, making it less observable to radar and other sensors than the F-16. The JSF is also being designed to be more reliable and easier to maintain than the F-16, meaning that it should be able to be deployed forward with 50-percent less support "tail." For avionics, the JSF will have an internal FLIR and targeting system, as well as an active, electronically steerable antenna (ESA) with radar modes that allow it to find moving and stationary ground targets. It will be able to carry either two 2,000-pound JDAMs (GBU-31), two WCMDs, or eight SDBs internally.

Our assumptions for the F-16C Block-60 are based on the aircraft being sold to the United Arab Emirates. It will have two conformal tanks, which will give it 30-percent more range than a F-16C Block-50. We assume it has the same avionics as the JSF with the same weapon carriage and survivability as the F-16C Block-50.

Table 4.5 summarizes our assumptions regarding the characteristics of the Air Force's fleet of heavy bombers. We assume that all the bombers have received programmed upgrades to their avionics and communication systems. In general, these upgrades will allow the bombers to locate, identify, and engage fixed and moving targets and to deliver the full range of air-to-ground weapons that will be available to USAF platforms in the coming decade.

FORCE MIX RECOMMENDATIONS

Fighter-Bomber Mix

In this section, we address the question of whether the Air Force should retire the B-1B force and use the savings to buy additional new stealthy fighters. The analysis reveals that, with adequate supplies of standoff weapons, a force with the B-1Bs is preferable to a force with additional new fighters. If the Air Force does not buy adequate supplies of standoff weapons, the opposite is true; in most scenarios, the additional stealthy fighters are preferable to the B-1Bs.

Figure 4.2 summarizes the findings of our campaign analysis for all four scenarios outlined previously, *given the currently programmed buy of JASSMs*. Each of the four quadrants of Figure 4.2 is labeled with the force mix option that yields the best outcome in the scenario depicted. For example, in the lower left quadrant (the defense of South Korea from an attack by the North), U.S. and Republic of Korea forces do best when B-1Bs are retained in the force. In each of the other scenarios, however, defending forces do better if all the B-1Bs are "traded in" for accelerated buys of F-22X (in the case of a Chinese attack on Korea), JSF (in the case of an Iranian attack on Saudi Arabia), or F-22E (in the case of a Russian attack on Poland).

The reasons for these results are clear: Since the B-1 depends on the standoff capability of the JASSM to survive in most air defense environments, retaining the B-1 force is preferable only when both the air-to-air and surface-to-air threats are low.[11]

---

[11]Because we believe that future commanders will be fairly risk averse, we did not allow the B-1B to fly when air superiority had not yet been achieved unless it had standoff weapons to launch.

Table 4.5

Heavy Bomber Characteristics

	B-2A	B-1B	B-52H
Annual O&S cost per aircraft (1998 M$)	34.0	12.0	9.5
Program	All are Block-30s	All are modernized	Retiring B-52Hs not considered because of SIOP role and carriage of unique weapons SIOP Role Unique weapons
Avionics	Radar upgrades (GMTI/GMTT) Link 16/SATCOM Joint MPS/in-flight replan	Radio/GPS/Link-16 GMTI/GMTT Computer upgrade Full DSUP Relative targeting system	Electro-optical viewing system Radio/GPS/Link-16 Ring laser gyro Survivability upgrades Enhanced mission management system and in-flight replan
Weapon integration	JASSM JSOW WCMD JDAM LOCASS SDB	JASSM JSOW WCMD JDAM LOCASS SDB	JASSM JSOW WCMD JDAM LOCASS SDB

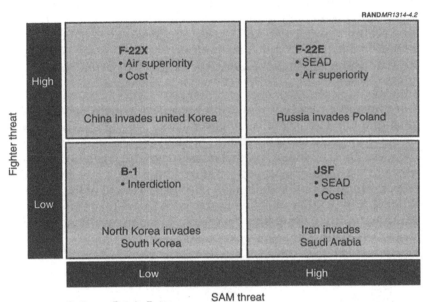

Figure 4.2—Preferred Force Mix Options: Currently Planned JASSM Buy

In scenarios with a high fighter threat and in which only a modest number of JASSMs are available to USAF aircraft, procuring additional F-22s is preferable to retaining the B-1s because more F-22s decrease the time required to achieve air superiority, which, in turn, allows nonstealthy aircraft to begin the interdiction campaign sooner. In scenarios with both high fighter and SAM threats, procuring the F-22E is preferable because its larger weapon bay allows it to carry more weapons for attacking enemy air defenses. In scenarios with a high SAM threat and a low fighter threat, procuring the JSF is preferable to retaining the B-1 because of the JSF's capability to destroy surface-based air defenses and because it is less expensive to buy and operate than the F-22.[12]

_____

[12]Recall that all the force mixes we examined impose equal costs over the 2000 to 2020 period.

The value of the B-1, in short, depends strongly on the number of standoff weapons in the inventory. The more standoff weapons in the inventory, particularly AGM-158s (JASSMs), the greater the value of the B-1s. Figure 4.3 shows the preferred aircraft for each of the four air defense environments *when the JASSM buy is increased from 2,400 to 9,600.*

With an increased buy of standoff weapons, retaining the B-1 is preferable in all air defense environments. The fighter force is still engaged in establishing air superiority and suppression of enemy air defenses (SEAD). The B-1, with its long range and large payload, is an efficient way to carry weapons to the enemy, as long as the standoff weapon supply is sufficient for effective employment of the bombers. Of the 9,600 JASSMs available in this set of force mix options, 4,800 are armed with the currently programmed unitary warhead. The rest dispense guided antiarmor weapons.

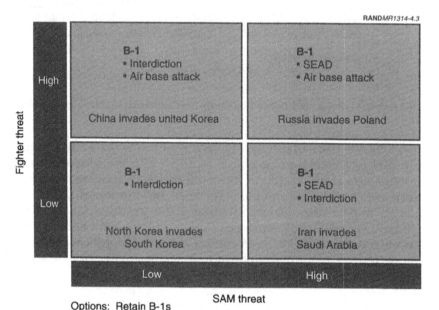

Figure 4.3—Preferred Force Mix Options: 9,600 JASSM Buy

In the scenarios we examined, the single most cost-effective action the Air Force can take to enhance its overall effectiveness is to buy more standoff weapons so that its nonstealthy platforms (particularly the bombers) can participate effectively in the early stages of the war.

To gain a sense of the robustness of different force mixes, we examined a range of cases for each of our four scenarios. Each of these cases assumed more-stressing conditions for the defender than those portrayed in the baseline scenarios. Specifically, for every scenario, we assessed the effects of the following eventualities on the allies' ability to defeat the attacking force (see Figure 4.4):

- Access to air bases in the theater is reduced ("limited basing").

- The enemy uses lethal chemical or biological agents or, in some cases, nuclear weapons to attack airfields and ports. The objective of these attacks is to impede the flow of U.S. forces into the theater and to destroy or reduce the operational tempo of the forces in the theater ("WMD").

- The enemy attacks with less warning than in the base case; that is, fewer U.S. forces are deployed to theater at the commencement of hostilities ("reduced warning").

- U.S. forces deploy more slowly to theater because 50-percent less airlift and sealift capacity is available. This condition might obtain if, for example, another large-scale operation were in progress in another region ("reduced lift").

Of course, the effect of all these assumptions is to make a successful defensive operation more difficult—an effect that is manifested in more territory lost, more time required to defeat the attack, and more allied losses. But some force mixes perform substantially better than others in these cases, and identifying the reasons for the differences provided insights about the relative strengths and weaknesses of each force mix.

Figure 4.4 portrays graphically the results of these more stressing assumptions on one measure of outcome, territory lost, and in one scenario, China invading Korea. In our analysis of this scenario, we assumed a defensive concept of operations in which the Republic of Korea (now unified) maintains two main defensive lines: one set just north of the current demilitarized zone and one set south of the demilitarized zone. If attacking forces were able to penetrate the southern positions, Seoul would be threatened and perhaps overrun. This, then, is the primary criterion determining the success or failure of the defense.

In Figure 4.4, the bars represent the point at which the enemy's advance is determined to have halted. The further "south" the bars reach, the worse the outcome from the standpoint of the defender. Bars that reach south near the lower line labeled "Fixed defenses" depict a likely failure of the defense. In each case, the black bars represent the threat army penetration with the baseline force mix. The gray bars represent trading the B-1s for F-22Xs, and the white bars represent trading B-1s for JSFs.

In all the stressing cases we examined in which the buy of JASSM did not increase, a force that traded the B-1B force for additional stealthy fighters performed better than the baseline force with B-1Bs. This result obtains because, without large stocks of standoff weapons in this scenario, the defending force cannot effectively employ the bombers without suffering unacceptable losses. This set of results is particularly noteworthy in light of the fact that each case adds conditions to the baseline scenario that, in principle, ought to favor long-range bombers. That is, they make it harder for shorter-range fighters to deploy to theater on time or to sustain a high tempo of operations once they reach the theater. Nevertheless, in all cases, modern fighters were preferable to B-1 bombers without additional standoff weapons.[13]

Figure 4.5 summarizes results from the same stressing cases, except now the AGM-158 (JASSM) weapon buy has increased to 9,600. Again, one-half of these weapons have unitary warheads, and the other half have guided antiarmor munitions. Under this assumption, retaining the B-1 is preferable to purchasing additional fighters in the baseline scenario and in all four of the scenario variations. A comparison of the results in Figure 4.5 with those of Figure 4.4 reveals the benefit of buying additional JASSMs. The additional JASSMs substantially improve the performance of all the force structures examined across all the scenario conditions examined.

---

[13]These cases were all run with the baseline AGM-158 (JASSM) weapon buy of 2,400.

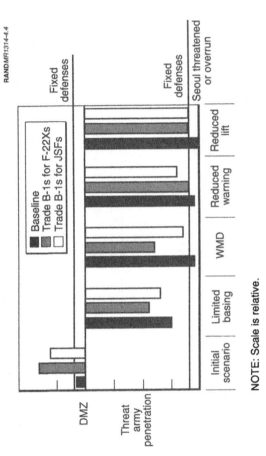

NOTE: Scale is relative.

Figure 4.4—Fighter Versus Bomber Trades in Stressing Scenarios: China Versus Korea, Current JASSM Buy

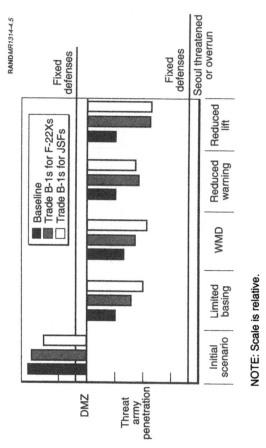

NOTE: Scale is relative.

**Figure 4.5—Fighter Versus Bomber Trades in Stressing Scenarios:
China Versus Korea, 9,600 JASSMs**

In conclusion, the preferred mix of fighters and bombers depends first and foremost on the size of the long-range standoff weapon stockpile. With large numbers of JASSMs, or a very-low-threat environment, retaining the B-1s is preferable to trading them in to buy additional stealthy fighters. With the current JASSM buy and either a strong SAM or air-to-air defense, additional stealthy fighters are preferable to the B-1s. If the air defense is primarily a SAM threat, additional JSFs are preferable. If the air defense is primarily air-to-air fighters, additional F-22s are preferable. If the air defense has both a high SAM threat and a high air-to-air threat, F-22Es are preferable. Finally, all the alternative force mix options benefit more from buying additional JASSMs than from changing only the mix of fighters and bombers.

Trades Among Fighters

The first issue to address when analyzing fighter mix options is whether the USAF should replace its current front-line air superiority fighter, the F-15C, with new models of the same design or with the stealthy F-22? The answer that emerged from our analysis was robust. Unless one can rule out the possibility of conflict with such adversaries as China or Russia—both states deploying substantial numbers of fourth-generation fighters—the capabilities the F-22 provides are essential to success in future conflicts. Without such capabilities, particularly the ability to take the first shot from beyond the range of the enemy's air-to-air radars, too much time and too many U.S. aircraft will be lost trying to defeat the enemy's fighter force. As a result, territory and ground forces will be lost to the enemy attack, resulting in the loss of the opening phase of the conflict. Because success in these scenarios was not possible without sizable numbers of F-22s, we did not examine options that reduced the currently planned buy of the F-22A.

Once the question of the F-22 versus the F-15C is settled, the primary issue becomes: Should the JSF force be reduced to fund two additional wings of an air-to-ground version of the F-22 (either the F-22E or F-22X)? All the cases summarized in this subsection assume that the B-1 fleet is retained and that 9,600 JASSMs are available. The conclusions for the trades among fighters do not depend on the JASSM buy.

Figure 4.6 summarizes the results of our assessment of the preferred fighter mix for our four scenarios. These results obtain when one assumes that no developmental weapons, specifically the SDB, are available to the force.

In air defense environments with a high fighter threat (i.e., those involving the projected forces of China or Russia), additional F-22s are preferable because of their ability to defeat advanced enemy aircraft. Assuming the presence of sizable numbers of advanced enemy fighters, in the low SAM threat environment (the northwestern box), the F-22X is preferable because of its lower cost relative to the F-22E and the reduced need to perform SEAD. For a high SAM threat, the F-22E, with its larger weapon bay, is preferable because of its ability to perform SEAD and interdiction while still contributing to air superiority. For a low fighter threat, retaining the planned JSF force is preferable because the planned quantity of F-22As is sufficient to handle the enemy fighter threat and because having larger numbers

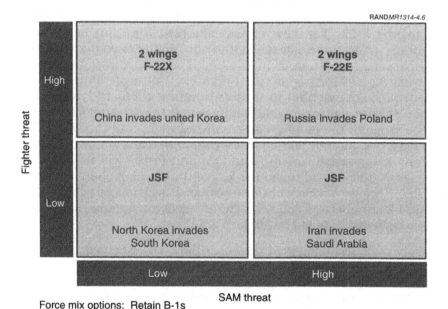

RANDMR1314-4.6

Force mix options: Retain B-1s
                    Reduce JSFs; buy two wings of F-22Xs or F-22Es

Figure 4.6—Fighter Force Mix Preferences: No Developmental Weapons

of JSFs available early in the conflict is useful for attacking ground targets.

These results are sensitive to assumptions about the air-to-ground weapons available to the fighter force. Table 4.6 describes the air-to-ground weapons now in development that we considered in our analysis. All these weapons can increase the capabilities of U.S. air forces. However, in our analysis, only three offered increases that were both sizable and platform-specific, thus altering the preferred aircraft force mix. In the previous section we showed the effects of a larger stockpile of JASSMs on the preferred mix of fighters and bombers. The other two weapons that affected the preferred mix are the SDB and the Low-Cost Autonomous Strike System (LOCASS).

The SDB and LOCASS benefit all aircraft but particularly help aircraft that carry weapons internally.[14] With SDBs and LOCASS, the JSF and

Table 4.6

Developmental Weapons Considered

Weapon	Assumptions
SDB	250 lbs, two versions: enhanced GPS (8 m) and laser radar (3 m)
LOCASS	Powered, 17-lb multimode warhead, laser radar seeker
Advanced dispensers	Low-Cost Dispenser (LODIS) or Advanced Dispenser System: high loadouts, internal or external carriage
Improved JDAM	3-m CEP
Energetic explosives	High energy content explosives (MK-82EE, equivalent to the standard MK-83)
Extended range systems	Switchblade, Longshot, and the Switchblade Wing Adapter Kit (SWAK): JDAM, SDB, MK-82
New warheads for JASSM	CEM, (2) LOCASS

[14]To avoid compromising their low-observable properties, stealthy aircraft, such as the F-22 and JSF, must carry their munitions internally. Nonstealthy aircraft can carry their weapons externally, although, of course, they are not as survivable.

F-22X will be twice as lethal per sortie as the F-15E is today.[15] That is, the SDB and LOCASS buy back the lethality lost due to the need to use an internal weapon bay. The F-22E, with the larger weapon bay, benefits the most from the SDB and LOCASS. The F-22E can carry twice the number of SDBs and LOCASS as the F-22X or JSF can.

Figure 4.7 shows the results for the preferred fighter mix for all four scenarios when the developmental weapons are available. With developmental weapons, the F-22E is preferable to either the JSF or the F-22X across all air defense environments because the increase in effectiveness (from carrying twice as many weapons) more than compensates for the increase in cost. Having fewer platforms with

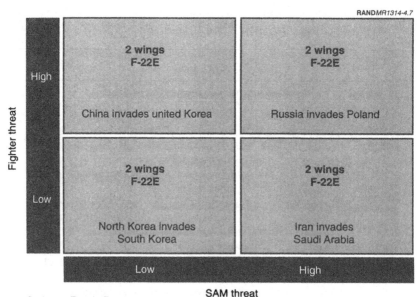

Figure 4.7—Fighter Force Mix Preferences: With Developmental Weapons

---

[15]This is a comparison of today's fighters and today's weapons against tomorrow's fighter and tomorrow's weapons. In the analysis, an F-15E carrying SDBs is more lethal than a JSF but still not as survivable.

more weapons yields more capability overall than does having more platforms carrying fewer weapons.

In conclusion, procuring two additional wings of F-22Es, (and paying for them by buying fewer JSFs) is always preferable when SDBs and LOCASS are available because the F-22E's large weapon bay allows it to carry more weapons that the other fighters. When developmental weapons are not available and when there is a large air threat and a small SAM threat, the F-22X is preferable to the JSF and F-22E because it is a better air superiority aircraft than the JSF and is less expensive than the F-22E. Only when there is no high air-to-air threat *and* when there are no developmental weapons is the JSF preferable to the F-22X and the F-22E.

Summary Force Mix Alternatives

Figure 4.8 presents a decision tree that summarizes our principal findings. External factors, shown in the diamond-shaped boxes, are the conditions that determine preferable force mix options.

Starting with the baseline fighter and bomber force, purchasing an adequate inventory of JASSMs to permit effective employment of the B-1 means that retention of the B-1 force is preferable to purchasing additional fighters. Continuing down the left-hand side of the figure, if SDBs and LOCASS are available, the Air Force should field two wings of F-22Es and reduce the JSF force to pay for them and for the required JASSMs, SDBs, and LOCASS. If SDBs and LOCASS do not become available and if a large air-to-air threat begins to emerge, an additional two fighter wings of F-22Xs are preferable. If it is decided that the emergence of a large air-to-air threat can be ruled out, the 1997 Quadrennial Defense Review force should be retained.

The right-hand side of the figure represents the decision processes when the inventory of JASSMs is not increased. Without additional JASSMs, the B-1 force will exhaust the JASSM inventory in a few days and cannot be safely utilized until air superiority is achieved. Therefore, in this case, we recommend retiring the B-1 force and using the savings to buy additional stealthy fighters and weapons. As before, if SDBs and LOCASS are available, the preferred fighter is the F-22E because of its large payload capability and effectiveness in the SEAD

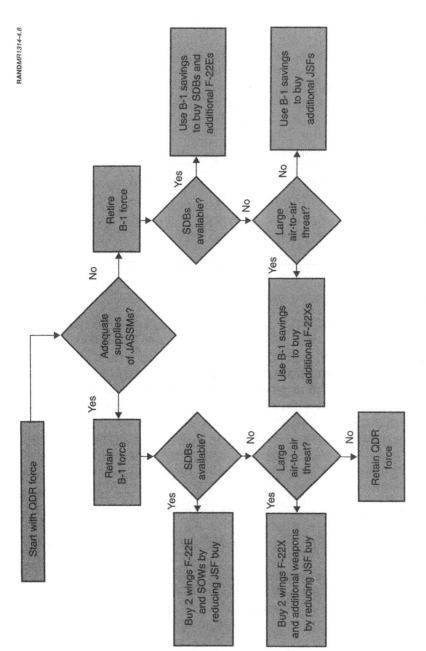

RAND*MR1314-4.8*

Figure 4.8—Force Mix Recommendations

mission. If SDBs and LOCASS are not available and if a large air-to-air threat is anticipated, the savings from retiring the B-1 should be used to buy additional F-22Xs. If it is decided that hedging against the emergence of a large air-to-air threat is not warranted, the B-1 savings should be used to buy additional JSFs.

With regard to weapons, buying additional JASSMs has a bigger influence on the outcome of the war than any changes to the mix of fighters and bombers. The single most cost-effective action the Air Force can take to enhance its overall effectiveness is to buy more standoff weapons to permit nonstealthy platforms to participate in the early stages of the war. Finally, SDBs and LOCASS are a great benefit to the F-22X, F-22E, and JSF because they restore the lethality lost due to the payload limitations inherent in using an internal weapon bay.

COST SENSITIVITIES

The above analysis outlined the preferred force mixes based on current estimates of aircraft costs. Given the unavoidable uncertainties associated with these estimates, it is important to assess the impact of potential cost growth in the F-22 and JSF programs on our force mix preferences.

The methodology for exploring this question is straightforward and is based on the assumption that a fixed amount of money will be available for each new aircraft program. Under this assumption, if the cost of an aircraft increases, the number of that aircraft bought must decrease proportionately. If the smaller buy of the new aircraft results in the need for a service life extension program for a current-generation aircraft, that cost is included. Also, the effect on unit cost of decreasing the annual rate for procuring the aircraft type is included. As costs rise, fewer aircraft are available until, finally, buying an alternative aircraft results in a more-capable force overall (based again on our theater-level simulation). For each case, we developed cost indifference lines to show the range of costs within which the decisionmaker is indifferent about which aircraft is preferable. The indifference line indicates how variations in flyaway cost and operations and support (O&S) costs for the aircraft under study affect the preferred aircraft mix.

Impact of Cost Growth in F-22X and F-22E Programs

As the previous section showed for scenarios that feature a large air-to-air threat (more than 250 fourth-generation fighters), the most capable force fields two wings of F-22Es or F-22Xs in place of the planned JSF procurement. Underlying this recommendation are the assumptions that the F-22X flyaway cost will be $87 million (1998 $), that the F-22E flyaway cost will be $95 million (1998 $), and that the O&S cost per aircraft will be $3.25 million (1998 $) per year for either the F-22X or F-22E. Figure 4.9 illustrates how sensitive this recommendation is to potential cost growth in the two programs. Along the bottom of the figure is the annual O&S cost for the F-22X or -E; along the vertical axis is the flyaway cost for the two aircraft. The lower region of the figure represents combinations of costs where two fighter wings of F-22Xs or F-22Es are preferable to the JSF.

For the current estimates of the cost to own and operate the F-22X or -E, we are in the "F-22 preferred" portion of Figure 4.9. Given an annual O&S cost per F-22X of $3.25 million, we continue to prefer the JSF unless the flyaway cost of the F-22X exceeds $95 million (that is, a cost increase of $8 million per aircraft), or about 9 percent. If the F-22X flyaway cost exceeds $95 million, we prefer the force with 591 JSFs (the baseline force) to the force with 144 F-22Xs. If the F-22E flyaway cost exceeds $110 million (a 16-percent cost growth), we prefer the force with JSFs. In addition, if the F-22X flyaway cost remains stable at $87 million per aircraft but the annual O&S cost exceeds $3.7 million, the F-22X is no longer preferred. And if the O&S cost of the F-22E exceeds $4.0 million, the F-22E is no longer preferred. Also depicted along the bottom of Figure 4.9 are the annual O&S cost for the F-15C and the expected O&S cost for the F-22A. As shown, if the annual O&S cost for the F-22X equals that for the F-15C, we are solidly in the "JSF preferred" portion of the figure. Conversely, if the annual O&S cost for the F-22E equals that for the F-15C, we are on the indifference line.

Impact of Cost Growth in JSF Program

In scenarios that do not have a large air-to-air threat, the JSF is preferable to additional F-22Xs or F-22Es. Underlying this recommendation is the assumption that the JSF's flyaway cost will be $30.4 million (1998 $) and that the O&S cost will be $1.9 million per year for

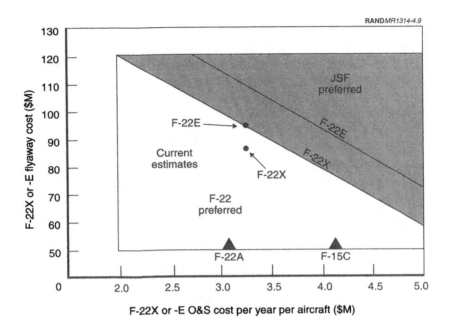

Figure 4.9—Impact of Cost Growth on Preferences for F-22X or -E

each aircraft. Figure 4.10 shows the effects of JSF cost growth on that recommendation.

If the JSF flyaway cost were to increase above $42 million, with no change in annual O&S costs, we would prefer the two F-22X fighter wings over the additional 353 JSFs. The JSF can sustain a 38-percent cost increase and still be preferable to the F-22X in scenarios lacking a large air-to-air threat. In addition, if the flyaway cost of the JSF were to remain stable but its annual O&S cost were to increase above $2.5 million, we would prefer the F-22X.

In scenarios that favor the F-22 derivatives, the JSF is preferable after only a 9- to 16-percent increase in F-22 flyaway cost. In scenarios that favor the JSF, the F-22 derivatives are preferable after a 38-percent cost increase in the JSF. At this point in its development, the JSF enjoys a much bigger cushion against cost increases. However, because the F-22A is much more mature than the JSF program, cost figures associated with the F-22 program are almost certainly more

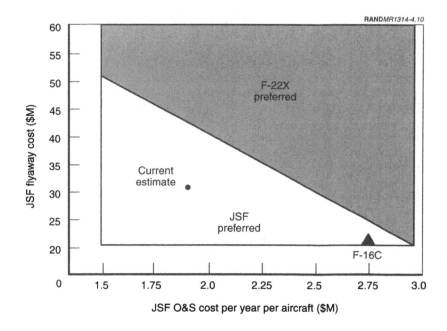

Figure 4.10—Impact of JSF Cost Growth on JSF Versus F-22 Decision

realistic than those associated with the JSF, prototypes of which have just begun to fly.

Figure 4.11 shows the effects of JSF cost growth in a comparison with the F-16C Block-60. The greater the air defense threat, the more preferable the JSF becomes. However, even for the high-air-defense threat, if the JSF's flyaway cost were to exceed $50 million, we would prefer the F-16 Block-60. Stated another way, at $50 million per JSF, an equal-cost force of F-16 Block-60s would be equally capable despite its lower survivability. Likewise, if the JSF's flyaway cost remains stable but if its annual O&S cost were to exceed what the Air Force pays today for the F-16C, we would again prefer the F-16 Block-60 to the JSF.

In conclusion, the cost assumptions underlying our recommendations are very important. To be preferable to the JSF, the F-22X must keep its flyaway cost below $95 million, and the F-22E must keep its flyaway cost below $110 million. To be preferable to either the F-22

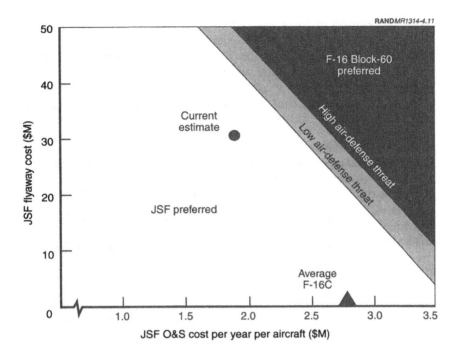

Figure 4.11—Impact of JSF Cost Growth on JSF Versus F-16C Block-60 Decision

derivatives or the F-16C Block-60, the JSF must keep its flyaway cost below $42 million. The JSF and F-22X must also hold their annual O&S costs to levels less than the aircraft they are replacing. Given that both the JSF and the F-22X are stealthy aircraft, whose exposed surfaces will likely demand a significant amount of maintenance, this could be a difficult challenge.

SSCs AND ONGOING DEPLOYMENTS

For the first 40 years of the Cold War, the size of the U.S. Air Force was predicated mainly on the demands of deterring and defeating Soviet aggression. The size and shape of Air Force's general-purpose forces were driven primarily by the need to defeat a large-scale Soviet combined-arms offensive in Central Europe. Since the end of the Cold War, the size of the Air Force has been driven primarily by the need to be able to fight and win two nearly simultaneous major regional conflicts. However, it has become increasingly clear that the post–Cold War security environment is placing heavy demands on the Air Force in peacetime as well. In fact, we believe that the Air Force needs approximately 20 fighter wings (the current force size) to sustain commitments that have been levied on it in peacetime and still be able to undertake deployments to SSCs.[16]

In most recent force structure analyses, SSCs have been considered to be lesser cases of major regional conflicts. It has been assumed that, if the United States fields forces that are adequate for prevailing in two nearly simultaneous regional conflicts, the forces will also be capable enough to meet peacetime demands, including succeeding in any SSC that might arise. This may be true for individual SSCs, but, as Figure 4.12 shows, the Air Force has consistently engaged in two to three SSCs, resulting in prolonged deployments.[17]

SSCs and ongoing deployments place different demands on the force structure than do major regional conflicts. Major regional conflicts require a large number of aircraft (10 to 12 fighter wings) for a relatively short time (one to three months). Ongoing deployments, such as Southern Watch, require a smaller number of forces (typically, less than two fighter wings), but may need them in theater indefinitely. Because political conditions preclude permanently stationing units (and families) in these theaters, aircraft and personnel are rotated into them on TDY, typically for 30 to 90 days and then are replaced.

Experience has shown that unit training and personnel retention suffer if personnel are required to spend more than about 120 days per year away from home on TDY. To examine how ongoing deployments influence force sizing, we examined the number and types of deployment operations or SSCs that can be supported with different sized forces, using the 120-day ceiling on annual TDY for the units' aircrews and support personnel.

---

[16]We include peacetime commitments among SSCs even though the former may not lead to conflict.

[17]The successful conclusion of Operation Allied Force in 1999 reduced the demands on the USAF for rotationally deployed forces to patrol the Balkans considerably, leaving Operations Southern Watch and Northern Watch as the primary determinants of overseas deployment–related operational tempo.

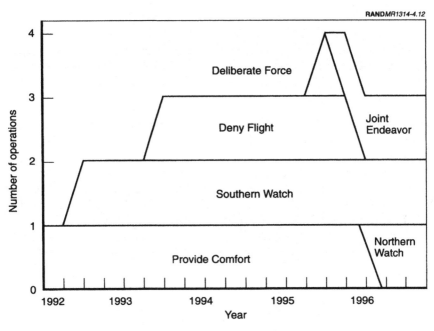

SOURCE: Thaler and Norton (1998).

Figure 4.12—Long-Term USAF Deployment Operations

We began by determining the force requirements for several different classes of small-scale operations. The three cases considered here are deterring a small- to medium-sized invasion, specifically a Basra-breakout scenario in Iraq; defending a no-fly zone; and enforcing an exclusion zone. We believe that these scenarios represent a reasonable range of potential long-term demands for the foreseeable future.[18]

---

[18]A fourth class of SSCs, the punitive strike, was examined but is not included here because it does not generally require deploying aircraft overseas for an extended period and, therefore, does not affect the required force size. Force structure requirements for SSCs are driven by the need to keep aircraft overseas for extended periods. Punitive strikes tend to be lesser-included cases of major theater wars. That is, if the Air Force can conduct a major regional war, it can also effectively conduct a punitive strike.

The first type of operation is deterring a small- to medium-sized invasion. The specific scenario, shown in Figure 4.13, assumes that Iraq has attempted a lightning strike to seize key objectives in Kuwait and Saudi Arabia. The scenario is set around 2015 under the assumption that Iraq has been able to rebuild a portion of its ground, air, and missile forces.

We assumed that, in recognition of the possibility of such an attack, the United States routinely deploys a force of 24 F-22As; 48 JSFs; 18 U.S. Marine Corps JSFs; one carrier battle group; and land-based support aircraft, such as the Joint Strategic Tracking and Radar System, the Airborne Warning and Control System, and aerial refuelers. Once hostilities commence, an aerospace expeditionary Force (AEF) will be deployed to supplement the prepositioned forces. In our analysis, the size and composition of this AEF varied from 33 combat aircraft (12 F-22As, 18 JSFs, and three B-1s) to more than 100. Note

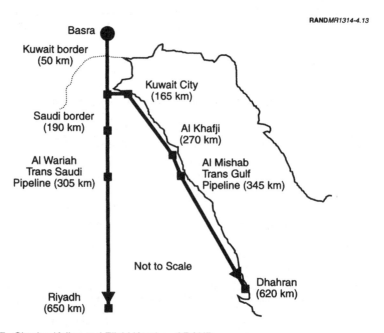

SOURCE: Charles Kelley and Eiichi Kamiya of RAND.

Figure 4.13—Basra Breakout Scenario

that AEFs also incorporate support aircraft for reconnaissance, command and control, and aerial refueling. We assume that the B-1s from the AEF enter combat on Day 2 of the conflict and that the fighters enter combat on Day 3.

The Iraqi goal is to overrun Kuwait, then continue into Saudi Arabia to disrupt the oil pipelines and, if possible, threaten Dhahran and destabilize the government in Riyadh. We assumed that the Iraqi army continues to advance until half of its combat vehicles have been destroyed. In our assessments, U.S. forces halting the Iraqi army before it overruns Kuwait City would be an allied victory because of the limited long-term effect the invasion would have on the oil supply and stability of Saudi Arabia. Alternatively, the Iraqi army overrunning the pipelines in Saudi Arabia and advancing on Dhahran would be an Iraqi victory because it would severely disrupt the supply of oil and, possibly, destabilize Saudi Arabia.

The distance the Iraqi army is able to penetrate into Kuwait and Saudi Arabia depends, among other things, on the size of the invading army and the forces available to attack the invading army. We examined Iraqi invasions consisting of three to six divisions.

Figure 4.14 summarizes the results of our assessment of the Basra-breakout scenario, again using the START model. The figure plots the maximum penetration distance for the invasion force (in this case, consisting of four divisions), given different sized predeployed forces and AEFs. The first column represents the baseline force of 72 prepositioned Air Force combat aircraft and 33 aircraft arriving as an expeditionary force. With the baseline force, the Iraqi army overruns Kuwait City and penetrates the Saudi Arabian border.

As aircraft (and munitions) are added to the AEF or the predeployed forces, the outcome improves from the standpoint of the defender. Adding 18 or 36 JSFs to the AEF does not substantially change the result. We have to add 54 JSFs to the AEF to hold the Iraqi army north of Kuwait City. However, adding as few as 18 JSFs to the predeployed forces halts the Iraqi army north of Kuwait City. The difference in effectiveness between adding the JSFs to the AEF and predeploying them arises because the AEF force does not arrive until Day 3 and because Kuwait City is overrun on Day 5.

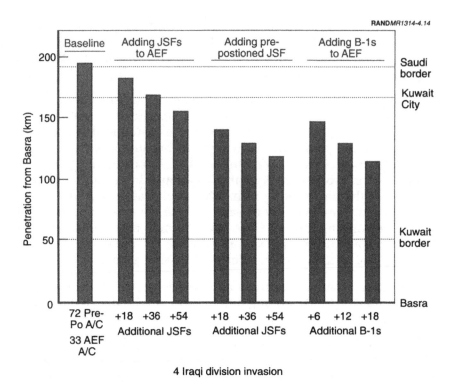

RAND*MR1314-4.14*

**Figure 4.14—The Basra Breakout: Results as a
Function of Defending Force Size**

Time, in short, equals land, at least in this scenario: A smaller force in place at the outset of hostilities is worth more than a larger force that arrives several days later. Unfortunately, adding JSFs to the predeployed force substantially increases the TDY rates for JSF units worldwide, as discussed later. Alternatively, if B-1 bombers were available from the continental United States (CONUS),[19] they could supply the necessary punch to halt the Iraqi army north of Kuwait City without requiring additional pilots to remain TDY with the prepositioned forces. Figure 4.14 shows that each B-1 bomber from

---

[19]The B-1s would fly their first mission from CONUS, recover in the theater, and continue flying missions from bases in the theater.

CONUS could, in effect, provide the same capabilities as three bomb-dropping JSFs predeployed to theater. For this finding to apply robustly to future contingencies, the B-1Bs would have to have access to sizable stocks of standoff weapons that can deliver capable antiarmor munitions (e.g., JASSM with LOCASS or, more marginally, JSOW with SFW). Such weapons allow the bombers to engage the vehicles associated with advancing ground forces early in the war, even before successful SEAD.

Force Requirements to Support Deployed Aircraft

The analysis Figure 4.14 summarizes shows that, in responding to short-notice aggression, aircraft predeployed overseas and ready for immediate action have greater utility than aircraft that must be deployed after the attack commences. This conclusion is strengthened by the need for forward-deployed aircraft to enable safe operation of reconnaissance and other aircraft essential to U.S. operations and to protect rear-area air bases, ports, munitions stocks, and other assets needed for the deployment and employment of other follow-on forces.[20] Too, forward-deployed forces have an incalculably higher deterrent value than forces in CONUS because they signal to potential aggressors an unambiguous U.S. intention to defend national interests in that region. Unfortunately, given unavoidable limits on TDY rates, keeping forces prepositioned in theater requires substantial force structure.

Table 4.7 shows calculations of the force structure required to support an aircraft overseas for one year.[21] As was noted above, experi-

---

[20]Of course, care must be taken to ensure that forward-deployed and reinforcing units are based survivably. As enemy forces gain the capacity to threaten fixed facilities with more-accurate missiles and precision-guided munitions, this will call for the construction of hardened aircraft shelters and other measures.

[21]We use aircrew and aircraft TDY interchangeably because we are concerned with the number of TDY days the entire force can generate. Recently, the Air Force has been increasing the crew ratios in Southern and Northern Watch. While this increases the sorties available for the individual conflict, it does not alter the total number of TDY days the entire force can generate. To increase that total, we need to increase the crew ratio across the entire force. This would create other problems, such as decreased readiness (fewer hours per year per pilot) or an increase in the number of hours each aircraft flies (using up the airframes faster), which are not addressed in this analysis.

Table 4.7

Force Structure Required to Keep One Aircraft TDY for One Year

	CONUS-Based Active Wing	Overseas-Based Active Wing	Reserve Wing
Pilot TDY limit per year (days)	120	120	50
TDY required for training and exercises (days)	−50	−70	−35
TDY days available for SSCs per year (days)	70	50	15
Percentage of force structure	×40	×20	×40
Days available for SSCs per year	28	10	6
Average days available for SSCs		28 + 10 + 6 = 44 days	
Number of aircraft required to keep one aircraft TDY for one year		$365 \div 44 = 8.3$	

SOURCE: Thaler and Norton (1998).

ence suggests that unit training and readiness, as well as retention of experienced personnel, suffer if people spend more than about 120 days per year TDY. Of these 120 days, personnel in CONUS-based wings spend an average of 50 days on TDY to special training events and exercises (such as Red Flag), which leaves a maximum of 70 days per year per person for deployments to overseas operations, such as Northern and Southern Watch. Personnel in most overseas-based wings must deploy from home station to accomplish important parts of their routine training syllabus. They have an average of only 50 days per year for deployments to operations.[22] Personnel in reserve and National Guard units spend an average of 35 days in training and exercises and, absent a call-up, have only 15 days per year during which they can deploy to SSCs. The CONUS-based active wings make up 40 percent (eight wings out of 20) of the force structure, while overseas-based wings make up 20 percent; the remaining 40 percent of the fighter force structure comprises reserve and guard wings. The weighted average of the number of days per year for CONUS-based, overseas-based, and reserve wings yields an average of 44 days per aircraft per year for deployments to routine SSCs. Since there are 365 days in the year, the force structure requires 8.3 aircraft to keep each aircraft deployed overseas.

Returning to Figure 4.14, each JSF we add to the deployed force requires eight aircraft in the total force structure. The B-1s flying from CONUS do not count against TDY limits and, therefore, can help to offset some of the demand routine overseas deployments make on force structure.

No-Fly and Exclusion Zones

We examined two other classes of SSCs: enforcing no-fly zones and enforcing exclusion zones. In a no-fly zone, the Air Force attempts to intercept any aircraft that flies within a designated area. In an exclusion zone, the Air Force attempts to intercept particular types of ground targets within a designated area—perhaps a border area or a demilitarized zone. To understand the potential demands of such

---

[22]Overseas-based wings are the aircraft deployed to Europe and Korea. Unit members are stationed at these bases for a year or more and are not considered TDY when at home station. In practice, units stationed in Korea have not been available for rotational deployments elsewhere.

missions, we must first estimate the number of aircraft that might be required to enforce a no-fly or exclusion zone.

Figure 4.15 illustrates the dimensions of a no-fly zone that might be imposed over Iraq. Aircraft fly out of a friendly airfield and sustain combat air patrols (CAPs) over the no-fly zone. Each CAP station has an intercept area that represents the area in which an aircraft on CAP can intercept an enemy aircraft. The size of each intercept area depends on a number of factors, including target flight time, target speed, interceptor speed, and requirements to identify the target (e.g., visually or by identification, friend or foe "squawk").

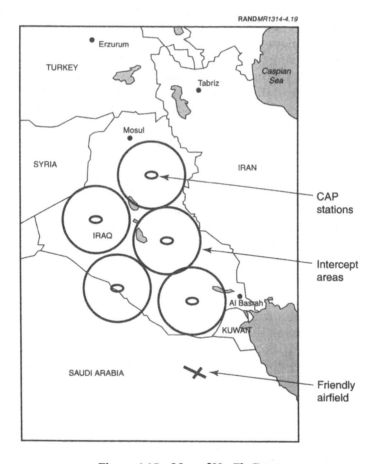

Figure 4.15—Map of No-Fly Zone

In computing the demands of enforcing a no-fly zone, we assumed that enemy aircraft attempt to fly only during daylight hours and that they attempt to remain airborne for 15 to 30 minutes. Figure 4.16 plots the number of aircraft required to patrol a no-fly zone, given both the optimistic and pessimistic assumptions for key factors.

Two points show the average number of air superiority aircraft deployed to Operations Southern Watch and Northern Watch. They fall midway between our optimistic and pessimistic assumptions. The triangles below the abscissa indicate the sizes of North Korea, all of Korea, and Iraq. Given a 12-hour operational day, covering North Korea with a no-fly zone would require less than a squadron of deployed aircraft. Covering all of Korea would require more than one squadron, but less than two squadrons. Covering all of Iraq would require a wing (three squadrons) of deployed aircraft. Not shown on the chart is Iran, which would require three wings of deployed aircraft.

Also shown in Figure 4.16 is the number of F-22As that a force of three fighter wings can provide on a long-term, sustainable basis to enforce a no-fly zone. Assuming that all three wings of F-22As are in the active forces, the force would need five F-22As for every one deployed abroad.[23] Therefore, three wings of F-22As (216 aircraft) can only support a sustained deployment of approximately 40 F-22As—less than two squadrons. Even dedicating all the F-22s to such deployments could not support a deployment the size of Southern Watch. This suggests that other portions of the force (such as JSFs) may have to play an important role in future no-fly zone enforcement operations.

Figure 4.17 shows the number of aircraft required to enforce an exclusion zone as a function of the size of the area to be patrolled. For exclusion zones, the Air Force must be able to attack any target on the ground within the zone within a particular time constraint. An example of this kind of target is an artillery piece being removed from an area of cantonment or an armored column moving into a demilitarized zone.

---

[23]For this calculation, we assume all the F-22s are in the active force and based in CONUS. From Table 3.2, each aircraft in a CONUS-based active wing has 70 TDY days available for SSCs per year, and 365 divided by 70 equals 5.2.

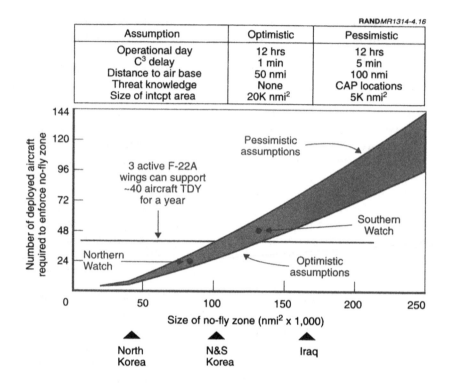

Figure 4.16—No-Fly Zone Force Requirements

The optimistic assumption is that the target is vulnerable to attack for 10 minutes, while the pessimistic assumption is that the target is vulnerable for only 5 minutes (including command, control, and communications delays).[24] Exclusion zones require between three and four times as many aircraft as no-fly zones because of the shorter exposure time and 24-hour operational day.[25]

---

[24]The amount of time the target is vulnerable is the amount of time the target is exposed minus the command, control, and communications time.

[25]For the no-fly zones, we assumed the threat air force only flew during the day (12 hours). For the exclusion zones, we assumed the ground threat could appear any time during the day or night (24 hours).

RAND*MR1314-4.17*

Assumption	Optimistic	Pessimistic
Operational day	24 hrs	24 hrs
Tgt exposure time	11 min	10 min
C^3 delay	1 min	5 min
Tgt vulnerable time	10 min	5 min
Distance to air base	50 nmi	50 nmi

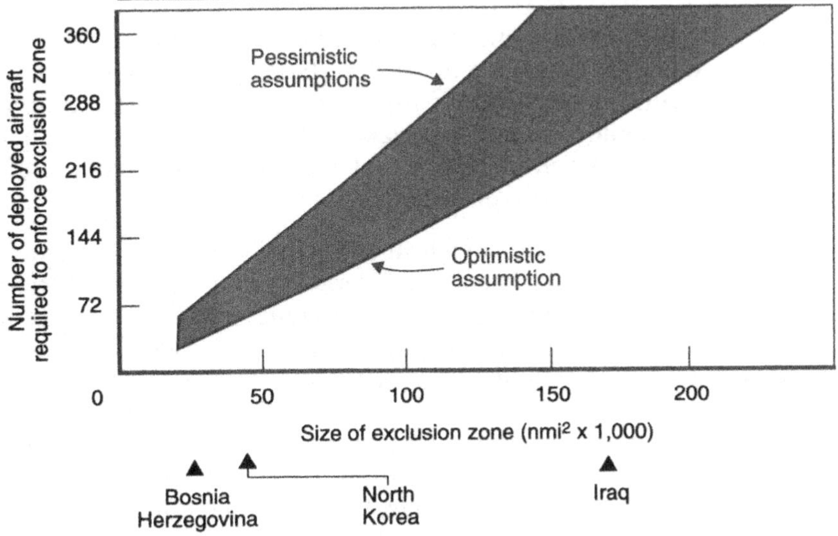

Figure 4.17—Exclusion Zone Force Requirements

Force Structure Requirements for Ongoing Deployments and SSCs

SSCs present three general classes of commitment:

- large commitments require a wing or more deployed and, therefore, a rotation base of eight to ten fighter wings

- medium-sized commitments require one to two squadrons deployed and, therefore, a rotation base of about three to five wings

- small commitments require a squadron or less deployed and, therefore, a rotation base of two wings or less.

Illustrations of each of these commitments appear in Table 4.8. The number of fighters and bombers deployed are taken from the foregoing analyses describing the Basra breakout, no-fly zones, and exclusion zones. The number of aircraft deployed is multiplied by 8 (see Table 4.7) to determine the force structure required for sustained support of each operation.

Figure 4.12 showed that, since the early 1990s, the Air Force has consistently deployed aircraft to three simultaneous SSCs. Figure 4.18 shows the number of fighter wings needed to support multiple simultaneous deployments. The black box at the bottom of each bar represents overseas commitments that will not release their aircraft for deployment operations elsewhere.

As a point of comparison, the leftmost bar in Figure 4.18 shows that the Air Force supported one "large" deployment (Southern Watch) and two "small" deployments (Northern Watch and Joint Guard) in 1997. This bar shows that it takes approximately 14 fighter wings to support these deployments on a sustained basis. Since the USAF today fields 20 fighter wings and since some other demands on force structure are not captured here,[26] this shows that the Air Force has been providing close to the maximum force it can to routine deployment operations. Historically, the commander in chief in the Pacific has not released the fighters deployed to South Korea to support SSCs because of the deleterious effect that the loss of these assets would have on the allied deterrent posture in Korea. It is reasonable to assume that there will continue to be areas where the risks are so high that commanders will not release aircraft deployed in their regions for SSCs.

The second bar in Figure 4.18 shows that conducting two large deployments simultaneously—in this case, a large no-fly zone and deterring a medium-sized invasion—could stretch the Air Force's 20-wing force structure beyond the TDY limits. The third bar shows that, if possible, operating bombers operating from their home stations to offset some of the deployed fighters in one of the operations could either reduce the required force structure by a fighter wing or reduce

---

[26]Most notably, the Air Force normally deploys several F-15Cs and combat search and rescue assets to Reykjavik, Iceland.

Table 4.8

Force Structure Requirements for SSCs

Class of Conflict	Specific SSC	Fighters and Bombers Deployed	Force Structure Required[a]	Size of Commitment
No-fly zone over large country	No-fly zone over Iraq	72	8	Large
Exclusion zone over a small country	Exclusion zone in North Korea	72	8	Large
Deter small- to medium-sized invasion	Halt four-division Basra breakout:			
	Fighters and three B-1s	141	11	Large
	Fighters and six B-1s	111	10	Large
No-fly zone over medium-sized country	No-fly zone over all of Korea	36	4	Medium
Exclusion zone over small areas	Exclusion zone over cantonment areas in Bosnia-Herzegovina	48	5	Medium
No-fly zone over small country	No-fly zone over North and South Korea or northern Iraq	24	2	Small

[a]Fighter wing equivalents.

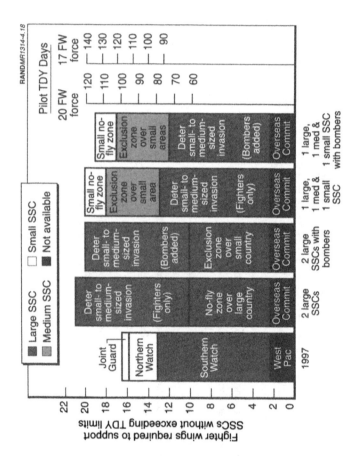

Figure 4.18—Force Structure Implications of Simultaneous SSCs

the TDY time by 5 percent. The fourth bar shows that an Air Force comprising 20 fighter wing equivalents could also support one "large" deployment and two "smaller" ones without exceeding TDY limits. The fifth bar shows, again, that using bombers to offset some of the fighter requirements in one of these operations could reduce the requirement for fighters by 5 percent.

Force Structure Implications of SSCs

During the 1990s, the Air Force was consistently involved in either two "large" or one "large" and two "small" deployment operations. If the Air Force is called on to continue this level of support, it will need to sustain approximately 20 fighter wings if it is to avoid exceeding personnel TDY limits. If Air Force budget levels remain about as they are today, fielding a force structure of this size will demand that the force mix contain a large number of aircraft that are relatively inexpensive to buy and operate, such as the JSF or F-16.

For SSCs that involve halting an invading army, the B-1s (if modernized with the right avionics and weapons), flying from CONUS, could help to provide the initial firepower needed for a successful defense. This capability will be important not only operationally but also as a way to help manage the rotational demands placed on the fighter force.

SUMMARY

A number of factors, including the shifting demands of U.S. strategies, new technological opportunities, and the unique competencies of U.S. military aviation units, have combined to make airpower the instrument of choice in a growing number of regions and situations. Indeed, it is becoming increasingly difficult to conceive of U.S. military forces succeeding in future combat operations without dominating the air (and space) dimensions of the battle. For these reasons, it is essential that the U.S. Air Force successfully modernize its aging fleet of combat aircraft.

Our analysis points to several insights that can help guide this effort. Chief among them are the following:

- The utility of the nonstealthy B-1 in future conflicts is highly dependent on the availability of an adequate supply of standoff weapons. With the currently planned inventory of JASSMs, retention of the B-1 force is preferable only in the least stressful threat environment. In the other three threat environments, retirement of the B-1 is preferable, with the savings devoted to procuring additional stealthy fighters. Alternatively, if the JASSM inventory is substantially increased, the B-1 makes valuable contributions to the joint campaign, and retaining it is preferable to buying additional fighters.

- The capabilities the F-22 provides are essential to the success of power-projection missions against adversaries whose forces are equipped with advanced air defenses. Without these capabilities—particularly the ability to take the first shot (or shots) in air-to-air engagements from beyond visual range—too much time and too many U.S. aircraft are lost trying to defeat the enemy's fighter force.

- If the USAF succeeds in developing a new generation of air-to-ground munitions deliverable by fighters—especially the SDB and the LOCASS—the best way to capitalize on the capabilities of these new munitions is to field two wings of F-22E aircraft and to reduce the number of JSFs purchased commensurately. The F-22E's survivability and large internal weapon bay make it an extremely capable aircraft whose contribution to the joint battle exceeds its relatively high cost.

- If SDBs and LOCASS are, for some reason, not available, the picture becomes more complicated. In environments with a high fighter threat (greater than 250 fourth-generation fighters), two additional wings of an air-to-ground variant of the F-22 are preferable to an equal cost number of JSFs. If the SAM threat is also high, F-22Es are preferable because of their larger weapon bay. If the SAM threat is low, two additional wings of F-22Xs are preferable because of their air-to-air capability and because they are less expensive than the F-22E. If the air-to-air threat is low, preserving the JSF force is preferable, independent of the SAM threat.

- The force preferences outlined above are based on the current cost estimates for each of the new fighter aircraft. If the F-22X flyaway cost exceeds $95 million (1998 $) or if the F-22E cost exceeds $110 million (1998 $), the original JSF force is preferred. Conversely, if the JSF flyaway cost exceeds $42 million (1998 $), adding the F-22 derivatives is preferable. In both cases, these results assume the O&S costs of these aircraft remain at their currently estimated levels. If they increase, the allowable increase in flyaway cost is reduced.

- Finally, if the JSF flyaway cost exceeds $50 million (1998 $), the F-16 Block-60 is preferable. Again, this is based on the JSF achieving about a 30-percent reduction in O&S cost relative to the F-16. If this does not occur, the flyaway cost where the JSF is preferable will be reduced.

REFERENCES

Cohen, William S., Secretary of Defense, *Report of the Quadrennial Defense Review*, Washington, D.C.: Department of Defense, May 1997. Online at http://www.defenselink.mil/pubs/qdr/ (as of December 12, 2001).

DoD—*See* U.S. Department of Defense.

Hosmer, Stephen T., *Psychological Effects of U.S. Air Operations in Four Wars 1941–1991: Lessons for U.S. Commanders*, Santa Monica, Calif.: RAND, MR-576-AF, 1996.

Joint Strike Fighter Joint Program Office, *Joint Strike Fighter Joint Interim Requirements Document, Version III (JIRD-III)*, Washington, D.C., 1998.

Thaler, David E., and Daniel M. Norton, *Air Force Operations Overseas in Peacetime: OPTEMPO and Force Structure Implications*, Santa Monica, Calif.: RAND, DB-237-AF, 1998.

U.S. Department of Defense, *Report of the Bottom-Up Review*, October 1993.

Chapter Five

SPACE CHALLENGES

Bob Preston and John Baker

This chapter identifies some of the key issues and challenges U.S. space activities will face over the next several decades, with emphasis on military and Air Force issues. Given the time horizon (and complexity of the subject), we paint this picture in broad strokes and focus on a few conditions whose consequences appear most serious for the Air Force.

The timing is opportune. After a dynamic period of expanding military use of space, of commercial space industry growth and turmoil, and of congressional scrutiny of national security space organizations culminating in OSD direction to make the Air Force its executive agent for space, the Air Force is on the threshold of significant change. The immediate changes are primarily organizational. The more-critical changes for the longer term may be those that result from a reexamination of the roles the Air Force plays in space.

To provide initial context (and highlight some key interdependencies), the chapter begins with a survey of current U.S. and international activities and capabilities. We then turn to a discussion of critical forcing functions and trends in threat and, finally, make some observations on future choices and actions.

CURRENT SPACE ACTIVITIES

For this discussion, it will be useful to divide U.S. space activities into three major sectors—civil, national security, and commercial—and to group varying activities in each. Falling into the civil sector are research and development activities, under the National Aeronautics and Space Administration (NASA), or operational activities, under either the Department of Transportation (DOT) or the Department of Commerce (DOC). National security includes both military and intelligence activities, although that distinction is more a historical artifact than a natural division. The commercial sector can be organized by kind of service or commodity: communications, remote sensing, launch, etc.

The Civil Space Sector

NASA. In the brief history of the U.S. space program, civil space has been more or less synonymous with NASA. The preponderance of government funding for civil space is still there. However, in recent years, responsibility for operational civil space has evolved significantly in other departments. At the same time, there has been increasing recent pressure on NASA for change.

Under Administrator Dan Goldin and the impetus of the National Performance Review, NASA completed significant restructuring of its workforce and reorientation of its program content, business practices, and relationship with other segments of U.S. space activity. As a consequence, NASA's relationships with military space activities are probably better now than they have been since NASA was formed.[1] Particularly notable is cooperation in technology investment and reusable spaceflight development. Significant examples of its technology cooperation opportunities are large, deployable optics and distributed, very small satellite technology.

However, NASA's existing commitments severely constrain its latitude as a partner, and will likely continue to do so. Of its roughly $13.5 billion annual budget authority, about 50 percent is committed to manned spaceflight (e.g., Space Shuttle program and the International Space Station) and 20 percent committed to infrastructure and civil service salaries. The last 30 percent must cover all the remaining enterprises: space science, earth science, and air and space technology. The magnitude alone of NASA's financial investment in manned

---

[1]The new NASA administrator, Sean O'Keefe, has expressed interest in closer cooperation between DoD and NASA in their space activities, particularly in the space launch area (Berger, 2002).

spaceflight should serve as a caution sign for others who might consider similar investments. Some of NASA's activities in manned spaceflight, however, hold some interest for the Air Force. First, NASA's exploration of the general development of human activity in space presents an opportunity. It serves as a valuable substitute for military investment in the area, particularly since such studies entail extremely large investments in overhead. No compelling reasons for military human presence in space have been identified and few interesting ones have been conjectured, so NASA's responsibility for developing and exploring human activity in space obviates the need for any significant Air Force investment in the area.

Second, NASA's work in reusable launch technology has more near term appeal, despite some setbacks. The troubled X-33, X-34, and X-37 experimental reusable launch vehicle (RLV) programs suggest some caution before investing substantially in large-scale vehicle prototypes.[2] Air Force experience with the Space Shuttle would also suggest caution about the extent, urgency, and nature of a military commitment to another civil RLV. In any event, NASA's own constraints may severely limit its alternatives, possibly to the point of extending the Shuttle fleet's life until technology and greater launch demand dramatically alter the outlook for reusable space launch.

Among the significant external influences on NASA's space launch future is congressional interest in encouraging private-sector financing, development, and operations. In presenting its acquisition strategy for the Evolved Expendable Launch Vehicle (EELV) the Air Force set a precedent for commitment to a commercial expendable launch market.[3] There might be additional opportunities for the Air Force to

---

[2]Confronted with technical setbacks and funding constraints, NASA terminated the X-33 and X-34 programs, which emphasized single-stage-to-orbit approaches to developing RLVs. To continue developing advanced launch vehicle technologies that could lead to a Space Shuttle replacement, NASA has initiated the Space Launch Initiative program for innovative launch architectures and new technologies for a second-generation RLV system (Cáceres, 2002; Berger and David, 2001).

[3]The EELV program seeks to reduce the cost of access to space significantly while improving the international competitiveness of the U.S. launch industry. DoD awarded contracts to encourage the development of two EELV families of vehicles (Lockheed Martin's Atlas V and Boeing's Delta IV) with medium- and heavy-class launch vehicles that could be used for government or commercial payloads (Knauf, Drake, and Portanova, 2001).

benefit from reusable launch, both in providing commercial launch services and in the adapting vehicle technology or designs for other military uses. However, the outcome of congressional encouragement of NASA to employ the private sector for reusable launch development will be sensitive to the emergence of launch market opportunities.

DOC and DOT. Beyond NASA, the growth of U.S. civil reliance on space has significantly expanded the involvement of other agencies, such as DOC and DOT, in managing space operations. Although civilian and military space activities often operated independently in the past, the current trend is toward more-collaborative efforts to avoid duplication, achieve budgetary savings, and fully exploit dual-use space technologies, such as the Global Positioning System (GPS), for a broad range of applications.

DOC has assumed important management and regulatory responsibilities concerning earth observation satellite systems. The department's National Oceanic and Atmospheric Administration (NOAA) has long been responsible for operating the nation's civilian weather satellites to support the national economy and to promote public safety. In 1994, an integrated project office was created involving NOAA, DoD, and NASA to develop, manage, acquire, and operate the National Polar-Orbiting Operational Environmental Satellite System (NPOESS). The NPOESS program is a unique effort to converge separate civilian and military meteorological satellite systems; it will procure and operate higher performance satellites starting about 2008. Merging the civilian and military programs involves several challenges (Johnson, 2001). First, the next generation of environmental monitoring satellite systems must have sensors and associated algorithms for satisfying distinctive civil and national security requirements, such as different data availability requirements. Second, the program must integrate distinct support infrastructures, including ground stations and program personnel. However, the successful convergence of the civil and military programs promises to modernize U.S. polar-orbiting weather satellites and to yield substantial cost savings over separate programs.

Along with NOAA's role in environmental satellite operations, DOC has a lead role in regulating commercial remote sensing enterprises. Its regulatory responsibilities derive from the Presidential Decision

Directive (PDD) 23 (White House, 1994). This policy statement on commercial remote sensing, which President William J. Clinton signed in March 1994, implemented earlier legislation governing U.S. land remote sensing (15 U.S.C. § 5621). Under PDD-23, the Secretary of Commerce has the authority to license the operations of U.S. private remote sensing space systems, while the Department of State leads the U.S. government reviews of any requests from U.S. firms to export remote sensing technologies or a complete imaging satellite system to another country.

In close consultation with other relevant agencies (e.g., Defense, State, Interior), NOAA reviews requests from U.S. firms for licenses to operate commercial remote sensing satellites. The firms must accept various conditions, such as government access to records of satellite taskings for imagery collection. In addition, the firms must agree to accept the government's right to impose "shutter controls" that could restrict how the firms can collect or distribute data when national security so demands. Despite such provisions, nearly a dozen U.S. firms have applied for and received licenses since 1994 to operate commercial observation satellites capable of collecting higher-resolution satellite images.

DOT's Federal Aviation Administration (FAA) is another domestic agency with important and growing responsibilities for regulating U.S. commercial space launch and reentry activities. Public safety is the FAA's principal concern in regulating such new commercial space launch systems as conventional expendable launch vehicles and newer air-launched and sea-launched vehicles, as well as even-more-innovative future reusable launch systems. Prompted by the prospect of dynamic growth in space launch, the FAA is moving to develop a Space and Air Transportation Management System to handle the growing overlap between aviation and space launch operations. The FAA's regulatory authority also covers licensing commercial launch site operations, including those at several projected nonfederal launch sites.

The Commercial Space Sector

The commercial space industry has been receiving a great deal of acclaim recently for its relative maturity. The U.S. industry has been able, for example, to convert earlier government investments into a substantial comparative advantage over international suppliers. The result was an ability to compete well in the growing market that increasing deregulation of telecommunications, remote sensing, and launch opened. The industry also displayed its maturity by attracting private financing for large-scale enterprises and executing development and deployment of the systems. Competition produced radically shortened product development cycles; rapid evolution of product lines through frequent technology insertion; and lean, agile satellite production in both individual and large-quantity procurements. By many measures of maturity—technology, product cycle, investment, revenue, launches, satellites—the commercial sector achieved parity with the national security and civil sectors.

The sector has not, however, achieved independence from them. Some business cases still rely on national security customers as anchor tenants, and the commercial space sector's success and projected growth have stimulated national security sector interest in opportunities for divestment or leverage, potentially increasing this dependence.[4] The industry as a whole still relies on government research and development at the technology level.

Moreover, despite intense interest and hopeful prospects, the U.S. commercial space sector is increasingly under fire. Although it could develop products more rapidly than either international competitors or the national security sector, it failed in notable cases to capture apparent market niches before terrestrial alternatives filled them. As a result, the financial community is less willing to support commercial space ventures, stranding many initiatives. More fundamentally, the trend of deregulation that created the opportunity for the industry's maturation has stalled in some applications and has completely reversed in international sales of space commodities. These difficulties span each of the key elements of the commercial space sector: communications, remote sensing, and launch.

Communication Satellites. Satellite communications has always been the backbone of the commercial space industry and, indeed, was for a long time the only space application that could be reason-

---

[4]For a quantitative review of commercial sector projections and discussion of opportunities for military space exploitation of a maturing commercial space industry, see DeKok and Preston (2000).

ably called commercial (McLucas, 1991).[5] It blossomed first as part of a larger and earlier trend of worldwide deregulation of telecommunications in general. U.S. commercial satellite manufacturers responded aggressively and well to these opportunities. The three established high-altitude communications satellite builders modernized and expanded their production capacity and shrank their order-to-launch product cycles from roughly three years to one. The net result of their competitive efforts was a substantial increase in U.S. comparative advantage in price and quality over international suppliers.

In addition, several new suppliers emerged, offering smaller, lower-altitude satellites to provide the services new spectrum allocations had enabled: distributed plant or fleet monitoring, terrestrial cellular phone service, and bandwidth-on-demand Internet service. The first few of these to space broke new ground in design, parts use, manufacturing, and testing; in the end, they were a technical tour de force. Unfortunately, the most spectacular technical accomplishments were also business failures, resulting in bankruptcy of all (Iridium, Globalstar, and ICO) of the first three attempts to supply cellular phone service from satellites (Barboza, 1999; Reuters, 1999).[6] Prospects for market niches big enough to support new large-scale space communications enterprises still look good for wideband data delivery. However, even those are sensitive to changes in the regulatory and competitive environment.

Commercial Remote Sensing. The successful launch of Space Imaging's Ikonos satellite in September 1999 was an important commercial milestone. Ikonos is the first commercial imaging satellite capable of collecting high-resolution images (about 1-m resolution) of the earth's surface. However, despite the technical success of Ikonos, the business viability of U.S. commercial remote sensing from space remains uncertain for several reasons, including an unpredictable market for satellite imagery products; stiff, subsidized

---

[5]A recent space commerce report for DOC notes that satellite communications is projected to account for 83 percent of the total space industry revenues, or about $430 billion, during 1996–2002 with space transportation (8 percent), GPS (9 percent), and remote sensing (less than 1 percent) accounting for the remainder (Futron, Inc., 2001).

[6]Globalstar suspended payment on its debts and began restructuring in negotiation with its creditors (Jeffery and Pascale, 2001; Pascale, 2001).

foreign competition; and U.S. policy difficulties in regulating the industry.

Space Imaging is only one of several firms that PDD-23 encouraged to proceed with developing commercial observation satellites. Although all these firms had high expectations for a rapidly growing market for their products, these expectations have yet to be realized. DOC (Futron, Inc., 2001) projected that the U.S. share of the global market will grow at a modest rate (see Table 5.1). The modest projected increases make the profitability of the reported original investment doubtful.[7]

Not surprisingly, therefore, American firms expect the U.S. government to be the anchor customer for their higher-resolution imagery, at least while the commercial market develops. The National Reconnaissance Office (NRO), as a part of its Future Imagery Architecture (FIA), has assumed the availability of this imagery.[8] The National

Table 5.1

Space Imagery Revenues

	Actual				Projected		
	1996	1997	1998	1999	2000	2001	2002
Revenues ($ millions)							
U.S.	28	32	38	43	50	63	83
Rest of World	74	88	101	111	123	134	148
Total	102	120	139	154	173	197	231
Market share (%)							
U.S. share	27.5	26.7	27.3	27.9	28.9	32.0	35.9
U.S. annual growth		14.3	18.8	13.2	16.3	26.0	31.7
Total annual growth		18.9	14.8	9.9	10.8	8.9	10.4

SOURCES: Adapted from Futron, Inc. (2001) and DOC.

---

[7] *The Washington Post* (Loeb, 2000) reported $700 million in 2000, compared with $278 million invested by 1996, based on a Space Imaging press release that a $25 million investment represented a 9-percent holding at the time.

[8] The congressional commission on NIMA reported that roughly half of the military's requirements for imagery of point and area targets in both peace and wartime were allocated to commercial satellites and military aircraft rather than FIA (NIMA Commission, 2001).

Imagery and Mapping Agency (NIMA) announced that it intended to award upwards of $1 billion in contracts for commercial imagery products and services over five years to American firms that can provide 1-m resolution imagery. This policy could give the intelligence community flexibility to offload some of its less-demanding requirements to commercial satellites so that national systems could concentrate on more-challenging problems. However, NIMA failed to budget funds for the purpose. And, as congressional commissions pointed out, there are fiscal disincentives for military users to purchase these products using their own funds (Kerrey and Goss, 2000; NIMA Commission, 2001).

Commercial Launch Capabilities. Commercial space launch, in contrast to communications and remote sensing, remains in its infancy. Developing a highly reliable space launch capability that is substantially less expensive than existing government-sponsored launches will require overcoming some major technological and operational hurdles. Operationally, launch rate demands must become high enough to sustain fleets of reusable vehicles. The specific technical hurdles, however, will depend on the system approach taken. Both aspects present real difficulties.

The demand for civil and national security satellite launches has been and is projected to remain steady. The only real growth has been in commercial communication satellite launches, but there are indications that this demand is decreasing (COMSTAC, 1999, 2001). While there is a major source of uncertainty in the demand for launches to low earth orbit, there is no indication that overall demand will grow enough to warrant commercial development of RLVs. Other potential sources of demand that might be high enough have yet to materialize.[9]

---

[9] The industry's best candidate for now is tourism. Short of that, NASA's need for space station logistic support could be a start. Other than those, a military need for substantial orbital replenishment might supply a future demand rate high enough to warrant reusable launch. For example, operating a constellation of space-based lasers sized for boost-phase missile defense would require replenishment launch capacity equivalent to a small satellite launch for each "shot" of laser reactant used in testing, training, or combat. Alternatively, a military demand for large volume, intercontinental range, and prompt strike could provide the basis for developing an RLV (Richter, 2001; Schneider and Sawyer, 2001).

It costs from $2,000 to 8,000 per pound to orbit a payload using current launch systems. Achieving anything approaching an order-of-magnitude cost reduction will require not only increasing launch rates but also new launch technologies. NASA's X-33 reusable launch technology program hoped to enable future costs around $1,000 per pound, if its advanced technology had been successful and if high enough launch rates were needed. In addition to the NASA program, a series of entrepreneurial commercial firms explored a range of smaller reusable vehicle concepts. Most of these, however, have been unable to make substantial progress. Declining demand associated with the high-profile failures of satellite mobile telephone services has diminished investor interest.

The Regulatory Environment. Regulation is a critical issue for any examination of the U.S. commercial space industry. The major source of concern is the problematic regulatory environment for export of space commodities. The environment for commercial remote sensing is similarly troubling.

Until the latter part of the first Bush administration, all space-related commodities were, by default, considered munitions items, subject to Department of State export controls. With the end of the Cold War, President George H. W. Bush directed a policy change to move as many space commodities as possible to DOC export controls.[10] The result moved most commercial communications satellites to more expeditious Commerce procedures, assisting the rapid growth of U.S. commercial communications satellite sales.

The resulting growth held the seeds of problems to come. To accommodate the demand for satellites, U.S. manufacturers needed to go offshore for additional launch capacity. Among the available international launch vehicles was a relatively immature Chinese Long March vehicle. After U.S. satellite makers assisted with troubleshooting of a few spectacular Long March failures, the U.S. government investigated the companies themselves for procedural improprieties and potential illegal technology transfers. These investigations were caught up with others looking into charges of illegal

---

[10]For some historic details on the liberalization of U.S. space export controls, see Preston (1994); for historical development of U.S. commercial communications satellites, see Chapter 3 in particular. Also see Acker (2001).

Chinese contributions to U.S. political campaigns and espionage at U.S. nuclear weapon laboratories (Cox, 1999).

In the end, Congress directed that satellites be put back under Department of State export control. The department, however, was no longer staffed to handle even the earlier, more leisurely, and smaller volume exports and, moreover, responded to the congressional mood by being extremely conservative. As a result, the abrupt transfer virtually halted export of U.S. satellites in 1999 and diminished the cycle-time advantages the lean, agile U.S. builders had achieved. Foreign customers, including long-standing U.S. allies, made it clear that they were unwilling to depend on U.S. satellite technology providers, given the new delays, complications, and licensing approval uncertainties the turbulence in the U.S. regulatory process had created (Pollack, 1999b). By 2000, the U.S. share of the geosynchronous satellites market had declined from 75 to 45 percent (Marquis, 2001; "U.S. Satellite Export Controls . . . ," 2001). Although U.S. satellite manufacturers rebounded in 2001 to capture a majority of the contracts for new geosynchronous communication satellites (de Selding, 2002a), the export control process remains a major challenge.

While the long-term effects for the communications satellite industry remain to be seen, the story is similar for remote sensing. Rather than encouraging U.S. companies to lead the market, U.S. licensing practice has been to constrain them to follow, particularly when advanced technologies, such as radar and spectral sensing, might confer a competitive advantage.[11] The regulatory difficulties in implementing the policy pose risks for the government's interest as well. Each application for a new type of imaging technology (e.g., radar, hyperspectral) has resulted in a protracted decisionmaking process within the U.S. government. In addition, the government has issued licenses to operate without having first developed clear rules and procedures for operational "shutter controls" that could with-

---

[11]Radar sensing provides an all-weather, day or night imaging capability. Depending on the particular radar, it also provides a sense of touch (in moisture, texture, or profile) and a remote position measurement. High-resolution spectral sensing provides the equivalent of a remote sense of smell, disclosing the presence or absence of materials in an element of a scene. For more detail on these emerging technologies and regulatory issues, see Preston (2001a, 2001b). For a market analysis and a tutorial on military use of remote sensing, see Chapter 2 and Appendix A of Preston (1994).

stand legal challenges on First Amendment grounds.[12] Without such rules and procedures, the risk is that commercial remote sensing will turn from a national security advantage into a liability.

The State of the Industry. Overall, the current state of the U.S. commercial space industry suggests a significant branch point between alternative futures. One extreme possibility is that heavy-handed regulation, neglect, and some bad luck could produce a future space industry like today's maritime industry. In this scenario, the government could still acquire critical security space capabilities from a small set of captive U.S. defense contractors to continue current uses of the medium while the bulk of the world's space industry—both capacity and innovation—would have moved offshore. In another extreme possibility, the international commercial space industry could resemble the computer industry, with substantial U.S. leadership and Moore's Law–like growth in capacity, capability, and innovation. In this scenario, U.S. national security space capabilities, with marginal investments to adapt and exploit the commercial sector, might make substantially greater contributions to U.S. security across all the applications for which the global reach, pervasive vantage, persistence, and immediacy space provides are advantages. At the moment, the way forward seems more likely to muddle along somewhere in the middle between these extremes. Some attention to the industry's state, regulatory issues, and market conditions could nudge things toward the brighter future or at least prevent stumbling into the unhappier alternative.

The National Security Space Sector

This sector includes both military (primarily procured and operated by the Air Force) and intelligence community (primarily procured and operated by NRO's space activities).

Military Space. Military space had a very public coming out party in the 1991 Persian Gulf War; since then, there has been steady progress at integrating the space capabilities developed originally for national

---

[12]The challenge of using "shutter controls" was highlighted by the George W. Bush administration's decision to limit access to U.S. commercial satellite imagery temporarily through an exclusive purchasing contract for Ikonos images of the Afghanistan theater rather to impose more-formal controls (Gordon, 2001).

intelligence and strategic nuclear deterrence into conventional military capabilities for virtually all U.S. military operations. The Gulf War was very much a come-as-you-are and make-it-up-as-you-go-along affair. Commercial GPS navigation receivers were bought in volume on a crash basis and were distributed widely. Satellite communications, both leased commercial transponders and military satellites (including retired, residual capacity), were reoriented to the theater. Mobile ground terminals were pushed to the theater and reallocated to smaller maneuver elements than they had been intended for. Couriers and airplanes disseminated the satellite imagery within the theater. National missile warning procedures were adapted and cobbled into existing intelligence broadcasts to provide theater warning and Patriot missile battery cueing. As usual, rapid innovation under the press of operational necessity brought quick progress.

Since the Gulf War, organizational changes have encouraged integration of space into conventional forces and widespread experimentation with and adaptation of space capabilities in new systems and operations.[13] Equally important is a change, both in the United States and globally, in the perception of space as an integral and essential element of military effectiveness. This change has affected the conceptual, pragmatic, and operational levels.

Conceptually, current and future opponents in a theater will understandably view any space capability contributing to the opposing military as part of the forces arrayed against it in theater. When the

---

[13]Examples are many. Advanced concept technology demonstrations for Operation Joint Endeavor in Bosnia experimented with battlefield communications using commercial, very-small-aperture satellite communications networks and commercial satellite broadcast equipment (Brewin, 1996). The first operational generation of high-data-rate global broadcast capabilities for intelligence data dissemination adapted from the commercial, direct-to-home television broadcast industry is now flying on the eighth, ninth, and tenth ultrahigh frequency follow-on satellites (Hughes Space and Communications Co., 1999). The Navy has adopted military and commercial satellite communications to connect the fleet in wide area networks for administrative, morale, and command and control communications. New generations of GPS-guided precision munitions and GPS-aided targeting demonstrated all-weather precision strike convincingly in Operation Allied Force in Serbia and Kosovo in 1999 (Brewin, 1996; Tirpak, 1999). By 2001, GPS-guided bombs were a dominant factor in the war in Afghanistan. Two-thirds of the bombs dropped in the first two months were precision munitions; of those, 64 percent were GPS-guided Joint Direct Attack Munitions (Clarke, 2001).

space capabilities represent an easier target than other critical nodes, we should expect interference with them. The natural consequence of space integration into military activity is a more hostile environment for space.

In pragmatic terms, the realization of space's operational role and utility has led to an international market (as with other weapons). International commercial space entities are supplying militarily useful products and services to U.S., allied, and opposing forces. Opponents that are not hampered by institutional habit or heritage systems with backward capabilities may be able to use the tools of commercial space—developed in competitive, information-industry markets—more effectively than the United States can use its heritage military systems. Opponents should be able to innovate more rapidly. Where both the United States and its opponents exploit a global commercial space environment to some degree, we will find a much broader, and more confusing, picture necessary for situational awareness.

Operationally, Air Force space systems (which make up the vast bulk of military space capabilities) are evolving from scheduled or static utilities to dynamically tasked systems that must operate with limited resources. This trend, a natural consequence of greater integration into military activity at the theater and tactical levels, can be found in almost all areas of military space activity:

- Missile warning sensors have historically provided continuous, global surveillance. Coming missile engagement sensors must be time-shared to track multiple, individual targets.

- Staring infrared battlespace characterization sensors must be tasked to the individual areas of interest at times of need.

- When, and if, space-based radar for moving-target detection joins the sensing mix, its power and aperture will have to be dynamically managed across the constellation of satellites and over varying global and theater needs for update rates, sensing modes, and target activities.

- GPS navigation signals have provided a global and ubiquitous utility for timing and location. As follow-on architectures provide the ability to manage signal power to users in limited areas for improved jamming resistance, it also will become a tasked system.

* Military communications satellites increasingly employ taskable spot beams for higher data rates or better jam resistance to selected areas and users.

* Weather observations from space have historically combined the scheduled (twice a day) detailed soundings of the atmosphere from low-altitude military satellites with more-or-less continuous, but coarser (continental scale), views from high-altitude civil satellites. Evolving architectures for environmental sensing may include atmosphere sounding using proliferated, low-altitude GPS occultation receivers for on-demand soundings and staring, narrow-field-of-view sensors for continuous fine detail from high-altitude satellites.[14]

Historically, Air Force space operations have primarily involved rear-area support activities[15]; the satellites and their payloads have largely flown themselves.[16] The shift to dynamically tasked systems is a fundamental change in the functions Air Force space operators perform and in the command, organizational, and informational relationships needed to perform these functions. Ironically, the Air Force's past efforts to make space more operational may have created

---

[14]GPS occultation receivers use the coherent dual frequencies of signals from a GPS satellite as it rises above or sets below the earth's horizon to make detailed refractivity measurements of the ionosphere and atmosphere in successive horizontal slices. Refractivity measurements give total electron content in the ionosphere and temperature and pressure in the atmosphere. The technique derives from planetary exploration. With the global presence of GPS satellites and a variety of proliferated, low-altitude satellite constellations available to host receivers, the technique enables global, on-demand sampling of the atmosphere. This is akin to being able to launch a weather balloon from the opponent's territory whenever and wherever desired.

[15]Primary activities have included providing range facilities, safety, and mission assurance monitoring of contracted satellite launches; providing network infrastructure for tracking and contacting satellites; monitoring maintenance activities; adjusting (e.g., conditioning batteries, changing power loads for eclipse conditions, and making station-keeping maneuvers); and troubleshooting.

[16]Many of the elements of these traditional Air Force space operations are becoming commercially available; the Air Force should be able to capitalize on this opportunity, as it already has for expendable vehicle space launch. This may mean some reduction of Air Force personnel needed for space operations.

obstacles to its ability to respond to this trend. These organizational challenges are discussed more fully later in this chapter.

Intelligence. In 1995, the intelligence community's part of the national security space sector began a new era of openness. Even so, the discussion in this section must remain quite general. The new openness has also brought newly adversarial oversight, and the legendary agility of this sector has been, to some extent, a casualty.

The new openness came at a time of consolidating organizational changes. NRO aligned itself according to its traditional product lines, rather than according to the organizations that supplied its personnel. This alignment was reinforced by an external orientation toward evolving national intelligence customers, such as NIMA, which are similarly organized along traditional intelligence community disciplines—"INTs"—such as imagery intelligence (IMINT), signals intelligence (SIGINT), and measurement and signature intelligence (MASINT).

These changes, while improving NRO's orientation toward traditional intelligence customers, have tended to preserve a focus on "national" customers, rather than on more-tactical military users of its products. Moreover, emphasizing the activities of the traditional intelligence production cycle inhibits alternative and more-dynamic views of information in military activity. For example, attacks against fleeting targets and adaptive opponents require timely means of controlling both sensor and weapon resources in response to the changing battlefield. Although the nature of the information and processing needed is similar, the traditional intelligence cycle embodies a more-structured, less-flexible, and relatively leisurely flow of information. The segregation of portions of the production cycle in different organizational elements adds delay to the flow and isolates the source of information from the user.

NRO's recent focus on national intelligence community products preserves these organizational boundaries and resource pool subdivisions. It preserves relationships in the production process, such as tasking authority and responsibility for information flows (for example, feedback of tasking commitments). This focus also sets the standard of improvement as marginal change in current processes. Opportunities for dramatic improvement through radical change are hard to articulate and hard to implement. They have no champion but many natural opponents.

These organizational changes have occurred in the context of successive, generational change of NRO satellite architectures. NRO is now embarking on its second post–Cold War intergenerational change across its product lines. With five- to ten-year product cycle commitments and similar constellation lifetimes, easy opportunities for further fundamental changes are likely a decade or more away—unless they occur sooner in conjunction with another organization's opportunities for mission, platform, or constellation change. For example, the recent award of NRO's FIA contract should freeze imagery-related opportunities, including potential advances in tactical use and new phenomenologies, through 2010 (Boeing, 1999). On the other hand, NRO's SIGINT architecture may provide opportunities for collaboration and innovation, for example, with the evolution of the GPS constellation (Hall, 1998).

Despite reinforcement of the national customer focus, NRO's capability to support military users has steadily improved. The new openness has brought formal military involvement in NRO requirements processes (themselves a recent innovation) and formal organizational overtures toward integrating Air Force and NRO space activities. These provide natural and lucrative opportunities for military space. The Navy's influence on and augmentation of NRO's operational electronic intelligence (OPELINT) architecture set the precedent for very lucrative and effective military leverage of national intelligence community architectures and funds.[17] The Air Force could balance or reverse the limitations described above through aggressive and consistent participation in NRO's requirements processes and pursuit of collaborative opportunities.

---

[17]OPELINT contrasts with technical electronic intelligence, which is concerned with the signal characteristics, modes, functions, associations, capabilities, limitations, vulnerabilities, and technology levels of foreign emitters. Using that technical background, OPELINT strives to provide timely warning of the current disposition of an enemy's electronic order of battle, including such information as the location, activity, and identity of ships, aircraft, and missile batteries.

World Players

Any discussion of U.S. space policies ignores international context at its peril. Internationally, a robust market and a competitive supplier community have increasingly combined to drive investment and innovation for almost all space applications. Here, we provide a few brief highlights on a few countries and broad trends.

Russia has historically been both a pioneer in space and our most serious competitor in the national security use of space. Despite severe economic, internal security, and political problems, Russia retains most of its Soviet-era industrial and technical space capabilities. Even as the rest of the Russian military endured cuts, the rocket forces continued to have relatively high priority for modernization. Although the rest of the Russian military has endured severe cuts in funding since the Soviet Union's dissolution, the rocket forces received priority funding until at least mid-2000, when President Putin was convinced to approve a more-balanced allocation of resources among the branches of the Russian armed forces (Williams, 2000; Danilov, 2000). Its space industry has been a source of some hard currency, particularly in launch vehicles and technology. Its exports encourage and its example typifies a distinctive Eastern school of space activity, which is generally more pragmatic than that of the West but quite effective.

The national space efforts of both China and India are also of the Eastern school, having roots in collaboration with the Soviet Union.[18] Both have achieved similar, substantial, independent capabilities for space launch, sensing, and communications. India has applied much fewer resources to the task but has made more progress in sensing. Although China has diluted its efforts significantly by pursuing a manned spaceflight program, preferring prestige over more-pragmatic or security needs, it has achieved substantial launch capability. Any of these Eastern-school countries could produce effective national security space capabilities for virtually any national security use and for either regional or global interests. The specific technology used and the specific approaches taken might be very different and might appear less elegant or lower technology than a

---

[18]For more detailed discussion of the Chinese and Indian space programs, see Harvey (1998) and ANSER (1999).

Western solution, but they should not be dismissed. They are likely to be economical and utilitarian.

There is an echo of that Eastern school in small countries that have been export customers of the three biggest players in the Eastern school and missile proliferators themselves. North Korea is probably the most immediately threatening of those, but less for its space capability than for its missiles and weapons of mass destruction. In this group, Iran may be a more interesting user of military space capabilities.

The Western school of space activity, in contrast, is modeled on U.S. practice. Canada's development of radar remote sensing puts it in this school. Europe has historically featured some national variations in capability and interests, with notable and distinct centers in France, Italy, Germany, and Great Britain. However, recent trends in European industrial consolidation and political unity suggest that viewing the continent's space capability as a whole is more appropriate. In that light, European space is technically capable of providing anything that the United States can provide, despite generally lacking the degree of sophistication in understanding applications that greater U.S. experience and past investment supply. Consensus on use and commitment of adequate resources to approach U.S. capabilities is also lacking. Europe is, though, a natural and generally compatible partner for U.S. national security space. It is also the most capable source of commercial competition for the U.S. space industry and the likely inheritor of the marketplace should U.S. industry be isolated.

At the other geographical extreme of the Western school is Japan. Its space industry is likewise technically capable of any use of space, although, as with Europe, it is behind in understanding the possibilities. While the Japanese constitution restricts its military to a very limited defensive charter, its concern for emerging threats has recently convinced it to develop its own imagery intelligence satellites. Israel distinctly echoes the Western style, much as some small countries do the Eastern school, and is emerging quickly as a potential competitor of U.S. commercial remote sensing.

The net result of this evolving international context is a multiplicity of sources from which foreign militaries can acquire useful space capabilities. Although performance and style may vary substantially over the different "schools" of space systems, even the less advanced can provide suitable capabilities.

MOTIVATIONS FOR CHANGE

The previous discussion of current U.S. and global space activities identified some issues that provide impetus for change in U.S. policies toward space (e.g., the drive toward inexpensive launch capabilities). Nonetheless, some specific aspects of the current situation and projections for the future will play central roles in U.S. decisionmaking. These fall into two broad classes: those that stem from internal factors or broad technical trends and those that stem from external threats.

Bureaucratic and Technological Forcing Functions

Politics. Our activities in space are human endeavors; as is often the case, politics can trump other considerations. Recent U.S. political activity could produce profound change for U.S. military space activities and, in particular, the Air Force. The first public airing of an ongoing political debate on military space was in a speech by Senator Bob Smith (R-N.H.) in November 1998, at the Fletcher School of Tufts University's Institute for Foreign Policy Analysis (Smith, 1998). Senator Smith challenged the Air Force commitment to the full exploitation of space for U.S. national security and asserted the need for "a sustained and substantial commitment of resources . . . vest[ed] . . . in a dedicated, politically powerful, independent advocate for spacepower." He discussed a variety of forms for such an independent advocate, ranging from a separate service to a space commander in chief with his own major force program and acquisition authority, similar to Special Operations Command.

While this speech arose naturally out of Senator Smith's long history of advocacy for missile defense, more was to come. In time, legislation reflected the broader political interest in the issue. The National Defense Authorization Act for Fiscal Year 2000 established a commission to examine U.S. management and organization for national security space (Congress, 1999). Among other issues,[19] the commis-

---

[19] The other issues included the use of space, interagency processes, relationships with the intelligence community, and military education for space.

sion was tasked to evaluate the range of organizational alternatives for space Senator Smith had mentioned.

Among its recommendations, the commission suggested making the Air Force DoD's executive agent for space and aligning NRO and Air Force space acquisition under a single Air Force official (Rumsfeld, 2001b). In the Secretary of Defense's response to Congress, he directed both of these recommendations (Rumsfeld, 2001a, 2001c).

Broader Security Issues. The broader U.S. security environment and strategy will also have a major effect on space policy. In general, the trend in the security environment is toward diversification—in numbers, in worldwide distribution, in the nature of U.S. interests, and in asymmetry with U.S. military capabilities (e.g., missile proliferation, antiaccess strategies). The natural U.S. strategy response to these diverse and dangerous environments has been to emphasize strategic agility and power projection (Shalikashvili, 1998). The U.S. response to the September 11, 2001, terrorist attacks, which has resulted in long-range military operations in the Afghanistan region, only highlights the importance of these crucial military capabilities. The prospect for counterterrorism operations, involving both direct U.S. military operations and indirect support to other countries and occurring almost anywhere, will continue.

Military activities in response to these conditions have increasingly stretched the U.S. military, demanding higher operational and deployment tempos. This is most obvious for high-demand, low-density assets, such as tankers and surveillance, command and control, and electronic warfare platforms. The most prominent Air Force response to this situation has been its reorganization around Expeditionary Aerospace Forces, intended to provide both rapid response where needed and some degree of predictability and stability for U.S.-based forces (Peters, 1999). Along with reorganization, the Air Force has emphasized the use of space as a potential enabler of "reachback" command and control capabilities in an attempt to improve the rapidity, effectiveness, and efficiency of its global reach.

Technological Opportunities. Finally, technological opportunities can enable new concepts of employment and, in and of themselves, motivate broader policy changes. In recent satellites, there has been a dual trend toward both small and large satellites. One of the more intriguing possibilities emerging in laboratories now is a result of guidance from the Air Force Scientific Advisory Board's New World Vistas Study,[20] which predicted clusters of many small satellites operating collaboratively. Air Force, NASA, and university laboratories have been developing the necessary technology. Such satellites hold the promise of having manufacturing and testing processes similar to those for high-volume electronic commodities, including the benefit of economy of scale.

In application, such clusters could perform like impossibly large monolithic satellites to provide sensing precision and resolution and surveillance area coverage. The distributed nature of clusters holds the promise of graceful degradation under failure and inherent resilience to attack. In the most optimistic view, these technologies might even enable ubiquitous, global satellite access at costs now associated with episodic access from combinations of high-demand, low-density air and traditional space platforms.[21]

Somewhat in contrast to the trend toward large clusters of very small collaborating satellites is another toward extremely large, deployable, precision structures in space. The first instance of this will likely be NASA's Next Generation Space Telescope, which will unfold in space as a very large mirror, on the scale of the largest terrestrial telescopes today (e.g., the Keck telescope). A few telescopes of this size (about six, depending on size and altitude) could provide continuous visibility of the earth. Such satellites could also provide images of distant satellites to identify their purpose and status. A few dozen, if successfully integrated with high-energy lasers,[22] could provide a layer of boost-phase missile defense.

----

[20]The Chief of Staff and the Secretary of the Air Force commissioned New World Vistas as a 50-year revisit of von Karman's pioneering study, Towards New Horizons, conducted for General Hap Arnold after World War II. The original study was the impetus for the founding of the Air Force Scientific Advisory Board (see U.S. Air Force Scientific Advisory Board, 1995).

[21]Another advantage of such "ubiquitous" platforms might also lie in architectures for real-time on-demand sensing of space and atmospheric weather at scales of measurement that could substantially mitigate the effects of weather on military operations. Among the effects would be improvement of the accuracy of geolocating the source of signals on the ground intercepted through the distorting effects of the ionosphere in space.

[22]Such integration is a challenge, something like connecting a jet engine operating at full throttle to an astronomer's telescope without causing it to vibrate while pointing it at distant stars.

Over the first decade of the 21st century, space communications may employ robust, very-high-data-rate, "intersatellite" links that will be based on the lasers and fiber-optic amplifiers used for terrestrial, fiber-optic communications. The networks would be substantially harder to intercept for uninvited eavesdropping; more important, they would allow more-capable and more-flexible sensing and communications architectures with processing, hardware, and human interaction located where most convenient. This could further reduce the need for low-density, high-demand command and control platforms, as well as the need for their logistic support footprint in theater.

Finally, marginal, evolutionary improvements in space launch technology will improve the economic prospects for RLVs—if demand for launch rates increases enough. This could cause an evolution in some space operational concepts, particularly replenishment and reconstitution. Applying this technology to long-range weapon delivery, such as a reusable ballistic missile or boost-glide missile, could change the effects of basing, access, and logistics lift issues on the interdiction and strategic strike missions of an air campaign and on the halt phase of a theater strategy.

Threat-Driven Considerations

Beyond bureaucratic or technical drivers, U.S. military space policy must account for actively hostile actors: those seeking to use ballistic missiles against the United States or its allies, those threatening U.S. security or forces from space, and those threatening U.S. space capabilities from any venue. The description here is necessarily speculative. The aim is not to cover the range of possible speculation but to stay close to a credible centroid of possibilities reasonably close to current capabilities.

Delivery of Weapons. Long-range ballistic and cruise missiles offer an effective means of delivering either conventional warheads or weapons of mass destruction. The increasing proliferation of missile technology and weapons of mass destruction is changing the U.S. political climate for missile defense, space weapons, and space technology. This climate change may create fundamental change in U.S. use of space in the long term. In the short term, it has already changed the government's regulatory climate for commercial space.

A spate of recent developments has heightened U.S. concern about the threat and consequences of missile proliferation. Examples that have generated concern include China's 1996 attempt to exert political pressure on Taiwan through the use of unarmed missile test launches into the Taiwan Strait and, perhaps most importantly, North Korea's test of a long-range, multistage ballistic missile in August 1998. The unannounced flight test seemed aimed at demonstrating Pyongyang's intent to acquire a missile capability that not only can threaten its Asian neighbors but also can hold the United States at risk. Other countries are making steady progress on developing longer-range ballistic missiles and have undertaken progressively longer-range missile tests (e.g., Iran, Pakistan, India).

At best, nonproliferation instruments, such as export controls, unilateral sanctions, and the multilateral Missile Technology Control Regime, can only delay missile proliferation efforts. Furthermore, the degree of advanced warning that the United States can expect on foreign missile developments is uncertain. In July 1998, the Rumsfeld commission on ballistic missile threats to the United States warned that there is a high risk of surprise because these countries are adept at using denial and deception to diminish the U.S. intelligence community's ability to make accurate forecasts of missile proliferation developments (Rumsfeld, 1998, pp. 21–22).

With growing concern over the threat posed by long-range missiles and weapons of mass destruction in the hands of potential opponents that have been heretofore comfortably distant and relatively impotent, the U.S. political climate toward missile defense has shifted. A pivotal event in this shift was the January 1999 commitment to funding for a limited national missile defense, with a deployment decision originally set for June 2000 (Cohen, 1999).[23] The second Bush administration further committed to "pursue a robust missile defense RDT&E [research, development, test, and evaluation] program to acquire the capabilities to deploy limited, but effective missile defenses as soon as possible." Finally, the Bush administration has declared its intention not to be constrained from pursuing

---

[23]Although the Clinton-era national missile defense program used space only to detect and track targets, evolving threats could provide the forcing function to move weapons to space as well.

missile defense technologies by the provisions of the existing Anti-Ballistic Missile Treaty.[24]

In the shorter term, this political climate of missile fears has had the unfortunate side effect of demonizing space. Missile and space technologies are fundamentally similar and related, differing only in details and degree. Not too surprisingly, exports of space commodities have become entangled with missile proliferation controls. Otherwise normal commercial activity becomes criminal. The industry's products are treated as munitions, and even legitimate activities have been associated with allegations of nuclear weapon espionage and external interference with U.S. political campaigns (Cox, 1999; Mintz, 1999; Schwartz, 1999). Remote sensing methods that are well known in the open scientific literature have become the subject of sensational allegations of espionage (Gerth and Risen, 1999; Liu and Wu, 1999).

Opposition Use of Space. Beyond missiles, opponents may use space-based capabilities against the United States or its allies. Most potential opponents in various kinds and levels of conflict will be able to derive substantial military utility from internationally available space commodities and services. At the strategic level of indications, warning, and intent, much of the kind of insight that the United States gained from its early reconnaissance satellites will be openly available from civil, scientific, and commercial remote sensing satellites. When other sources and indications (e.g., the Cable News Network [CNN], scientific communications) draw attention, quite detailed insight will be available at reasonable prices from high-resolution commercial remote sensing satellites.

At the theater level, opponents should still be able to see movement and placement of large maneuver elements and logistics stores using civil earth-resources satellites to observe the effects of traffic on the terrain. As high-spectral-resolution sensing from air and space develops a market, U.S. attempts at camouflage, concealment, deception, and denial will have to be increasingly comprehensive, consistent, and sophisticated if they are to succeed in surprising

---

[24]On December 13, 2001, President Bush announced that he had given Russia formal notice that the United States was withdrawing from the ABM Treaty, in accordance with treaty procedures requiring a six-month prior notification (White House, 2001).

opponents.[25] As a corollary, the United States should expect opponents to develop increasing skills in this area.

At the tactical level, opponents will have access to secure command and control using services or products adapted from business communications. Very-small-aperture satellite terminal networks will allow high-data-rate exchanges between maneuver elements; large platforms; and air defense, artillery, and missile batteries. Commercial trends toward increasing reuse of frequencies with smaller spot beams will make intercepting or interfering with the communications increasingly difficult if the jammer or listener is not increasingly close to either the satellite or its user. Handheld terminals will allow moderate data rates and secure voice contact down to small units, although they will be relatively susceptible to downlink jamming.

These commercially and publicly available capabilities will be within the reach not just of conventional forces and the larger nations able to field them but also of unconventional forces and terrorists with resources comparable to the operating budgets of small business offices.

Moreover, for opponents with the resources to acquire space systems of their own, either indigenous or imported or adapted from commercial vendors, the capabilities listed above may be augmented with others. Having more satellites can improve situational awareness by providing more-frequent updates (this seems to be the intent of the Israeli commercial imaging program). Imaging radar and SIGINT satellites would be less susceptible to bad weather and darkness. Two or three moderate-aperture SIGINT satellites in highly elliptical orbits would give a midlatitude country with regional ambitions long-dwell access to its area of interest. They would permit detection, identification (of type and activity), and geolocation (not precise enough for targeting but adequate for warning and characterizing activity) of fixed and mobile emitters. Larger constel-

---

[25]In the future (and at a price), extremely precise terrain and structure profiles (and change detection) will be available from interferometric radar. After earth resources sensors have detected gross indications of troop and supply movement, commercial higher–spatial resolution sensors will confirm the broad area picture with samples that will permit checking the source and details of terrain scarring for attempts at deception.

lations or clusters of satellites with long baselines between receivers could provide greater precision.[26]

For almost any opponent, but especially for those with the resources to augment internationally available civil navigation satellites, all-weather precision ordnance will be available comparable to the accuracy the United States achieved initially in Serbia in 1999 and dramatically in operations in Afghanistan in 2001. This accuracy will extend to ballistic and cruise missiles as well as to bombs.

Denial of U.S. Use of Space. In addition to some leveling of advantage from using space themselves, some opponents will attempt to deny or disrupt U.S. space-based capabilities. Almost any state-level opponent will have some ability to jam, dazzle, or deceive some U.S. and most commercial sensing from space. Similarly, the same class of opponents will have some ability to jam communications satellite uplinks or downlinks. Commercial systems, wideband services, and small mobile users will be the most vulnerable. Opponents will also be able to engage in the cat-and-mouse game of electronic countermeasures and counter-countermeasures against military use of satellite navigation signals.

Opponents with moderately developed ballistic missile technology could develop such missiles as direct-ascent antisatellite interceptors against low-altitude satellites.[27] Larger, more-developed opponents could have more-effective and selective weapons against low-altitude satellites in ground-based high-energy lasers. For example, Russia, China, or India could reasonably have them sometime in the first three decades of the 21st century.

Opponents with access to space could place various kinds of weapons or intelligence-gathering means in orbit and target them

---

[26]For example, France has flown two small developmental satellites for SIGINT and plans to fly them in a cluster for greater precision (Brunet, 2001; Grasset, 2000).

[27]Those with nuclear weapons could choose to use them as warheads, either to increase the probability of killing an individual target or to shorten the lives of unshielded and older satellites by temporarily increasing the radiation environment in the Van Allen belts. The effectiveness of the latter use would depend on the latitudes the weapons could access and the state of the satellites in orbit at the time. Aside from uncertain effect, the utility of this use of nuclear weapons seems relatively low compared with other uses, and it entails a large amount of collateral damage to other satellites.

against space systems. Many of these could pose surveillance challenges. For example, some could be very small satellites operating in close proximity to U.S. satellites. Several universities around the world have been developing just such satellites. The surveillance challenge arises both from the small size of the satellite and the difficulty of resolving two satellites in close proximity from a distance. The surveillance architecture may thus need to include putting proximity sensors on high-value satellites.[28]

Continuing U.S. Advantages. On balance, with these advances in opponent use of space, the United States could still expect to enjoy advantages, if it provides reasonably for the threats to its own space capabilities and continues to improve them. The United States should, for example, enjoy the following advantages:

• situational awareness over wider areas, thanks to having more vantage points

• greater precision than opponents in locating targets from space, to the point of handoff to a terrestrial weapon platform

• active (e.g., radar) sensing from space, sufficient for tracking more-challenging moving targets (ground and air) and over much broader areas

• greater resilience to interference, adverse weather, and attempted deception through a greater diversity of sensing means and phenomenologies.

---

[28]The volume of space that needs to be searched for these and similar satellites presents an additional surveillance challenge if they are deployed from a space object that has already been cataloged at a time when that object is not being tracked. New orbital objects are normally cataloged when a satellite is launched from earth, when an existing object breaks up, or when one of a small number of surveillance "fences" reveals an uncataloged object. Downward-looking missile warning sensors can readily detect the high energy of a satellite launch. Without seeing the actual breakup, routine tracking of known objects can lead to detection of a breakup by revealing multiple objects in proximity to the original satellite's described orbit. However, if a hostile satellite is not launched from earth, the event could very well go undetected until the satellite has done its mischief or has become unresolvable because of its proximity to another space object. Such a satellite could instead deploy from an apparently inert body, such as a rocket that had previously delivered another satellite to orbit, or from another satellite (which would remain in place, serving as something of a decoy). This surveillance challenge would require an upward-looking system capable of detecting deployment events in space without external cues.

These advantages would accrue from a technological edge, decades of experience in application, and supplying a greater scale of resources than other nations.

FUTURE CHOICES

The preceding sections have examined both the relative strength and dynamism of U.S. space capabilities and the motivations the nation may have for making changes to its military space activities. Change can take many forms; here, we focus on three realms of choice for the country and the Air Force for their future in space: policy, enterprise, and organization.

Policy

Within the realm of space policy, the Air Force helps shape national policy through the interagency process, through legislative liaison, and through activities that inform or influence public debate. With other implementing agencies, the Air Force also shapes policy through the details of execution and through the context that its operations and programming decisions create. For example, a decision to acquire GPS-guided munitions may emerge from a simple desire to benefit from satellite navigation. But that decision in turn influences the larger policy context through the related needs to promulgate GPS as an international standard and utility and to supply it as a free good. Doing so discourages independent, competing capabilities that might not be as useful to the U.S. military.[29]

While the Air Force is not the sole determiner of defense space policy, the service's competence, commitment of people, and role as steward of roughly half the national security community's space resources make its voice significant de facto. The tone and carrying power of that voice depend to some degree on the state of the Air Force's intimate, but sometimes ambivalent, relationship with NRO, the other major steward of national security space. Recognizing the limits of Air Force influence, we can still usefully identify looming policy issues and opportunities.

---

[29]A good example is the European Union's difficulties in attracting private investment for its Galileo project, which seeks to use a combination of public and private sector investment to finance an alternative satellite navigation capability (de Selding, 2002b).

Three issues in particular have significant consequences that the Air Force might find of concern:

- assertion of space as an economic center of gravity
- space weapons
- industrial policy.

Space and the Economy. U.S. Space Command's long-range plan asserts that U.S. use of space and investment in space infrastructure will constitute an economic center of gravity and will, therefore, be a vulnerability needing military protection (Estes, 1998). The assertion invokes a saying from the maritime and mercantilist era that the "flag follows trade." While the maritime simile may seem awkward and the assertion may seem premature under scrutiny, current national security policy for space hints at this assertion.

Realization or acceptance of the assertion's prediction would have significant effects on the structure and operations of the national security space force. At the level of force structure, the space equivalent of convoy escort is unlikely in the near term, and reprisals elsewhere would likely be a more cost-effective deterrent than a defense in space would be. As our earlier discussion of space surveillance suggests, the degree and kind of space situational awareness would need to change substantially, as would the contributing abilities of overhead intelligence systems. At the operational level, important changes would also be necessary, particularly in military relationships and information-sharing with U.S. commercial space entities.

Space Weapons. The issue of space weapons (either applied from space against terrestrial targets or applied from anywhere against space targets) is emotional, domestically and internationally. The emotion is evident on both sides of the issue, as the exaggerated fears about the consequences of such weapons and the attributions of their value demonstrate.[30] The issue looms in U.S. public debate in conjunction with missile defense. The U.S. Space Command's long-range plan presumes that the policy issue will be resolved to allow

---

[30]For a comprehensive introduction to the issues of space weapons in terrestrial conflict, see Preston et al. (2002).

U.S. preparations for conflict involving space weapons of all types but circumspectly avoids advocacy (Estes, 1998). National space policy has always reserved U.S. prerogatives to defend its interests in space and to deny adversaries advantage from use of space. While some space-based weapons might find interesting cost-effectiveness niches in some operational contexts, most should neither be greatly desired nor greatly feared.

On the other hand, weapons of various sorts targeted against space systems are an almost certain near-term outcome of U.S. military dependence on space. Air Force interests in space weapon policy should lie principally in avoiding or mitigating surprise from others' decisions to deploy them. In particular, given the relative ease of interfering with some space capabilities using readily available terrestrial means, the Air Force should organize and train aggressively for operations against space-using and space-interfering opponents.

A more extreme case, which is topical because of congressional interest, is space-based missile defenses. If or when the nation decides to deploy space-based defenses against ballistic missiles in boost phase, the implications for Air Force force structure and operations would be profound in both scale and consequences.[31] However, the operational deployment of space-based laser missile defenses seems unlikely in the next 30 years. Space-based interceptors are a readily available technology, but they are helpful primarily as a midcourse layer of a multilayer defense if opponents find it too easy to saturate a limited terminal defense. They are of limited utility against short-range theater missiles or the boost phase of long-range missiles.

Policy Toward the Nation's Space Industry. Although stated policy is to encourage and support the nation's space industry, particularly the maturation of the commercial sector, implementation of that policy across the government is uneven, even counterproductive, in

---

[31]A carrier battle group, including its needs for guided missile cruisers, antisubmarine warfare escorts, airborne early warning aircraft, fleet oilers, etc., is an apt metaphor for the scope and scale of surveillance, defense, and logistics support that a space-based laser would need. An Air Force with a space-based laser constellation might resemble a two-dozen carrier battle group navy (although with much less manpower). The effects on the rest of the Air Force would naturally be profound, even if the lasers did not reach into the atmosphere to change our notions of air superiority.

places. The Air Force has little direct opportunity to set that policy and limited opportunity to influence its implementation. Where the service has latitude, it has been at the forefront of implementing the policy, for example, in acquisition strategy. This is natural; the Air Force stands to benefit more than most from improved efficiency and new capabilities enabled by a mature commercial space industry. The Air Force might seek a more vocal role in influencing policy and implementation by others. It should perhaps also consider exit strategies limiting its liability if current industry trends do not stabilize or improve.

Enterprise

The second realm of choice available to the Air Force in defining its future in space is largely within its power to choose: its choice of enterprise or role. Historically, the Air Force has been the service most encumbered by resource and personnel commitments to space and the least dependent on or effective in employing space for military purpose. That has been changing radically. The Air Force continues to be the most encumbered by stewardship but has been rapidly integrating (and therefore depending on) space capabilities to support expeditionary operations and to conduct all-weather precision strike operations.

The Air Force's role as a supplier of space capabilities may change as more space services and commodities become available more economically from civil, converged, or commercial sources. This does not, however, mean that the Air Force should expect to supply fewer such services; rather, the level of demand is likely to be sustained or to expand, thanks to new and expanded capabilities in areas for which the commercial sector does not have a viable business case. These areas include surveillance, warning, targeting (including navigation), protected communications, and weapons. To find the resources for an expanded level of supply, the Air Force may find it desirable to partner or integrate architectures with NRO; this would be easier to implement with the realignment of NRO and Air Force space acquisition under a single Air Force official.

As the Air Force's use of space matures and expands, it can consider an evolution from operations to command. In this context, command is not meant as the relay of instructions to satellites but rather the accountability for allocation of limited resources to military purposes, for example, power, boresight dwell time, receiver tuning, revisit rate, target designation, weapon release, and expendable consumption. The change in enterprise is from operator to employer. The ability to make the change is a consequence of the increasing ability of satellites to "fly" themselves. The pressure to make the change will come from the evolving nature of Air Force space operations that we noted in our earlier discussion of the trend toward tasked systems over scheduled utilities.

The last choice of enterprise the Air Force has to make in space is in the nature of its role as defender. We noted earlier the call in U.S. Space Command's long-range plan to protect economic interests in space (Estes, 1998). Presumably, protecting the interests and infrastructure will be balanced with protecting other economic interests and critical terrestrial infrastructure. Protecting space interests and infrastructure may include some element of military deterrence. It will likely include a dominant contribution by legal, diplomatic, and law enforcement activity. It should likely not require substantial change in military force structure, although it would certainly require substantial improvement in space situational awareness. However, independent of space's economic value, increasing military dependence on space, including commercial space, will require a greater role in protecting its utility. The choices available to the Air Force are in the level of threat and conflict, in the scope or extent of satellites and services protected, and in the approach to its role as defender.

Organization

A natural corollary of choice in enterprise is choice in organization. A discussion of Air Force organization requires some historical context, and several factors have driven the current organization. First (and by happenstance), the Air Force operational space command was formed around a nucleus of nuclear deterrence and warning. The operational culture rightly demanded caution and predictability over initiative and responsiveness. This happenstance was reinforced by the merger of space and missile operational career fields. Not surprisingly in that operational culture, the missileers rose rapidly to the top. The culture and the people may prove an obstacle to tactically useful space capabilities.

Second, the natural organizational model for normalizing space operations was air operations. As a consequence, the focus is on platforms rather than missions. This has had the curious, and ostensibly harmless, effect of enshrining space maintenance activity as operations. The harm becomes apparent to the degree that space operational requirements and priorities for funding are phrased in terms of platform characteristics and maintenance that resemble those of air operations rather than in terms of the end use and usability of a system's capabilities.

Third, in distinguishing itself from its past, the newly operational Air Force space community distanced itself from its historical roots in the research and development and intelligence communities. An arm's-length relationship with research and development creates obstacles to rapid innovation and isolates the creativity of research and development from the real needs of operation and mission.

A similar disdain for intelligence, or at least a lack of interest in it, isolates space operations from a crucial source of situational awareness necessary for tactical use of space. In a double irony, isolation from NRO has isolated Air Force space operations from a space operational culture that has long dealt with shared, taskable capabilities engaged in multiple missions—albeit only recently with increasing military missions and tactical uses. The Air Force also isolated itself from NRO's cradle-to-grave approach to space operations that encompassed technology, acquisition, and operation of all elements contributing to the mission's accomplishment and employment of their products.

Finally, and perhaps the biggest source of obstacles, isolating Air Force space operations organizationally from its air operations impedes communication between elements that must be integrated to bring space to the theater. This poses resource issues that should be integrated as separate or even competing needs. The result is competing advocacy of air and space platforms or independent allocation of budget cuts to air and space resource pools when the question should often be: "What is the right mix of combined air and space capabilities for the mission tasks at hand?"

With this context, the Air Force faces two serious organizational issues in the near term: space integration and the relationship between intelligence and operations. With regard to integration, trends in Air Force space systems and a choice of the "employment" enterprise would both suggest the need to further integrate space into combat operations rather than segregate it from air combat and combat support commands. Increased outsourcing of space launch and satellite operations and infrastructure should make integration an easier choice. The risk in integration is that a prematurely integrated organization might lack the maturity in its understanding of space employment to make wise choices in allocating combined resource pools. It might err in either direction—conservatively toward familiar air or enthusiastically toward novel space. Conversely, the potential rewards of integration are more-effective use of air and space and more-efficient use of limited overall resources. Because the risks are tied to resource advocacy, the Air Force might wish to integrate operation and employment (training and organizing) before resource, or at least before modernization, decisions (equipping).

The Air Force faces a fundamental question with regard to its organization and relationship to the intelligence community: the nature of the relationship between operational military space and operational intelligence space activities. Even before the Space Commission's report (Rumsfeld, 1998), the Air Staff had been tentatively exploring opportunities for integrating Air Force and NRO planning activities. So far, the Space Commission's recommendations and DoD implementation have addressed alignment of only the acquisition activities of Air Force and NRO space systems, but the Air Force and NRO should go beyond this to explore more-significant organizational changes, such as consolidation of launch and satellite operations activities. Consolidation of today's installation and maintenance satellite operations could occur in conjunction with outsourcing decisions. Alternative organizational models could also address the opportunity for a more effective and efficient relationship between military space and intelligence space operations.

WAYS AHEAD

This concluding section contains some observations on ways ahead for the Air Force. Where they sound definite, it is only an attempt to keep the observations succinct. These observations should be viewed in the same broad terms as the sketches of trends and possibilities in the preceding material. We suggest that the Air Force could

- Engage to shape opportunity and develop partnerships with the U.S. space industry—employ it as a means of shaping international space capabilities and U.S. advantage. The price may seem high, but aerial alternatives will be increasingly at risk to air defenses and limited in preconflict employment.

- Expect space to enable ubiquitous awareness tactically throughout the depth of the theater and flexibly around the world.

- Stay above the debate on space weapons, informing and calibrating the expectations of both sides while developing suitable technology against eventual need. When the weapons are clearly needed, the debate should subside.

- Prepare to defend critical space capabilities and interests and to attack those of opponents. This is not a call for space fleets, although some such forces may be needed eventually. Rather, it is a prescription for situational awareness incorporating space and theater perspectives coupled with responses employing the full range of means currently available in joint military operations.

- When defining enterprise, consider command—the employment of space in joint military missions—a priority compared with its historical emphasis on the installation and maintenance of satellites.

- Organize to integrate—space with other joint forces and military space with intelligence space architecture, acquisition, and operations.

Given its new role as the DoD's executive agent for space and new direction to reorganize national security space, the Air Force faces pivotal opportunities. We hope these observations and suggestions shed some useful light on the ways ahead.

REFERENCES

Acker, Russell, "A Vane Blown with All Winds: Technology Export Policy, Economic Competitiveness and National Security," in Ray A. Williamson, ed., *Dual-Purpose Space Technologies: Opportunities and Challenges for U.S. Policymaking*, Washington, D.C.:

George Washington University, Space Policy Institute, July 2001, pp. 151–189.

ANSER Corporation, *India, Its Space Program, and Opportunities for Collaboration with NASA: An Overview*, Arlington, Va., January 14, 1999.

Barboza, David, "Executive to Raise at Least $1.2 Billion to Aid ICO," *New York Times*, November 2, 1999, p. C3.

Bender, Bryan, "USAF Looks to Phase Out Laser-Guided Bombs," *Jane's Defence Weekly*, November 17, 1999.

Berger, Brian, "NASA Chief Calls for Closer Cooperation with Pentagon," *Space News*, January 28, 2002, p. 8.

Berger, Brian, and Leonard David, "Goldin Defends SLI, Encourages Shuttle Privatization," *Space News*, November 5, 2001, p. 9.

Boeing, "Boeing Team Wins Future Imagery Architecture Competition," press release, Seal Beach, Calif.: Boeing, September 3, 1999.

Brewin, Bob, "Bosnian Mission Demands Most Complex Network Ever," *Federal Computer Week*, April 29, 1996.

_____, "IT Gives Navy Global C2 Capabilities," *Federal Computer Week*, May 19, 1997.

Brunet, Patrick, "France Unveils New SIGINT Satellites," *Journal of Electronic Defense*, July 2001, p. 32.

Carnegie Endowment for International Peace, Non-Proliferation Project, Administration Missile Defense Papers, 2001. Online at http://www.ceip.org/files/projects/npp/resources/EmbassyCable NMD_copy.htm (as of December 12, 2001).

Cáceres, Marco Antonio, "Hopes Fading for Most RLVs," *Aviation Week & Space Technology*, January 14, 2002, p. 142.

Clarke, Victoria, Assistant Secretary of Defense, Public Affairs, quoted in Bloomberg.com, "Boeing Co. JDAM Most Widely Used Precision Bomb in Afghanistan," Bloomberg.com, December 11, 2001.

Cohen, William S., Secretary of Defense, and Gen. Henry H. Shelton, Chairman of the Joint Chiefs of Staff, DoD News Briefing, Subject:

Missile Defense, Washington, D.C., January 20, 1999. Online at www.defenselink.mil/news/Jan1999/t01201999_t0120md.html (as of December 13, 2001).

Commercial Space Transportation Advisory Committee, *1999 Commercial Space Transportation Forecasts*, Washington, D.C.: Federal Aviation Administration, Associate Administrator for Commercial Space Transportation and the Commercial Space Transportation Advisory Committee, 1999.

_____, *2001 Commercial Space Transportation Forecasts*, Washington, D.C.: Federal Aviation Administration's Associate Administrator for Commercial Space Transportation and the Commercial Space Transportation Advisory Committee, 2001.

COMSTAC—*See* Commercial Space Transportation Advisory Committee.

Cox, Christopher, *U.S. National Security and Military/Commercial Concerns with the People's Republic of China*, Washington, D.C.: U.S. House of Representatives, Select Committee on U.S. National Security and Military/Commercial Concerns with the People's Republic of China, 1999.

Danilov, Igor, "Russia: Analyst Views Prospects for Rocket Forces After Security Council Talks," Moscow Interfax report, August 14, 2000, Foreign Broadcast Information Service (FBIS-SOV-2000-0814).

DeKok, Roger, and Bob Preston, "Acquisition of Space Power for the New Millennium," in Peter L. Hays, James M. Smith, Alan R. Van Tassel, and Guy M. Walsh, eds., *Spacepower for a New Millennium: Space and U.S. National Security*, New York: McGraw-Hill, U.S. Air Force Academy, Institute for National Security Studies, 2000.

de Selding, Peter B., "U.S. Satellite Builders Dominate Market in 2001," *Space News*, January 21, 2002a, p. 3.

_____, "Europe Offers Last-Ditch Effort to Save Galileo," *Defense News*, February 4–10, 2002b, p. 19.

Estes, Howell M., *USSPACECOM Long Range Plan*, Colorado Springs, Colo.: Headquarters, U.S. Space Command, 1998.

Fogleman, Ronald R., and Sheila E. Widnall, *Global Engagement: A Vision for the 21st Century*, Washington, D.C.: U.S. Air Force, 1997.

Futron, Inc., *Trends in Space Commerce*, Washington, D.C.: Department of Commerce, 2001.

Gerth, Jeff, and James Risen, "Reports Show Scientist Gave U.S. Radar Secrets to China," *New York Times*, May 10, 1999, p. A1.

Gordon, Michael R., "Pentagon Corners Output of Special Afghan Images," *New York Times*, October 19, 2001, p. B2.

Grasset, Bernard, *French National Assembly Report on Sigint, Comint*, Vol. III: *Defense—Space, Communications, and Intelligence*, Paris: French National Assembly Presidency, Committee on National Defense and the Armed Forces, Opinion No. 2627, translated excerpt, Foreign Broadcast Information Service, 2000.

Hall, Keith R., Presentation to the Senate Armed Services Committee Strategic Force Subcommittee, Washington, D.C.: National Reconnaissance Office, 1998.

Harvey, Brian, *The Chinese Space Programme*, New York: John Wiley & Sons, Inc., 1998.

Hastings, Daniel, *"Doable" Space Quick Look*, Washington, D.C.: Office of the Chief Scientist of the Air Force, 1998.

Hughes Space and Communications Co., "Hughes, Navy Sign Contract for 11th UHF Follow-on Satellite," El Segundo, Calif.: Hughes Space and Communications Co., 1999.

ICO, New ICO markets, 2000. Online at http://www.ico.com/products/index.htm (as of July 30, 2001).

The Independent Commission on the National Imagery and Map ping Agency, *The Information Edge: Imagery Intelligence and Geospatial Information*, Washington, D.C., 2001.

Jeffery, Mac, and Michael Pascale, Globalstar Telecommunications Appoints Restructuring Officer, press release, July 18, 2001. Online at http://www.globalstar.com/EditWebNews/205.html (as of December 13, 2001).

Kerrey, J. Robert, and Porter Goss, *Report of the National Commission for the Review of the National Reconnaissance Office*, Washington, D.C.: National Commission for the Review of the National Reconnaissance Office, 2000.

Knauf, J., L. Drake, and P. Portanova, "Evolved Expendable Launch Vehicle System: The Next Step in Affordable Space Transportation," paper presented at the 52nd International Astronautical Congress, Toulouse, France, October 1–5, 2001.

Liu, Antony K., and Sunny Y. Wu, "Submarine Wake Detected in Raw RADARSAT SCANSAR Image? NASA Goddard Spaceflight Center," October 4, 1999. Online at http://kaon.gsfc.nasa.gov/kaon_root.html (as of October 12, 1999).

Loeb, Vernon, "Sharp Eye in the Sky Lets Nations Spy—For a Price," *Washington Post*, May 10, 2000, p. A3.

Marquis, Christopher, "Some Lawmakers Urging U.S. to Speed Exports of Satellites," *New York Times International*, July 9, 2001, p. A7.

McLucas, John L., *Space Commerce*, Cambridge, Mass.: Harvard University Press, 1991.

Mintz, John, "Grand Jury Probes Boeing's Rocket Deal with Russia, Ukraine," *Washington Post*, March 2, 1999, p. A8.

NIMA Commission—*See* The Independent Commission on the National Imagery and Mapping Agency.

Pascale, Michael, "Globalstar Acts to Assure Continued Funds for Operations," Globalstar, January 16, 2001. Online at http://www.globalstar.com/EditWebNews/179.html (as of August 13, 2001).

Pearlstein, Steven, "Canada Balks at New U.S. Policy: Ally Prepared to Turn to Europe for Military Technology," *Washington Post*, August 12, 1999, p. E2.

Peters, F. Whitten, "Today's Challenges—Tomorrow's Vision: Positioning the EAF for the 21st Century," presented to the Foundation Forum at the Air Force Association's Air Warfare Symposium in Orlando, Fla., February 5, 1999. Online at http://www.aef.org/symposia/honpeter.html (as of December 20, 2001).

Pollack, Andrew, "Despite Its Promise, Satellite Industry Grapples with Myriad Woes," *New York Times*, August 4, 1999a. Online at http://www.nytimes.com/library/financial/080499space-biz.html (as of April 11, 2002).

_____, "Export Rules Are Said to Be a Threat to Satellite Industry," *New York Times*, August 1, 1999b, p. A14.

Preston, Bob, *Plowshares and Power: The Military Use of Civil Space*, Washington, D.C.: National Defense University Press, National Defense University Institute for National Strategic Studies, 1994.

_____, "Emerging Technologies," in John C. Baker, Kevin M. O'Connell, and Ray A. Williamson, eds., *Commercial Observation Satellites: At the Leading Edge of Global Transparency*, Washington D.C.: RAND and American Society for Photogrammetry and Remote Sensing, 2001a.

_____, "Regulatory Landscape," in John C. Baker, Kevin M. O'Connell, and Ray A. Williamson, eds., *Commercial Observation Satellites: At the Leading Edge of Global Transparency*, Washington D.C.: RAND and American Society for Photogrammetry and Remote Sensing, 2001b.

Preston, Bob, Dana J. Johnson, Sean Edwards, Mike Miller, and Calvin Shipbaugh, *Space Weapons, Earth Wars*, Santa Monica, Calif.: RAND, MR-1209-AF, 2002.

Reuters, "Iridium Declares Bankruptcy," *New York Times*, August 14, 1999.

Richter, Paul, "U.S. Looking at Spacecraft as Bomber," *Los Angeles Times*, July 28, 2001, p. A1.

Rumsfeld, Donald, *Executive Summary: Report of the Commission to Assess the Ballistic Missile Threat to the United States*, Washington, D.C., 1998.

_____, letter from the Secretary of Defense to the Honorable John Warner, Chairman, Committee on Armed Services, May 8, 2001a. Available from http://www.defenselink.mil/news/May2001/d20010508space.pdf (as of July 31, 2001)

_____, *Report of the Commission to Assess United States National Security, Space Management and Organization*, Washington, D.C.: Commission to Assess United States National Security Space Management and Organization, 2001b.

_____, Secretary Rumsfeld Announces Major National Security Space Management and Organizational Initiative, press release, May 8, 2001c. Online at http://www.defenselink.mil/news/May2001/b05082001_bt201-01.html (as of March 25, 2002).

Ryan, Michael E., and F. Whitten Peters, *America's Air Force Vision 2020: Global Vigilance, Reach and Power*, Washington, D.C.: U.S. Air Force, 2000.

Schneider, Greg, and Kathy Sawyer, "New Mission for Lockheed Space Plane?" *Washington Post*, April 13, 2001, p. E1.

Schwartz, Stephen I., "A Very Convenient Scandal," *Bulletin of the Atomic Scientists*, May–June 1999, p. 34.

Scoffield, Heather, "Ottawa to Cut U.S. Out of Satellite Project: Manley Takes Radarsat Business to Europe," *The Globe and Mail* (Canada), August 11, 1999, p. A.1.

Shalikashvili, John M., *National Military Strategy: A Military Strategy for a New Era*, Washington, D.C.: Defense Technical Information Center, 1998. Online at http://www.dtic.mil/jcs/nms/(as of December 13, 2001).

Smith, Bob, Senator (R-N.H.), "The Challenge of Spacepower," paper presented at the Fletcher School/Institute for Foreign Policy Analysis Annual Conference, Cambridge, Mass., November 18, 1998.

Standard & Poor's, "S&P Cuts Hughes Electronics," press release, New York, 1999.

Tirpak, John A., "With Stealth in the Balkans," *Air Force Magazine*, October 1999, p. 23.

Trimble, Paula Shaki, "U.S. Air Force Study Finds No Commercial 'Pot of Gold,'" *Space News*, November 8, 1999, p. 4.

U.S. Air Force Scientific Advisory Board, New World Vistas: Air and Space Power for the 21st Century, 1995. Online at http://www.sab.hq.af.mil/archives/reports/1995/NWV/afrtnwv.htm (as of February 27, 2002).

U.S. Congress, National Defense Authorization Act for Fiscal Year 2000, S. 1059, Title XVI, Subtitle C—Commission to Assess United States National Security Space Management and Organization. Washington, D.C.: 106th Congress, 1999.

"U.S. Satellite Export Controls Said to Cost $1.2 Billion, 1,000 Jobs," *Aerospace Daily*, Vol. 197, No. 24, February 7, 2001, p. 201.

The White House, Office of the Press Secretary, statement on the sale of space imagery, Washington, D.C., March 10, 1994. Online at http://clinton6.nara.gov/1994/03/ (as of December 20, 2001).

_____, "President Discusses National Missile Defense: Remarks by the President on National Missile Defense," press release, December 13, 2001. Online at http://www.whitehouse.gov/news/releases/2001/12/20011213-4.html (as of March 25, 2002).

Williams, Daniel, "Putin Moves to Resolve Military Dispute," *Washington Post*, August 12, 2000, p. A10.

Chapter Six

U.S. MILITARY OPPORTUNITIES: INFORMATION-WARFARE CONCEPTS OF OPERATION

Brian Nichiporuk

INTRODUCTION

Information warfare is often seen as a new threat, a tool for adversaries to use against the U.S. homeland or U.S. forces. Numerous stories about break-ins at Pentagon computers, disabled satellites, and downed phone networks have focused the attention of the public and the national security community on the need for information-warfare defense. The possibility that these new information-warfare tools could threaten America's ability to project power or to realize its national interests is real and deserves analytical attention and public awareness. However, information warfare creates more than just vulnerability—it may also mean many new opportunities for the U.S. military. New information-warfare tools and techniques hold the potential for the United States to achieve its national security objectives using cheaper, more-efficient, and less-lethal methods.

Although these potential opportunities are a frequent topic of research and discussion within the defense analysis community, they have not received much attention beyond very specialized pockets of that community. This topic garners little outside attention, largely because the literature on information-warfare opportunities falls into one of two distinct categories: (1) broad policy and strategic-implications work and (2) highly technical feasibility studies. Research in the former is often too general to be of specific use to military planners. Research in the latter is often highly classified and compartmentalized. This chapter seeks to bridge the gap between the two by providing an operational-level view of how a set of offensive information-warfare concepts of operation (CONOps) could expand the U.S. Air Force's capabilities to fight future wars (in the

2010–2015 time frame). It seeks to answer the question: How might the Air Force expand its doctrinal thinking about the systematic use of offensive information warfare to improve performance?[1]

What Do We Mean by "Information Warfare"?

One of the major features of information-warfare research is the pot-pourri of different definitions for the term *information warfare*. Without engaging in that debate, this chapter will simply define information warfare as the process of protecting one's own sources of battlefield information and, at the same time, seeking to deny, degrade, corrupt, or destroy the enemy's sources of battlefield information. This is taken to include six preexisting subareas that have only recently been grouped together under the heading of information warfare: operational security, electronic warfare (EW), psychological operations (PSYOPs), deception, physical attack on information processes, and information attack on information pro-cesses.[2] Since operational security is all about defensive information warfare, it is not as important to us here as the other five subareas. Therefore, offensive information warfare consists of the aggregation of EW, PSYOPs, deception, physical attack, and information attack.

EW encompasses the traditional concepts of jamming and spoofing radars and radio communication links. The Air Force's now-retired EF-111 aircraft and the Navy's EA-6B aircraft are good examples of traditional EW platforms. PSYOPs are all about using information dissemination to weaken the enemy's morale and, ultimately, to break his will to resist. Classical PSYOP techniques include the air dropping of propaganda leaflets and using airborne loudspeakers that broadcast demands for surrender to enemy troops. Deception involves the employment of physical or electronic means to camou-flage one's own force posture in theater. Deploying dummy aircraft on the tarmac of a major air base or broadcasting radio situation

---

[1]Note that this chapter previously appeared in an earlier volume in the Strategic Appraisal series. It has been updated and is reprinted here because of its relevance to the present topic. The author would like to thank RAND colleagues Alan Vick, Martin Libicki, Jeremy Shapiro, and Zalmay Khalilzad for their insightful comments on earlier drafts of this chapter.

[2]This grouping is derived from Fig. 8 in Hutcherson (1994), p. 22, as well as Joint Staff (1996), Ch. 2.

reports in the clear from "phantom" or nonexistent units are two instances of deception that have been used in the past.[3] Physical attack is simply the act of physically damaging or destroying an adversary's means of collecting, processing, and organizing information. This includes means as diverse as using aircraft to deliver dumb iron bombs to destroy a corps-level command bunker and using a high-powered ground-based laser to cripple an enemy communications satellite permanently. Finally, information attack involves the use of computer technology to electronically shut down, degrade, corrupt, or destroy an enemy's information systems in theater. Viruses, logic bombs, and sniffers are but three of the "information munitions" that experts in this area commonly discuss.

Many authors tend to equate offensive information warfare with information attack. However, for a true appreciation of the breadth of offensive information warfare, it is really necessary to consider all five elements. Indeed, as we shall see later, a rich mix of all five gives the best chance for success in the information campaign.

The Importance of Offensive Information Warfare

Offensive information warfare is not a "new" way of attacking one's adversary. To be sure, some of the current tools and technologies in this area are novel, but the goals of offensive information warfare today bear striking resemblance to those of the "military deception" campaigns of wars past. In short, while the means for offensive information warfare have changed, the ends have remained similar to those of yesterday.

Broadly speaking, the goals of an offensive information-warfare campaign are to deny, corrupt, degrade, or destroy the enemy's sources of information on the battlefield. Doing so successfully, while maintaining the operational security of your own information sources, is the key to achieving "information superiority"—that is, the ability to see the battlefield while your opponent cannot.[4] In

---

[3]Deception may appear to be a purely defensive tool at first glance; however, deception operations have been used throughout history as integral parts of information campaigns that were heavily offensive. Therefore, in this analysis, deception will be considered to be part of offensive information warfare.

[4]For an overview of how important information superiority in general will be to the United States in future conflicts, see Joint Chiefs of Staff (1996).

today's era of smart weapons and compressed decision cycles, there can be little doubt that the acquisition of information superiority in conventional warfare goes a long way toward achieving final victory.

History provides multiple examples of previous uses of "old-fashioned" offensive information warfare.[5] In the Revolutionary War, American agents supposedly inserted forged documents into British diplomatic pouches as a way of convincing the British that George Washington's army was far larger than it actually was. During World War I, the U.S. Army in France executed an important deception operation called the "Belfort Ruse" before a major attack on St. Mihiel. The Western Allies in World War II accomplished what was perhaps one of the largest "information warfare" successes in history when they fabricated the Calais invasion force in 1944, fooling some German leaders (including Hitler) into believing that the invasion of Northwest Europe would come at Calais, which is well to the north of the actual Allied landing sites in Normandy.[6] All of these historical examples involved the types of tactics that an early 21st century defense analyst would place in the category of offensive information warfare.

Despite the fact that information-warfare campaigns have occurred before, it is now possible to say with confidence that information-warfare campaigns are a relatively more important part of conventional wars than they have been in the past. The increased importance of information-warfare campaigns to the United States in general and the Air Force in particular is due to a combination of technological, doctrinal, and force-structure factors. First, the growth in information technologies is making offensive information warfare a more potent instrument against enemy militaries. As such, offensive information warfare offers new possibilities and options to the regional commanders in chief (CINCs) when they prepare their war plans. As part of a recognition of these new options, U.S. military doctrine is moving away from the platform-centric warfare of the Cold War toward a new concept of network-centric warfare (Cebrowski and Garstka, 1998). In network-centric warfare, information superiority is an essential ingredient of success. Finally,

---

[5]The historical examples provided here are drawn from Hutcherson (1994), pp. 23–24.

[6]For a succinct account of the Allied deception campaign before D-Day in 1944, see Ambrose (1994), pp. 80–83.

America's shrinking conventional force structure demands innovative solutions to emerging problems. As the number of U.S. wings, divisions, and combatant ships declines, U.S. commanders will increasingly rely upon advanced information technology and computer-savvy soldiers to gain the upper hand against adversaries in conventional warfare.

In recognition of this fact, this chapter will develop and elaborate four CONOPs that rely on offensive information warfare. Each CONOP is designed to counter some of the new (and not so new) asymmetric strategies that U.S. opponents are likely to use in future regional conflicts. The first section therefore discusses how the use of such strategies by regional adversaries could make the tasks of the Air Force more difficult. A rich literature on asymmetric strategies already exists, so the discussion here will be heavily derivative. Nonetheless, to set the stage for the proposed CONOPs, this section will lay out the types of asymmetric strategies posing the greatest threat to the United States. The second section presents and evaluates four offensive information-warfare CONOPs that appear to be promising countermeasures to this menu of asymmetric options. The chapter concludes with a third section that evaluates the utility of each of the CONOPs presented.

EMERGING ASYMMETRIC STRATEGIES

The lopsided American victory in Desert Storm and the successful NATO eviction of Serb forces from Kosovo in 1999 both featured clear displays of the vast margin of superiority the U.S. Air Force holds over any conceivable adversary. Most analysts agree therefore that, in future wars, hostile regional powers will use asymmetric options to counter the U.S. advantage in airpower. To organize our thinking about the contributions that offensive information warfare–oriented CONOPs could make toward defeating these asymmetric strategies, we need to begin by listing and categorizing the different strategies. As was noted earlier, a rich literature on asymmetric strategies has developed over the past few years.[7] The work on

---

[7]Two examples of this literature are unpublished manuscripts by Marcy Agmon et al. and by Kenneth Watman, both of RAND.

asymmetric strategies has revealed three types of enemy options the United States needs to be concerned about:

- increasing capabilities in selected niche areas
- enemy strategies that target key U.S. vulnerabilities
- creation of political constraints that hinder U.S. force deployments.

Increasing Niche Capabilities

Regional powers could achieve significant niche capabilities in a number of areas. However, the two that present the greatest cause for alarm are surely the acquisition of weapons of mass destruction (WMD) and improvements in command, control, communications, computers, intelligence, surveillance, and reconnaissance (C⁴ISR) networks.

Enhanced WMD Inventories and Delivery Systems. Several regional powers have stockpiles of biological and chemical weapons, along with the means to deliver them. Making this already difficult problem even more complicated, some regional powers (e.g., Iran, North Korea) may soon come into possession of what can be termed a mature small nuclear arsenal. This would be an arsenal of at least five or six secure and deliverable nuclear weapons supported by a reliable command and control and early-warning network.

The possession of a mature arsenal of nuclear, chemical, or biological weapons by a hostile regional power could restrict airpower's freedom of action.[8] It would be relatively easy for the leadership of that regional power to interpret many types of air strikes that U.S. Air Force planners would regard as strictly "conventional"—such as attacks on air defenses, command and control systems, or mobile missile launchers—as attempts to destroy, or at least degrade, its modest nuclear deterrent.

It is difficult to predict the reactions of small leadership groups in closed states, such as Iran, Iraq, and North Korea, to U.S. air opera-

---

[8]This concern may become one of more than academic significance to U.S. military planners if President George W. Bush decides to expand the war on terrorism to include military strikes against the "New Axis" nations—Iran, Iraq, and North Korea.

tions that threaten their deterrent. Clearly, if the enemy leadership comes to perceive a U.S. conventional air campaign as part of a thinly veiled counterforce plan, the risk that the adversary will escalate to nuclear use increases.[9] The adversary's homeland might evolve into a kind of sanctuary in which large masses of U.S. combat aircraft and cruise missiles could not operate freely because of concerns about escalation to WMD.[10] We will see later on that offensive information-warfare tools, working in concert with small packets of strike aircraft, could be a mechanism for both regaining some operational freedom and reducing the risks of escalation in a sanctuary-type environment. Offensive information-warfare tools can achieve this purpose because they can temporarily degrade or disrupt elements of an adversary's early-warning and air-defense systems without permanently destroying them. This reduces the chances that a U.S. air-defense suppression campaign will be interpreted as veiled counterforce.

The emergence of a homeland sanctuary in wartime would have concrete implications for Air Force planners and operators. Specifically, the enemy's leadership, national command and control, and internal security networks would all become harder to target. Supply and communications for enemy ground forces could not be disrupted on a regular basis, and a large chunk of the enemy's industrial warmaking capacity (including electric power generation and telecommunications capacity) would be essentially off limits to the orthodox offensive use of airpower.

U.S. leaders could choose not to let the enemy establish a homeland airspace sanctuary. If the U.S. leadership is not highly risk-averse, it could deal with WMD in other ways besides offensive information warfare. The United States could, for example, threaten massive nuclear retaliation for any adversary use of WMD and then proceed to carry out an air campaign against the enemy homeland under the assumption that the threat of escalation dominance by the superior U.S. nuclear arsenal cancels out the enemy's nuclear capability. Another option would be to mount a conventional counterforce campaign aimed at destroying the enemy's WMD capabilities and

---

[9]For a discussion of related issues, see Wilkening and Watman (1995).

[10] This sanctuary concept was first proposed by RAND colleague Alan Vick in internal discussions in late 1996.

delivery systems (ballistic missiles, cruise missiles, and fighter-bombers) before they could be employed. Yet a third alternative would be to hurriedly develop and deploy effective theater missile defense systems that would have the effect of reducing an adversary's expectations of the damage he could inflict should he attempt to use WMD against U.S. allies or forces in the field.[11] A future U.S. president could well select any of these approaches. However, in the event that the national leadership is highly risk-averse in a future major theater war (MTW), it behooves the Air Force to plan to deal with scenarios in which much of an enemy's homeland is off limits to sustained aerial attack.

Improved C⁴ISR Capabilities. The information revolution that is now sweeping the world will create more opportunities for regional powers to access advanced space-based communications and reconnaissance systems. Much of this increased opportunity will result from having relatively easy access to multinational commercial assets; some will come from being granted access to dedicated military satellites owned by major powers that could become hostile to U.S. interests (e.g., China, Russia, India); and yet a smaller amount will be due to the development and exploitation of indigenous capabilities.

The proliferation of space-based military and commercial capabilities for both imagery and communications will offer tremendous opportunities for regional powers to increase their capabilities, bringing them closer to those of the United States. The greatest concern in terms of space-based imagery is the proliferation of foreign systems with resolutions equal to or below 5 m. This threshold is critical because 5 m is the level at which one can discern large, soft military targets—such as ports, air bases, and defense ministry buildings—in a theater with enough accuracy to target them specifically using cruise or ballistic missiles, especially if these weapons are Global Positioning System (GPS)–guided.[12] By 2002, France, Israel, India, and Russia will have deployed commercial or military systems

---

[11]For a review of the recent course of the U.S. missile defense program, see Graham (2001).

[12]See Air Force Space Command (1996), p. 24. Recent RAND analysis has made some quantitative assessments concerning the impact of GPS guidance upon the cruise and ballistic missile accuracies likely to be achieved by the militaries of hostile regional powers. See Pace et al. (1995), especially pp. 45–91.

capable of 5-m accuracy (Air Force Space Command, 1996, p. 24; Stoney, 1997). However, the available evidence suggests that no midsized regional power will be able to build its own spaceborne imagery satellites by 2010.

Growth in communication satellites will be more explosive than for imagery systems. There are plans for a whole host of new commercial space communication systems in both low earth (LEO) and geosynchronous orbits (Keffer, 1996). Some of the planned geosynchronous systems will exploit the Ka-band and will use cross-links between satellites to minimize the need for ground stations. Experts predict that, by the end of this decade, there will be two or three new global Ka-band geosynchronous systems and at least one or two "Big LEO" global constellations. Some of these systems will bring massive capacity increases into the world market. As an example, the Hughes Spaceways Ka-band geosynchronous system is projected to have a capacity of 88 Gb/s. This can be compared to the current total Department of Defense requirement for satellite-communication capacity, which is a mere 12 Gb/s (U.S. Space Command, 1997, p. 4-14). At least five such Ka-band systems have been planned for the near future.

Important advances are also occurring in transoceanic fiber-optic technology. Satellite communications may be the optimal solution for mobile military users, but fiber-optic connectivity is probably the most efficient communication option for fixed military users in rear areas. Research into such areas as wave division multiplexing promises to produce per-fiber capacities of up to 160 Gb/s.[13] The number of transoceanic fiber-optic lines is increasing as well, with many large new projects, such as the FLAG line from England to Japan, now entering service.

The upshot of this proliferation of highly capable commercial imagery and satellite and terrestrial communications is that it will be easier in the future for hostile regional powers to have access to the type of C4ISR architectures that only the most advanced militaries could access a few years ago. This applies for both voice and data transmissions. The sheer number of available redundant commercial

---

[13]For an overview of technological developments in the field of undersea fiber-optic lines, see Submarine Systems International (1997).

routes and links will make it almost impossible for the United States to deny service to the adversary on a large scale for a long period of time—because too many communication "choke points" would need to be destroyed, disrupted, or corrupted. However, large-scale service denial for short periods during a theater campaign may still be possible, and such denials would indeed have military significance.

Increased access to overhead imagery will allow regional powers to monitor U.S. and allied force deployments both into and within a theater with greater fidelity than was possible before. The greatest military impact of this new capability is the availability of accurate and timely targeting data to aid in the planning of rapid ballistic- and cruise-missile strikes against air bases, port facilities, and logistics stockpiles being used by U.S. forces in the region.[14] Increased access to highly capable communication systems will lend regional powers the potential for much more timely control of their forces in theater. Decision cycles for these militaries could decrease dramatically. Furthermore, the ability to access large, new international communication networks could facilitate a regional power's offensive information warfare against the Department of Defense's worldwide command and control systems.

Enemy Strategies That Target Key U.S. Vulnerabilities

Another asymmetric option available to regional adversaries of the United States is the use of strategies that threaten key U.S. vulnerabilities and centers of gravity. Such strategies and tactics would be most effective in conjunction with the improved capabilities discussed above, but they could also pose a threat if used on their own. Three strategy types merit consideration: short-warning attacks, antiaccess operations, and deep-strike operations. Each will be covered briefly below.

Short-Warning Attacks. The first strategy that could be used would be a so-called standing-start attack, in which the U.S. intelligence community has little warning of an impending attack. Such an attack would take place before any major U.S. deployment to the region had begun.

---

[14]See Stillion and Orletsky (1999), Chapters Two and Three.

A short-warning attack would force the Air Force either to fight with major early disadvantages or to take time to build up its strength in the theater, thus letting the regional adversary make some initial territorial gains. This would be a difficult decision to make. If the President and the Secretary of Defense elected to commit combat aircraft immediately to battle against a standing-start attack, the Air Force might have to operate initially without its normal complement of critical enabling assets, such as tankers, the Airborne Warning and Control System, the Joint Surveillance and Target Attack Radar System, jamming aircraft, and dedicated air-defense suppression aircraft. In such a situation, the Air Force also could find itself at a heavy numerical disadvantage in early air-to-air engagements.

The upshot is that the Air Force could suffer significant losses in the early phase of a standing-start attack, especially if the opponent possessed advanced surface-to-air missile systems.[15] Risks would also be involved if the President and the Secretary of Defense chose to delay their response until U.S. forces were fully deployed. Serious political implications could result from the territorial losses that a local U.S. ally would almost certainly suffer in a delayed-response scenario. While most conceivable short-warning attack scenarios would not result in an ultimate U.S. defeat, they would almost certainly all extract a greater price in terms of blood and treasure.

Antiaccess Operations. Perhaps the cardinal mistake the Iraqis made during Desert Shield and Desert Storm was the six months of unhindered deployment and buildup time they gave to coalition forces before the January 1991 commencement of hostilities. During the height of a deployment of U.S. forces into a theater during an MTW contingency, future regional adversaries will have greater opportunities to avoid the error the Iraqis made and to mount strike operations designed to hinder U.S. access to critical points in the battlespace. This would likely be done through the use of missiles, unmanned aerial vehicles (UAVs), mines, and aircraft to damage and/or shut

---

[15]Of special concern here are the advanced "double digit" Russian-made surface-to-air missiles, like the SA-10. Russia is marketing these systems to several countries that could become military adversaries of the United States. China, for example, already deploys a version of the SA-10.

down both aerial and sea ports of debarkation in the region so as to cut down the throughput capacity of such facilities.[16]

Although the Air Force is attempting to diminish the threat of antiaccess operations by shaping itself into an expeditionary force with enhanced force-protection capabilities, the realm of offensive information warfare should also offer possibilities for mitigating the antiaccess threat.

Deep-Strike Operations. The final threat in the area of strategies and tactics has to do with deep-strike operations that a regional adversary could mount during the counteroffensive stage of an MTW, the phase during which U.S. forces would be fully assembled in theater and attempting to roll back any initial gains that the adversary had made. During this phase of an MTW, the logistical demands of major ground and air offensive operations will compel the United States to amass large stockpiles of fuel, ammunition, and spare parts throughout the rear areas of the theater. Major scripted offensive operations would also force U.S. air bases in the region to operate at a high tempo, possibly with little room for slack. These realities would create tempting targets for an adversary's remaining cruise and theater ballistic missiles. While the aforementioned antiaccess operations would concentrate on disrupting and delaying a U.S. deployment into theater, the goal of deep-strike operations would be to slow down and prolong a U.S. counteroffensive so as to keep U.S. forces off balance and to inflict greater casualties, possibly breaking down the U.S. national will to continue the campaign. Likely targets for adversary cruise and ballistic missiles in the deep-strike campaign would include ammunition and fuel storage sites throughout the theater, air bases, the theater air operations center, early-warning radars, anti–tactical ballistic missile batteries, ports, troop concentrations, and headquarters.

Political Constraints on U.S. Force Deployments

Not all of the troubling asymmetric options available to a regional adversary involve military means. Indeed, some of the most potent options may be political and diplomatic. There are a variety of diplomatic tactics available to a smart regional adversary for the

---

[16]Stillion and Orletsky (1999), Chapters Two and Three.

purpose of complicating U.S. military deployments. The goal of such tactics would be to intimidate potential or existing U.S. allies to back out of political coalitions or at least to deny the use of their air bases to U.S. forces. The blunt approach for an adversary would be to attempt direct coercion against a U.S. ally by threatening that ally's cities with WMD attacks from theater-range delivery vehicles. More-nuanced political strategies could include furnishing support to opposition groups in allied countries and encouraging them to foment civil unrest during a crisis. An alternative approach for regional adversaries would be to emphasize carrots over sticks by promising substantial political and/or economic rewards to their neighbors for keeping U.S. airpower off their soil.

Denial of U.S. access to theater bases would most likely force the Air Force to adopt a standoff approach to combat—that is, conducting air operations from bases outside the immediate theater. The Joint Forces Air Component Commander (JFACC) would face a number of penalties as a result of the need to pursue a standoff CONOP, including lower sortie rates for strike and counterair operations, greater demands on the tanker fleet, reduced chances of rescuing downed aircrews, increased pressure on heavy bombers and cruise missiles to hit deep fixed targets because of a lack of alternative delivery vehicles, a substantial degradation in the capability to hold critical mobile targets at risk, increased difficulty in supporting U.S. and allied ground forces, increased aircrew fatigue, and greater maintenance turnaround times. Indeed, early reports indicate that the fall 2001 air campaign against the Taliban and Al Qaeda in Afghanistan faced exactly these kinds of problems while it was being run largely from distant bases.

A recent RAND study examined the operational effects of using a standoff strategy in response to an adversary's employment of chemical or biological weapons against close-in air bases. The study found that a 600-mile standoff range in Southwest Asia reduces the Air Force's sortie rate by approximately 25 percent; in Northeast Asia, a 500-mile standoff range reduces the sortie rate by roughly 30 percent.[17] Such reductions could result in a substantially longer and bloodier conflict than would otherwise be necessary.

---

[17]Chow et al. (1998), pp. 66–78.

DEVELOPING OPERATIONAL CONCEPTS FOR FUTURE OFFENSIVE INFORMATION WARFARE

Now that we have identified the major asymmetric options available to regional adversaries, we can begin to think about the role of offensive information warfare in improving U.S. chances of dealing successfully with such challenges. Figure 6.1 maps each of the asymmetric options outlined above to a CONOP using offensive information warfare that provides a possible way to negate the enemy's strategy. The following subsections will discuss each of the potential offensive information-warfare CONOPs in detail. Here, we will only provide a preview.

Short-warning attacks can perhaps best be dealt with through effective regional deterrence strategies. The "information-based deterrence" CONOP attempts to expand upon previous notions of deterrence by using an array of information technologies to affect an opponent's perception of the overall political and military situation in his region during peacetime or during a crisis.

WMD possession and base denial are grouped together because both strategies have to do with an adversary striving to decrease the Air Force's freedom and capability to operate over his homeland on a sustained basis. We attempt to address these through a CONOP enti-

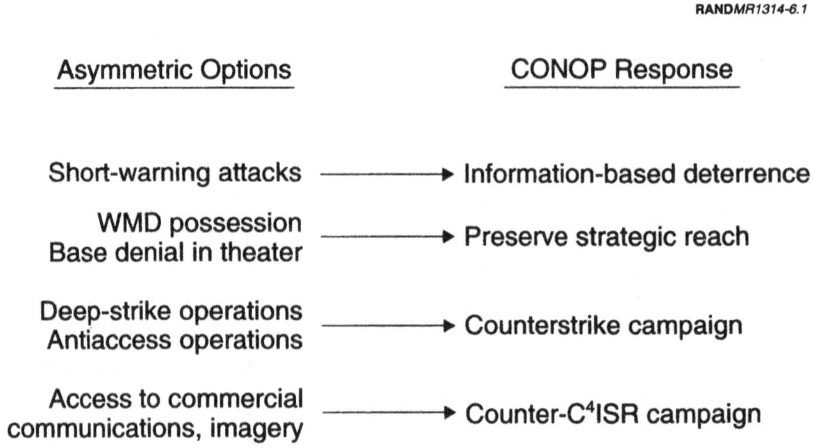

RAND*MR1314-6.1*

Asymmetric Options	CONOP Response
Short-warning attacks ⟶	Information-based deterrence
WMD possession / Base denial in theater ⟶	Preserve strategic reach
Deep-strike operations / Antiaccess operations ⟶	Counterstrike campaign
Access to commercial communications, imagery ⟶	Counter-C^{4}ISR campaign

Figure 6.1—Adversary Asymmetric Options and Potential U.S. CONOPs

tled "preserve strategic reach," which seeks to use offensive information warfare to suppress enemy air defenses to facilitate conventional strategic air attacks upon selected targets in the enemy homeland.

Next, there is the risk that the enemy may mount antiaccess and deep-strike operations. Both these asymmetric options involve the use of relatively new technologies (e.g., GPS-based targeting) to strike at U.S. and allied rear areas. The "counterstrike campaign" CONOP is a possible remedy. Counterstrike also uses a variety of offensive information-warfare tools, this time to disrupt an enemy's strike-planning and execution functions.

Finally, the increasing ability of regional powers to exploit space-borne communications and imagery assets is mapped against a CONOP called the "counter-C⁴ISR campaign." Counter-C⁴ISR uses a variety of offensive information-warfare tools to disrupt an enemy's ability to collect and process information gained from overhead assets.

Information-Based Deterrence

The overarching goal of the information-based deterrence CONOP is successfully manipulating the attitude of a potential adversary during peacetime or a crisis to prevent him from ever attacking an ally. Such efforts have been made in the past, but technological growth is creating opportunities to increase the power of information campaigns aimed at either long- or short-term deterrence.

Information-based deterrence strives to sow doubt in the mind of a potential adversary about the likely outcome of his aggression. This can be done in three ways: turning international opinion against the aggressor, altering his perception of the military correlation of forces in theater, and fostering instability in his country. Information-based deterrence does not require a pure strategy; it can include a combination of two or three options, depending on the circumstances. Although these three mechanisms could also be used during wartime itself as a means of coercing an enemy, history demonstrates that wartime coercion is much more difficult than is deterrence. Therefore, the U.S. military would have a better chance of success with these mechanisms using them within the context of a deterrence effort.

Three cautions are important when discussing perception-shaping strategies against other states. First, in recent times, technology has often outpaced international norms and standards. We still do not have a clear sense of which types of perception-shaping activities will be construed as legitimate peacetime behavior and which as casus belli by international organizations and institutions. Therefore, to reduce the risk of inadvertent escalation, it will be necessary to rethink our doctrine for perception shaping periodically in accordance with developing international norms and standards.

Second, perception-shaping activities carry a constant threat of "blowback": Operations designed to manage the opponent's perceptions may end up distorting our own perceptions to an equal or even greater extent. For example, while it may be advantageous to convince the enemy that U.S. forces are more capable than they actually are, it would be less helpful to convince oneself of that fiction. Yet, because of the need for consistency and secrecy to accomplish perception-shaping objectives, these two effects are, in practice, not completely separable.

Third, deterrence of any sort relies on convincing the adversary not to act. While our actions can affect the adversary's calculus, we must always be prepared for deterrence to fail. For our purposes, this reality means that information-based deterrence is not the complete solution for short-warning attacks. Other means must be developed to cope with the possible failure of information-based deterrence.

Turning International Opinion Against the Aggressor. A major U.S. strength lies in its ability to create wide coalitions against potential enemies that can isolate opponents from external support, both material and moral. Such coalitions reinforce U.S. combat power, reduce enemy access to critical supplies, and provide a greater legitimacy for U.S. action—a legitimacy that solidifies domestic U.S. support for deterrence actions and war, if that becomes necessary. Repeatedly, U.S. leaders have stressed that U.S. forces will only engage in a multilateral context. An increased likelihood of such a coalition will therefore have a deterrent effect on potential foes.

Creating and maintaining such coalitions require that international opinion views U.S. foes as aggressors with little regard for international law or human rights. Optimally, such a coalition would be sustained by a continuous and long-standing information campaign.

However, new conflicts and enemies can arise, and U.S. leaders must be prepared to cut such coalitions nearly from whole cloth. Particularly in the case of a short- or no-warning attack (such as occurred on September 11, 2001, in New York and Washington, D.C.), prospective foes will take great care to hide their intentions until shortly before hostilities break out. Preventing and responding to short-warning attack or responding to a no-warning attack therefore may necessitate a rapid-reaction information campaign that is prepared to foster the appropriate climate of international opinion.

In the short period before the outbreak of hostilities, a rapid-reaction information campaign has two basic parts. First, television and radio broadcasts of accurate information from U.S. sources should show enemy intent and preparations for attack. Second, the information campaign should include television and radio broadcasts that demonstrate both U.S. friendly intent and allied military prowess. These broadcasts should go to enemies and prospective allies, as well as to domestic audiences.

Both before and after hostilities break out, short-warning attacks often provide ready material for images that will outrage and inflame international opinion against the aggressor. Such attacks will typically require such preparations as the loading and unloading of trains, massing of supplies and forces, and shock attacks by rapidly moving forces. Capturing these preparations and attacks on video or satellite imagery will demonstrate the enemy's aggressive intent, give the lie to any pretext they might have established for invasion, and serve to catalyze international opinion and support for a broad coalition to oppose aggression. There is some historical precedent for the use of simple images as a tool for marshaling international opinion against an aggressor. During the Cuban Missile Crisis, for example, aerial reconnaissance photos of Soviet missile sites in Cuba helped strengthen the U.S. position at the United Nations.

Increasingly, the independent media can be counted on to capture and broadcast this information. However, some countries still maintain fairly effective control over even foreign media outlets operating in their territory. To avoid leaving such things to the intrepid action of individual reporters, information deterrence could be aided by a rapid-reaction information force that can quickly establish video surveillance of potentially hostile territory. UAVs equipped with video equipment and command planes capable of gathering, editing, and instantaneously disseminating that coverage are the essential features of such a force. If the risks of such aircraft being shot down or identified over enemy territory are too great, one could even use satellite imagery to provide evidence of mass graves, burned-out villages, etc. At the same time, U.S. leaders would need to work hard to counter enemy propaganda campaigns by tirelessly presenting the major elements of the true situation on a wide spectrum of information outlets (Internet, television, radio, etc.). Video coverage of the battle area will help expose any enemy attempt to portray U.S. actions in a deceptive light. Great effort needs to be expended to provide counterevidence for the inevitable enemy propaganda campaign. Good video images can make ruses—such as the Iraqi attempt to portray a U.S. attack on a military target as an attack on a facility that served solely as a baby-milk factory—nearly impossible.

It is critical that such a campaign maintain a consistency of message and purpose throughout its broadcasts. This kind of campaign should make no effort whatsoever to deceive or manipulate the international media, concentrating instead on the simple goal of using modern technology to highlight the hard physical evidence of a rogue state's or nonstate actor's aggressive intentions. Deceptive techniques are unnecessary and counterproductive in this context. In an age of numerous media outlets and largely unconstrained information flows, it makes little sense to risk American credibility as an honest international citizen by manufacturing video images.

Altering the Enemy's Perception of the Correlation of Forces. When leaders of the various ex-Yugoslavian factions met at Dayton, Ohio, to divide zones of control in Bosnia, the U.S. military provided satellite imagery to assist in the demarcation of borders. To the shock of the participants, the imagery demonstrated a knowledge of terrain and force dispositions far in excess of what the participants had previously believed possible. Indeed, the imagery showed a three-dimensional picture of the contested areas that demonstrated a more detailed knowledge of the participants' own forces than the parties themselves possessed. All sides now understood that the U.S. military could see virtually anything on the battlefield; the implicit threat was that it could destroy anything it could see.[18]

---

[18]For accounts of this episode, see Watters (1996); Libicki (1997), Ch. 3; and Nye and Owens (1996), p. 32.

While such knowledge was useful in negotiating the peace, it might be even more useful in deterring a future MTW. The U.S. military, particularly the Air Force, excels at simulation, which it uses to train its troops in conditions as realistic as possible. These simulations can similarly be used to demonstrate U.S. combat power, without actually employing it. Indeed, in the summer of 1998, NATO carried out simulated air raids over Kosovo as a deterrent to further Serb repression in that province. Realistic simulations have a tremendous capacity to impress an enemy's population, frighten his soldiers, and radically alter the enemy's assessment of U.S. military power. Such simulations would include images of U.S. military equipment operating, simulated attacks on significant military targets, and broadcasts of past U.S. combat successes. While these simulations and replays will certainly give insight on actual U.S. combat capabilities, they need not always reflect actual U.S. capabilities or intentions. As noted, if a simulation is realistic enough, it will create such a strong image in the mind of the adversary that it will irrevocably alter his perception of the correlation of forces, regardless of actual U.S. capabilities.

Fostering Instability. Every society contains divisions simmering below an often calm exterior. Nondemocratic societies, especially, contain latent tensions that, if exploited, can severely limit a country's ability to engage in offensive military action. Offensive information warfare offers many new covert means to exploit the tensions because it increases the opportunity to communicate directly with constituent parts of the adversary's society. As the information revolution increases the number and types of communication channels within any society, the opportunities to introduce false data into communication links between constituent parts of the society also increase.

It should be noted that fostering instability is the riskiest of the three methods of information deterrence. It should only be used against nations that pose a particularly grave military threat. It is also the most difficult method to implement as it requires a detailed understanding of the target society and the cultural context in which such action will be received. Inappropriate or clumsy efforts to foster instability may well create unity in an otherwise divisive polity by providing evidence of an external threat to the nation. The United States should not attempt this type of tactic unless it possesses an experienced cadre of intelligence analysts who have proven themselves to have an extremely high degree of cultural understanding and sophistication with respect to the target state.

Nonetheless, by exploiting cleavages among the government, the population, and the military (the so-called Clausewitzian trinity), it could sometimes be possible to convince the adversary leadership that its hold on power is fragile and that it thus cannot afford any type of military contest, let alone one with the United States. There are many ways such an approach could be pursued. False messages inserted into national communication networks could be used to create mistrust between the civilian and military leaderships by spreading rumors of military coup plots or planned purges against the officer corps. Support to nongovernmental organizations operating on the Internet could be used to spread popular disenchantment with government policies and to foster public protests against the regime. The United States could also use media organizations in the target country and its neighbors as a lever to influence public opinion in the adversary state and turn that opinion against its own government's policies. Finally, the low-technology approach should not be forgotten: Leaflets dropped from U.S. aircraft were extremely effective during the Gulf War in convincing Iraqi troops to surrender (Hosmer, 1996).

Some nations will be far less vulnerable to such measures than others by virtue of their closed political systems. However, as time goes on and as international connectivity increases, there will be fewer and fewer nondemocratic nations that can be sanguine about their ability to insulate themselves against the effects of a well-coordinated information strategy exploiting the mass media, the Internet, and proprietary communication networks.

Preserving Strategic Reach

The prospect of facing regional adversaries with mature nuclear arsenals raises questions about how much freedom the Air Force will have to conduct parallel warfare against the enemy homeland without substantially increasing the risk of escalation. The prospect of having local allies deny the U.S. Air Force the use of bases in the theater means that it may be prohibitively expensive and dangerous to employ air assets over the enemy's territory for the reasons outlined above.

The "preserving strategic reach" CONOP is intended as a response to these emerging challenges. The chief mechanism of preserving strategic reach is the periodic use of offensive information-warfare means to degrade the enemy's integrated air-defense system (IADS).[19] The significant degradation of the enemy IADS would allow the U.S. Air Force to operate over enemy territory in reasonable safety and, given well-chosen targets, with much less fear of nuclear escalation. As with information-based deterrence, however, preserving strategic reach may not be the final solution to these problems. Using offensive information warfare will not eliminate the possibility of escalation and will not completely make up for the loss of theater bases. Nonetheless, it represents an important part of the response to these relatively new challenges.

More-conventional operations to suppress enemy air defenses that use physical attacks, such as those mounted by F-16s equipped with High-Speed Anti-Radiation Missiles, contain a risk of escalation when used against an adversary with WMD. The adversary leadership may not be able to distinguish such an operation from an attempt to destroy nuclear warning and command and control systems in preparation for a counterforce attack designed to eliminate the adversary's nuclear deterrent. Such operations also require putting friendly forces at risk and are particularly difficult and dangerous to launch from a standoff posture. Offensive information-warfare operations contain a much smaller risk of escalation because they need not involve physical attacks on command and control systems and because they can be done covertly. They do not put friendly forces at risk and are not affected by the loss of theater bases, because they can be launched just as easily from outside the theater.

Any IADS contains information systems and information-based processes that are essential for its operations and that are lucrative targets for offensive information warfare. Schematically, an IADS consists of one or a few air-defense headquarters connected by communication links to sensors, such as early-warning radars, EW sensors, or aircraft like those for the Airborne Warning and Control System. Each headquarters also communicates with and controls a

---

[19]An IADS includes surface-to-air missiles, antiaircraft artillery, air-superiority fighters, and the communication and sensor infrastructure that connects them.

variety of antiaircraft weapons, such as fire-control radars, missile launchers, and air-superiority fighters.

Without physical destruction, offensive information warfare can attack an IADS at three points. First, offensive information warfare can attack the system's sensors, either degrading their ability to gather information or feeding them false data. Second, offensive information warfare can degrade or plant false information in the communication links between headquarters and the sensors or shooters. Third, offensive information warfare can degrade or deceive the information processes that compile the sensor information, interpret it for human decisionmakers, and assign particular weapons to targets.

The centralized nature of this system implies that the air headquarters is a critical choke point, the disabling of which will render the entire system useless without the need to disable every sensor and weapon. This point should not be taken too far, however. An IADS can be configured to work in several modes from centralized to fully autonomous. The characteristics of these systems in each different mode should be well understood because information munitions that prove effective against an IADS operating in centralized mode may be ineffective against the same set of surface-to-air missile batteries and sensors operating in autonomous mode. Indeed, in autonomous mode, the local air-defense headquarters may not even be that significant to the overall function of the system.

The decision about which part of the system to attack therefore depends on the reason for using offensive information warfare to bring down the IADS. If escalation to WMD is a concern, the emphasis should be on allowing the enemy to believe that the IADS is still functioning even as one has severely degraded its effectiveness. While attacks on sensors and communications can be useful under such circumstances, attacks on the information processes themselves are probably most useful under such circumstances because errors in such processes are difficult to trace, badly understood, and widely expected in the normal course of operation. Such processes can be degraded by means of various information munitions (viruses, worms, logic bombs, etc.) prepositioned or inserted into the enemy air-defense computers. This degradation could cause the IADS to fail to assign targets, assign targets to inappropriate weapons, lose orders to weapons, misinterpret sensor data, or mis-target surface-to-air missile batteries. If cleverly applied, these weapons can go undetected, and any errors in IADS information processes will be attributed to operational errors. The key difficulty in such an attack is timing. The information munitions must "go off" only just before the air strike, or the degradation in the IADS is likely to be detected and corrected. Timing such information munitions is a tricky problem. Viruses and worms travel at an unpredictable rate, and logic bombs are difficult to trigger remotely. One must also keep in mind that some threat air-defense systems will contain bounds-checking features that ensure the system does not malfunction in certain drastic ways (such as assigning targets to inappropriate weapons); these bounds-checking features could present clues that an information attack was in progress to an alert and well-trained air-defense commander.

If the offensive information-warfare attack is meant to allow the United States to operate from a standoff posture, the information attack need not be so unobtrusive in its methods. In this case, the most lucrative targets are the extremities of the system: the sensors and the antiaircraft weapons. Offensive information warfare would attempt to disable the sensors or weapons temporarily in particular nodes of the IADS in closely timed coordination with strike missions routed to pass through the resulting geographic gaps. Such temporary effects could be achieved by overloading sensors with false data, jamming communications via EW, inserting false data into communication streams, or conducting perception-shaping campaigns via broadcast or leaflets that threatened operators who turned on their radars or acquired allied targets.

Two caveats are in order. First, as we have already seen, timing is crucial for realizing the full potential of the preserving strategic reach CONOP. Precise timing of effects will be difficult to achieve and will require Air Force planners with considerable skill. Second, there is the issue of reliable damage assessment for offensive information-warfare attacks against IADS. How do you know if your attack has done its job and if it is safe for manned aircraft to fly through the area? This information-warfare battle damage assessment (BDA) problem could become larger the more frequently this particular CONOP is used. A cunning enemy, once he sees a pattern developing, may set traps by intentionally shutting down the radars in an air-defense sector during an offensive information-warfare attack and then luring American aircraft into an ambush. Once again, the only solution here is to support research into technologies that might make information-warfare BDA a more accurate science.

Preserving strategic reach should only be used if the following three conditions are met. First, U.S. policymakers need to have made a clear decision that other approaches to reducing the significance of the adversary's WMD have less potential. These other approaches include deterrence through the threat of massive retaliation; deterrence through the threat of escalation dominance; and a conventional counterforce campaign aimed at destroying WMD ordnance, delivery vehicles, and storage sites through the use of precision-guided munitions and the use of effective theater missile defenses to reduce adversary expectations about the ultimate results of any WMD attacks with theater ballistic missiles. In many cases, the other approaches could be more appropriate to the situation at hand than the cautious strategy embodied in preserving strategic reach.

Second, it will be critical for other components of the unified geographic command to be fully aware of the JFACC's concept of offensive information warfare and also to be prepared to coordinate actions if necessary. The importance of sharing information across organizational boundaries must not be underestimated when planning for offensive information warfare.

Last, national-level authorities must be made aware of the risks of enemy retaliation against the U.S. National Information Infrastructure in response to U.S. offensive information-warfare attacks against enemy IADS. These authorities should take appropriate precautionary measures. Addressing these vulnerabilities would give the United States more freedom of action to use offensive information warfare in MTWs.

Counterstrike

The purpose of the counterstrike CONOP is to keep the enemy from mounting antiaccess operations against U.S. power-projection capabilities and deep-strike operations designed to target U.S. logistics bases critical for sustaining U.S. air operations. Offensive information-warfare operations provide a new capability in this regard because they offer an opportunity to attack the enemy's rear areas and affect his capacity for antiaccess and deep-strike operations even before U.S. forces have deployed in strength to the theater. Through remote attacks on the enemy's planning and assessment processes, offensive information warfare denies him the use of a homeland sanctuary from the very beginning of the deployment.

It should also be noted, however, that offensive information warfare in this context is intended to be used in conjunction with conventional attacks on enemy strike assets. Offensive information warfare will enhance the effectiveness of conventional counterstrike operations, especially early in the battle, before all forces have deployed, but it will not replace the traditional missions.

At the most basic level, offensive information warfare is useful for this purpose because strike operations are highly information intensive. Successful strike operations require detailed planning, careful coordination, and reliable data on target locations. When viewed as an information process, strike operations can be seen as iterative, with three stages: planning, execution, and BDA. Offensive information warfare will aim to disrupt or defeat all three stages of that process. For this purpose, the United States would deploy an information-warfare squadron to the theater to support the JFACC. This squadron would also have access to numerous information-warfare centers based in the continental United States that can provide analysis and expertise, such as the Army's Land Information Warfare Activity or the Air Force Information Warfare Center.

Strike Planning. Strike planning is the process of allocating and coordinating scarce attack assets (cruise missiles, ballistic missiles, strike aircraft) to inflict the greatest possible damage on the target's operational capacity. To be employed efficiently, such a planning process will need to have access to mountains of data on U.S. capabilities, force-deployment plans, orders of battle, air defenses, and target locations. Planners will also need more prosaic data, such as terrain and navigation information and weather reports.

Offensive information-warfare operations can deny or corrupt all these data sources by attacking information systems. The adversary will use a variety of information systems to collect, process, and disseminate the data. Most of the imagery, weather, and navigation data necessary to pinpoint fixed U.S. targets will be collected from commercial satellites, as well as from such open sources as commercial maps and media outlets. Terrestrial and satellite communications will be used to disseminate raw data to planners, strike plans to the assigned units, and mission plans to the individual shooters. Finally, the creation of any complex strike plan will involve software to evaluate the mountain of data involved, produce mission plans, and transfer data to weapon systems.

There are many methods for denying and corrupting the data. Some are quite conventional, such as limiting access to critical facilities, camouflaging ship and aircraft movements, or periodically moving high-value assets and air defenses. Other methods are more novel and potentially more effective. Physical destruction or electronic jamming of critical junctures in the strike-planning process—satellites, satellite downlink stations, and mission-planning centers—will be particularly effective. Information munitions could also be implanted in the enemy's strike-planning system to render it inoperable at critical moments.

We should keep in mind that planners are adaptive and will find workarounds to the problems of missing data, downed systems, and destroyed communication links. A more subtle method, then, might be to corrupt the mission-planning process, thus causing the enemy to squander scarce strike resources. This can be done primarily by introducing false data on U.S. and allied force disposition, terrain, and even weather data into the enemy's striking-planning process. There are many potential points of access, the most promising being via falsified, corrupted, or hijacked satellite downlinks. Since the enemy may collect some targeting data on large, fixed targets long before a strike, another option is to implant errors in the enemy's mission-planning hardware or software to cause subtle errors in the strike-planning process.

Strike Execution. Strike execution is the process of carrying out the strike plan. For information systems, strike execution depends primarily on a dense system of communication links and navigational aids. This includes communication links—from command and control units to aircraft and missile launchers—used to make changes in the mission plan or report intelligence gained from the mission, communications between aircraft used to synchronize the attack, and navigational data acquired from satellite systems, such as GPS or GLONASS.[20]

Once again, each of these links is potentially vulnerable to destruction or disruption. Most vulnerable are the navigational systems. Without this type of navigational data, enemy cruise missiles and even strike aircraft will be far less accurate. U.S. forces can easily turn off or jam GPS and can jam GLONASS in local areas. Of course, this may also adversely affect the U.S. capacity to operate. Even if access to GPS were limited to U.S. and allied forces, the enemy might well be able to jam U.S. access to GPS in retaliation. Any degradation of satellite navigation systems must always be assessed in a relative perspective. Given that there is a good possibility that the loss of GPS would affect U.S. forces more than it would enemy forces, a more effective measure might be to introduce errors into the GPS or GLONASS signal at critical periods during strike execution. Alternatively, the United States could use information munitions to attack the information processes that load targeting and navigational data into enemy cruise missiles and strike aircraft. Both attacks could introduce navigational errors that should cause strike aircraft to attack erroneous targets or cause cruise missiles to collide with terrain features.

Similarly, the United States could hope to introduce false data into communication links between strike assets and command and control centers or between the strike aircraft themselves. Such false data could generate false targets or give false target updates to strike assets already in the air. Unfortunately, the growing use and sophistication of encryption techniques makes such insertion increasingly difficult, so jamming these links may soon become the only option. Nonetheless, aircraft or cruise missiles that are unable to receive information from their command and control centers will be unable to adjust their mission plans to reflect real-time changes in target disposition and will be unable to function as forward sensors.

New offensive information warfare or related methods may also soon be available for destroying the platforms themselves. One can well imagine having the technology available to generate bogus electronic signals that would prevent arming or prematurely activate warheads

---

[20]GPS is a U.S. satellite navigation system. GLONASS is a similar Russian system.

on inbound cruise missiles and fighter aircraft. Another interesting possibility is the use of high-altitude electromagnetic pulse weapons based on airborne platforms to disable navigation, flight control, target acquisition, and fire-control systems on inbound aircraft and cruise missiles, rendering them all but useless as offensive weapons or causing them to crash.

Battle Damage Assessment. BDA, the least examined part of the strike process, is critical for a successful overall strike plan. Given the scarcity and price of sophisticated strike assets, it is vital to know which defensive systems have been disabled and which targets have been destroyed in order to allocate strike assets efficiently and safely. BDA has several information sources, including the strike asset itself; open-source media outlets; human intelligence agents; and remote-sensing platforms, such as satellites, UAVs, and surveillance aircraft.

Once again, all of these data sources can potentially be degraded or destroyed by offensive information-warfare operations. Although jamming and physical destruction of communications and satellite downlinks will be very useful measures in this regard, BDA also presents ample opportunity for deception. False damage signatures may fool strike aircraft into thinking they have hit their target. Careful camouflage can fool satellite imagery into believing that targets have been destroyed or, conversely, remain unharmed.

Perhaps the most promising method of complicating the enemy's BDA process is by tainting open sources. In the future, much BDA may be done through the media or through human agents reporting openly available information from within the target zone via e-mail, cell phones, etc. The problem of open-source BDA of strike operations will no doubt persist and, indeed, may increase as more of the population gains access to cell phones and Internet e-mail. Offensive information-warfare operations can turn this intelligence drain into an asset by planting false information on battle damage into open sources. False damage reports and even false video images of bomb damage (or, conversely, of false images of still-operating facilities) can greatly complicate enemy planning. In contrast to information deterrence, if this activity damages the credibility of the media and human agents as sources of BDA, so much the better. It is again important to be aware of the risk of blowback: Special BDA spoofing efforts against the enemy may well fool some planners and operators on the U.S. side who are not familiar with these programs.

The Utility of Counterstrike. The importance of counterstrike will vary with the current phase of conflict. Pentagon planners have divided MTWs into three phases: halt, stabilize (or "buildup and pound"), and rollback. The counterstrike CONOP would be the most useful during the halt phase of a stressful MTW, in which the enemy has physical forward momentum on the ground and a numerical advantage in the air. Normal U.S. air and missile defenses may not be fully deployed, creating opportunities for the opposing side to mount deep-strike and antiaccess attacks against American and allied rear areas.

The usefulness of counterstrike drops steeply as one enters the buildup-and-pound and counterattack phases of an MTW. During the last two phases, the Air Force will presumably have established a comfortable level of air superiority; anti–tactical ballistic missile systems, such as Theater High-Altitude Area Defense and Patriot, will be fully deployed throughout the battlespace; and dispersal and decoy arrangements will be in place at the Air Force's main operating bases. In such an environment, counterstrike would probably be unnecessary, and it would be far better to devote offensive information-warfare resources to other purposes.

Counter-C⁴ISR

The "counter-C⁴ISR campaign" CONOP involves using a mix of offensive information-warfare tools and techniques to attack adversary sensors and communication assets across the board at critical "transition points" during a campaign. The goal of the counter-C⁴ISR campaign is to reduce the enemy's battlespace awareness at key junctures by degrading his ability to collect and process information from space, airborne, and ground-based C⁴ISR assets.

Earlier in this chapter, it was noted that the size and capacity of global commercial satellite and terrestrial communication networks is increasing at a rapid rate. By the 2010–2015 time frame, there will be so many redundant communication paths and links in existence that it will be impractical to achieve large, sustained reductions in the enemy's C⁴ISR capacity across the board from the tactical level up through national-level command and control. In fact, available evidence suggests that the United States was not even able to achieve this goal completely against Iraq during Operation Desert Storm.

However, this reality does not render counter-C⁴ISR futile. Instead, it suggests that U.S. operations should focus on degrading key choke points of the enemy's C⁴ISR system at critical moments in the campaign, rather than on an attempt to destroy all enemy communications.

Increased access to commercial space-based imagery, communication, and navigation systems will greatly enhance the enemy's C⁴ISR capacity but may also make the systems more vulnerable to offensive information warfare. Use of space-based systems will introduce choke points into the enemy's C⁴ISR system. Satellites themselves become critical nodes that, if disabled, would drastically reduce enemy C⁴ISR capacity. Commercial communication satellites require downlink stations and locally dense communication networks that are also vulnerable to physical and information attack. Imaging satellites also require downlink stations and an imagery analysis center to read, interpret, and disseminate the information gleaned from satellites. Reception of GPS navigation, while more dispersed, will often require differential GPS transmitters to achieve needed accuracy. A well-timed attack that disables or degrades these systems may well leave the enemy worse off, especially for short but critical periods of time, than if he had never grown accustomed to their significant advantages over terrestrial communications, surveillance, and navigation.

Counter-C⁴ISR involves many of the same means and mechanisms as the two previous CONOPs, "preserving strategic reach" and "counterstrike." There is a fundamental difference in their goals, however. Counter-C⁴ISR is designed to have a decisive effect on the outcome of a campaign, while the other CONOPs would have more-limited objectives. Counter-C⁴ISR is a potential war winner; preserving strategic reach and counterstrike are not. This distinction means that, even more than the other CONOPs presented, counter-C⁴ISR requires tight integration with the regional CINC's overall campaign plan. Indeed, counter-C⁴ISR will depend for its success on careful synchronization with more-conventional attack assets. This implies placing responsibility for the C⁴ISR campaign firmly in the hands of the JFACC, rather than creating a special information-warfare component commander, so that the principle of centralized control with decentralized execution can be maintained. Creating a special information-warfare component command may sound appealing at first blush, but would probably be unwise, because it would only add another layer of command and control that could slow down U.S. and allied decision cycles. Enemy forces will use C4ISR systems to anticipate U.S. force movements and order force movements in response. They will use satellite reconnaissance to show large force movements or attack preparations, such as the movement of large amounts of supplies and weapon platforms. They will use commercial communication satellites, as well as dedicated terrestrial communications, to receive human intelligence on such movements and to order counterpreparations and strike missions against massing U.S. forces and supplies. They will use satellite navigation to provide targeting information to cruise missiles and strike aircraft on the position of key U.S. forces and supplies. Finally, they will use advanced software and computer systems to program targeting information into cruise missiles. Again, the intent of counter-C4ISR is to deny the enemy these sources of information and information processes, not always or everywhere, but just at the critical moments and places where they are most needed.

This implies that the leading edge of the rollback phase offensive will be a coordinated offensive information-warfare attack on these information systems, including physical attacks. The exact nature of that attack would depend on the CINC's overall campaign plan. As an example, however, it will be helpful to consider how the CINC might have attempted to achieve a surprise flanking attack, such as U.S. forces accomplished during Desert Storm, despite the presence of a sophisticated enemy C4ISR system.

The first, and probably most difficult, task would be to allow the large force movement necessary to accomplish such a flanking maneuver to go undetected. This would require first jamming or disabling any commercial imagery satellites capable of providing images of the staging area. As these systems will be assets of neutral nations, this may require a certain delicacy of approach. One possibility is nonlethal attacks from ground-based antisatellite (ASAT) lasers that could only temporarily blind a satellite. A less-controversial method would be precision weapon attacks on the enemy's imagery analysis centers or the satellite's communication links with enemy command centers.

Unless such movements take place in trackless deserts, they are also likely to be detected by enemy agents on the ground. This unpleasant reality implies that the movement must be accompanied by efforts to cut off the enemy's external communications. The means to accomplish this task include local jamming of commercial communication satellites from mobile transponders, precision-guided weapon attacks on satellite communication downlink stations, and information munitions implanted in key communication switches that control communications with the downlink station. Once again, the enemy is likely to be able to reconstitute these systems in the space of days or even hours, so timing is critical.

Although counter-C⁴ISR may not be able to prevent, for a sufficient period, enemy detection of a movement of the same scope as the famous Desert Storm "left hook," it can also help in stymieing enemy responses. First, attacks on communication links will make it difficult for enemy commanders to receive satellite imagery and targeting data or to give and receive orders to respond to U.S. movements. Second, as with the counterstrike CONOP, the United States can hope to deny, degrade, or corrupt enemy access to satellite navigational aids and information processes that control enemy targeting systems. U.S. forces on the move, once detected, will present tempting targets for enemy cruise missiles. If their navigational and targeting systems can be degraded at the critical moment, however, they will present little danger to U.S. forces.

Over the long term, counter-C⁴ISR would benefit operationally from the deployment of space weaponry. Such systems as space-based co-orbital jammers, lasers, and obscurants would increase the chances of success for this CONOP. However, the price in terms of arms-race risks and military opportunity costs could be steep, and it is not clear that the price would be worth paying.

There are tangible arms-race risks to consider when thinking about the deployment of space-based ASAT weaponry. Other nations with significant scientific, industrial, and technical wherewithal could respond by deploying their own such systems to threaten U.S. satellites. The result of this could be a net negative for the United States, since the U.S. military's main advantage in future wars will come from its superior ability to collect, process, and act upon large amounts of data in very compressed time cycles. Conflicts in space resulting in the destruction or degradation of U.S. communication or imagery satellites would hurt American military capability more than they would a regional adversary's, even if the United States inflicted more damage on enemy space capabilities than it suffered itself. Furthermore, the deployment of space-based ASAT systems could invite adversaries to use asymmetric options to negate U.S. space capabilities-options that could include terrorist acts against U.S. commercial and military satellite ground stations worldwide. Another possibility would be retaliatory jamming against U.S.-owned satellites, including the GPS navigation satellites.

All in all, the operational advantages afforded to the Air Force in terms of being able to better execute the counter-C⁴ISR campaign look to be outweighed by the many potential disadvantages created by space-based ASAT deployments. The counter-C⁴ISR campaign will still likely be effective with only ground-based ASAT assets and would even have some use if employed without any ASAT weaponry at all. However, it would be wise for the United States to continue a research and development program into space-based ASAT technologies and also to be prepared to deploy an operational system if another nation shows signs of getting ready unilaterally to place an ASAT system in space.

COMPARING THE FOUR CONOPS

Table 6.1 is a crude attempt to summarize the strengths and weaknesses of the four CONOPs that have been presented. Our four offensive information-warfare CONOPs are listed vertically along the column at the far left. Each of the four is then assessed against five metrics: the risk of escalation that the CONOP carries with it, the

Table 6.1

Comparing the Four CONOPs

	Escalation Risk	Long-Term Relevance	Against Medium Powers	Against Large Powers	Military Decisive-ness
Information-based deterrence	Medium	Yes	Yes	Yes	Yes
Preserving strategic reach	Low	Yes	Yes	Yes	No
Counterstrike	Low	Yes	Yes	Maybe	Maybe
Counter-C⁴ISR	Medium	Maybe	Yes	Maybe	Maybe

ability of the CONOP to remain relevant as the revolution in information technology continues, the usefulness of the CONOP against medium-sized powers (such as North Korea and Iraq), the usefulness of the CONOP against larger powers (such as Russia), and the potential of the CONOP to be militarily decisive in and of itself.

First, in terms of escalation risk, none of the CONOPs described presents an extreme escalation risk. Information-based deterrence, because it will take place before any hostilities have broken out, may contain some escalation risks, especially if it involves fostering instability in the target state. If the offensive information-warfare campaign is discovered and if that campaign is considered tantamount to an act of aggression, it may provoke an adversary to conventional retaliation. Because international norms on how to treat information attacks are still evolving, it is difficult to say how any adversary might react to this provocation. Indeed, some Russian writings, for example, declare that Russia would interpret an offensive information-warfare campaign against its homeland as being tantamount to physical attack. While this is probably hyperbole, it does point out the need to be especially careful in using offensive information warfare against states that perceive themselves to be in a position of ever-increasing weakness. Preserving strategic reach was specifically designed to minimize escalation risk. Counter-C^{4}ISR, however, presents some risk of nuclear escalation, if the enemy has built a small nuclear arsenal. Because this CONOP involves overt attacks on the enemy's command and control networks and because these networks will likely also be used for control of the nuclear arsenal, there is some risk that the enemy will regard these CONOPs as preparatory to a counterforce first strike and respond by escalation. Preserving strategic reach was specifically intended to solve this problem.

As for long-term relevance, all the CONOPs except counter-C^{4}ISR should remain viable options for the foreseeable future. Counter-C^{4}ISR may become obsolete if international connectivity continues to increase at its current exponential rate. Under such circumstances, the density of the enemy's communication and surveillance networks will limit the number of choke points in the system and, consequently, the possibility of seriously degrading the system even for short periods of time. Counterstrike could lose some of its relevance if the sophistication and proliferation of digital signature technology reach a point where even regional powers could prevent the insertion of false information into the strike-planning and execution processes.

All of the CONOPs will have utility against medium-sized powers. Indeed, they were designed with such powers in mind. Against larger powers, the utility of counterstrike and counter-C⁴ISR will greatly diminish. Information deterrence does not depend on the size of the adversary, while preserving strategic reach can still be used to allow U.S. air assets to operate safely over particular areas of a larger adversary. However, the size and density of the communication networks of a large power would make it extremely difficult to create even the short communications blackout required for counter-C⁴ISR without taking drastic steps, such as the use of a high-altitude nuclear burst for electromagnetic pulse effects. Counterstrike may not be an efficient use of resources against an adversary who has very large numbers of cruise missiles and fighter aircraft available during the halt phase. Against such an adversary, it may be better to combine passive defenses with offensive counterair operations to deal with the threat of antiaccess and deep-strike attacks.

Finally, we arrive at military decisiveness. Only information-based deterrence has the potential to be militarily decisive, because it can dissuade an adversary from even starting a conflict. Preserving strategic reach is certainly not decisive, because its whole purpose is distraction, not decisiveness. Counter-C⁴ISR could be decisive under certain circumstances but not in others. Ultimately, counter-C⁴ISR creates the conditions under which other means of warfare press decisive operations against the opponent; counter-C⁴ISR is an enabler of decisive operations rather than a component of them. Counterstrike falls into the same category. The only scenario in which counterstrike could become decisive would be against an adversary who staked all his hopes on deep-strike and antiaccess operations against U.S. rear areas during the halt phase and had no backup plan in case the attacks failed. In such a scenario, counterstrike could act to fend off the attacks and thus implicitly compel the adversary to sue for peace. In virtually all other instances, counterstrike does not offer an opportunity for a decisive outcome in and of itself.

REFERENCES

Agmon, Marcy, et al., *Thwarting the Superpower: How the Smart Adversary Might Use Political Weapons to Offset U.S. Military Power*, Santa Monica, Calif.: RAND, unpublished manuscript.

Air Force Space Command, "Space Capabilities Integration," briefing slides, July 12, 1996.

Ambrose, Stephen E., *D-Day, June 6, 1944: The Climactic Battle of World War II*, New York: Simon & Schuster, 1994.

Cebrowski, Vice Admiral Arthur K., and John J. Garstka, "Network-Centric Warfare: Its Origin and Future," *Naval Proceedings*, January 1998, pp. 28–35.

Chow, Brian, Gregory S. Jones, Irving Lachow, John Stillion, Dean Wilkening, and Howell Yee, *Air Force Operations in a Chemical and Biological Environment*, Santa Monica, Calif.: RAND, DB-189/1-AF, 1998.

Graham, Bradley, *Hit to Kill: The New Battle over Shielding America from Missile Attack*, New York: PublicAffairs, LLC, 2001.

Hosmer, Stephen T., *Psychological Effects of U.S. Air Operations in Four Wars 1941–1991: Lessons for U.S. Commanders*, Santa Monica, Calif.: RAND, MR-576-AF, 1996.

Hutcherson, Norman B., *Command and Control Warfare: Putting Another Tool in the War-Fighter's Data Base*, Maxwell Air Force Base, Ala.: Air University Press, September 1994.

Joint Chiefs of Staff, *Joint Vision 2010*, Washington, D.C., July 1996.

Joint Staff, *Joint Doctrine for Command and Control Warfare*, Joint Publication 3-13.1, February 7, 1996.

Keffer, John W., "Trends in Commercial Satellite Communications Systems and Implications for MILSATCOM," briefing slides, November 20, 1996.

Libicki, Martin, *Defending Cyberspace and Other Metaphors*, Washington, D.C.: National Defense University Books, 1997.

Nye, Joseph, and William Owens, "America's Information Edge," *Foreign Affairs*, Vol. 75, No. 2, March/April 1996, pp. 20–36.

Pace, Scott, et al., *The Global Positioning System: Assessing National Policies*, Santa Monica, Calif.: RAND, MR-614-OSTP, 1995.

Stillion, John, and David T. Orletsky, *Airbase Vulnerability to Conventional Cruise-Missile and Ballistic-Missile Attacks: Technology, Scenarios, and U.S. Air Force Responses*, Santa Monica, Calif.: RAND, MR-1028-AF, 1999.

Stoney, William, "Land Imaging Satellites Planned to Be Operating in the Year 2000," data sheet, Mitretek, July 22, 1997.

Submarine Systems International, Inc., "Global Undersea Networks for Government Applications," briefing slides, July 1997.

U.S. Space Command, "Department of Defense Advanced Satellite Communications Capstone Requirements Document," June 23, 1997.

Watman, Kenneth, *Asymmetric Strategies for MRCs*, Santa Monica, Calif.: RAND, unpublished manuscript.

Watters, Ethan, "Virtual War and Peace," *Wired*, 4.03, March 1996, p. 49.

Wilkening, Dean, and Kenneth Watman, *Nuclear Deterrence in a Regional Context*, Santa Monica, Calif.: RAND, MR-500-A/AF, 1995.

Chapter Seven

NUCLEAR WEAPONS AND U.S. NATIONAL SECURITY STRATEGY FOR A NEW CENTURY

Glenn Buchan

During his campaign for the presidency, George W. Bush promised a fundamental reevaluation of U.S. nuclear strategy to make it more relevant to the contemporary world. Early in his administration, he announced his intention to reduce the size of U.S. nuclear forces quickly and unilaterally without necessarily relying on cumbersome arms control negotiations to produce such a result (Bush, 2001). He also took the dramatic step of announcing the formal withdrawal of the United States from the 1972 Antiballistic Missile (ABM) Treaty, a U.S.–Soviet agreement that had been central to the orthodox views of nuclear strategy that had evolved during the Cold War. Finally, he announced that U.S. nuclear strategy would no longer focus on a potential nuclear threat from Russia. Rather, the most urgent problem was dealing with rogue states or terrorists that might acquire weapons of mass destruction.

In early 2002, the Bush administration announced the completion of its initial Nuclear Posture Review (NPR), which was its first formal attempt to craft a new nuclear policy. As of mid-March 2002, the NPR itself had not been made public. However, senior DoD officials have briefed the press on its basic findings (Crouch, 2002), and some of the more controversial proposals from the classified version have been leaked to the press (Richter, 2002; Arkin, 2002; Gordon, 2002).

This could be the first shot in a new national debate on U.S. nuclear weapon policy. The objective of this chapter is to review the basic issues involved in crafting such a policy and help evaluate the various policy options that might be available.

WHY A REEVALUATION OF U.S. NUCLEAR POLICY IS NEEDED, AND WHY PEOPLE SHOULD CARE

Such a reevaluation is long overdue. During the Cold War, there was never agreement in the United States on the most basic philosophical underpinnings of nuclear strategy. The notion of deliberately leaving one's society vulnerable to the most destructive weapons ever conceived and relying instead on threats of retaliation in kind to deter an enemy from attacking hardly inspires confidence even among those who considered it the least unattractive option available. The alternative of developing and operating nuclear forces according to narrowly defined military effectiveness criteria as if a nuclear warhead was "just another weapon" appeared to be at best wasteful and at worst likely to increase the chances that a nuclear war might occur and the destructiveness of such a war if it were to occur. Attempts to "finesse" the logical inconsistencies and moral ambiguities of such policies with elegant conceptual strategies (e.g., the "no cities" doctrine) never really proved persuasive enough to affect practical policies. In fact, even the larger debate between "deterrers" and "warfighters," not to mention the nuclear abolitionists who found the whole discussion to be morally abhorrent, might arguably have had only a modest impact on what nuclear forces the United States chose to build and deploy and the plans it made to use them.

The end of the Cold War appeared to offer the possibility of, if not actually resolving the underlying logical conflicts and moral ambiguities of Cold War U.S. nuclear strategy, at least providing the possibility of eliminating some of the more dangerous practical manifestations of past U.S. nuclear strategy—e.g., massive nuclear arsenals, hair-trigger alert levels.[1] Unfortunately, it did not happen. After an initial flurry of activity involving both formal arms control and a series of reciprocal unilateral initiatives between President George H. W. Bush and Russian leaders Mikhail Gorbachev and Boris Yeltsin in which nuclear forces were reduced dramatically and alert levels of

---

[1]President Bush's use of the "Looking Glass" metaphor in his speech was somewhat unfortunate as an example of the "peace dividend" that ending the Cold War produced. Reducing the alert levels of command and control systems is not necessarily reassuring, even in times of reduced tensions.

some elements of the force were reduced, paralysis appeared to set in.

Questions involving U.S. nuclear forces clearly moved to the "back burner" in discussions of critical national security issues and battles for funds and attention. There was a widespread view that nuclear issues no longer matter much for the United States. At the very least, there appeared to be no clearly articulated view of why the United States still needed nuclear forces (if, indeed, it did), what the forces need to be able to do, and what criteria an effective U.S. nuclear force needs to meet. The result has been a U.S. nuclear policy that is a combination of stasis and neglect at a time when there are opportunities for new directions and when active proposals for change are already on the table. A prerequisite to a more sensible approach to U.S. nuclear policy is the kind of basic reexamination of nuclear strategy that President George W. Bush has promised.

During the Clinton administration, U.S. nuclear policy seemed set on autopilot, guided mainly by the decisions made in the entirely different context of the Cold War. The 1994 NPR that it conducted amounted to an endorsement of "business as usual." (See Nolan, 1999a; Nolan, 1999b; and Lodal, 2001, p. 81.) The United States had a declaratory policy on nuclear weapons during this period. Speaking for the Clinton administration, then–Assistant Secretary of Defense Edward L. Warner III laid it out for Congress in April 1999 (Warner, 1999). Some of the key points Secretary Warner made were the following:

- Nuclear weapons remain a vital part of U.S. national security policy.

- The United States cannot eliminate nuclear weapons or even reduce forces to very low levels (e.g., a few hundred weapons) because it needs a hedge against a resurgent Russian nuclear threat, needs to deter China, and needs to deter rogue nations that develop weapons of mass destruction and long-range delivery systems.

While this statement had some interesting elements (e.g., the explicit expectation that U.S. nuclear weapons might help deter rogue states from using chemical or biological weapons), it was at once too vague and too specific to make a compelling case for what kind of nuclear capabilities the United States needs in the current world and why. It was too vague because it does not explain what explicit strategy the United States would rely on to deal with, for example, a revanchist Russia. Nor did it explain why lower levels of nuclear forces could not achieve the ends that Secretary Warner described. It was too specific because it appeared to limit the role of U.S. nuclear weapons in non-nuclear situations to deterring use of biological or chemical weapons. Essentially, this policy amounted to "treading water" while the United States figured out what it wanted to do with nuclear weapons in the future.

Ironically, stasis was not entirely bad, at least for a while. Absent a compelling reason to change directions and a clear vision of the future, standing pat with an overall strategy and basic force structure that have seemed adequate in the past may have been sensible. However, such stasis cannot last indefinitely, if for no other reason than that decisions and actions will be required just to keep the current force up and running. For example, the Bush administration has already moved to eliminate the force of Peacekeeper intercontinental ballistic missiles (ICBMs) because, as Secretary of Defense Donald Rumsfeld put it, "there's no money in the budget for the next five years to keep it; there's no money to retire it; and there's a law that says you can't retire it" (Dao, 2001).

The fact that more practical decisions, such as Rumsfeld's, are unavoidable, and some could require spending significant amounts of money, provides an opportunity for the United States to rethink its basic nuclear policy and address fundamental issues.

Considering the destructive power of nuclear weapons, the stakes could hardly be higher—e.g., national survival, destruction on an unprecedented scale. Thus, even if nuclear weapons are to remain "out of sight," they can hardly afford to be "out of mind." That is why the revamping of U.S. nuclear strategy that President Bush has promised deserves the most serious attention. A fundamental review of U.S. nuclear policy must address the following kinds of basic questions:

- What military and political value does the U.S. expect its nuclear forces to provide?

- What does that imply about the kinds of nuclear weapon systems that the United States needs?

- What sorts of nuclear forces are required?
- How should they be operated?
- What role should formal arms control play?

Answering these questions adequately will require considerable analysis, and the Bush administration will certainly not suffer from a shortage of advice, most of it presumably unsolicited.[2] The new NPR is its first formal attempt to define a position. To begin to address these issues, I will first discuss the historical legacy of how U.S. nuclear forces and strategy have evolved. Then, I will review the basic advantages and disadvantages of nuclear weapons. Next, I will examine the generic spectrum of strategic options involving nuclear weapons that are available to the United States. Finally, I will reevaluate the possible applications of nuclear weapons in the new, more complex, post–Cold War context.

THE HISTORICAL CONTEXT: THE LEGACY, LESSONS, AND CONSTRAINTS

Today's U.S. nuclear force structure was shaped by decisions made in the 1950s. U.S. nuclear strategy was shaped by analysis done and political decisions made in the 1950s and early 1960s. That alone suggests that a reevaluation of U.S. nuclear policy is long overdue, given the end of the Cold War over a decade ago. Still, not everything has changed, and the lessons of the Cold War experience with nuclear weapons may be of some use in considering future strategic options.

The Cold War Legacy

From the inception and first use of the atomic bomb to the end of the Cold War, nuclear weapons held center stage in U.S. foreign policy. Indeed, during the Cold War, the nuclear confrontation between the United States and the Soviet Union was the central reality in world politics.

---

[2]See, for example, Lodal (2001), McNamara and Blight (2001), Blair et al. (2001), McKinsie et al. (2001), Payne et al. (2001), and Buchan (2001), not to mention all the other proposals that have been put forth over the last decade or so, for example, Blair (1995), Butler (1998), Feiveson (1999), Defense Science Board (1998), Turner (1997), and Buchan (1994).

The Cold War as it evolved posed an unprecedented challenge to U.S. national security. From the outset, the Soviet Union appeared to be an aspiring great power with a massive conventional army that threatened war-weary U.S. allies in Western Europe and elsewhere, expansionist goals, a hostile ideology, and—in the early Cold War days, at least—a monster running the country.[3] As the Cold War progressed, the Soviet Union developed nuclear weapons of its own—thus ending the U.S. nuclear monopoly—and long-range delivery systems. That development gave the Soviet Union the capability to threaten the U.S. homeland with unprecedented destruction. Meanwhile, successful communist revolutions in China and Korea and insurgencies elsewhere in the world created the specter of a global communist threat to U.S. interests.

The United States relied primarily on its nuclear weapons to solve this array of security problems. It developed a capacity to launch massive nuclear attacks on the Soviet Union to defeat its warmaking capability and destroy it as a society. The initial U.S. emphasis was on launching a preemptive first strike against the Soviet Union in a time of crisis, and there was apparently at least a flirtation among some in the strategic planning community with the notion of preventive war—launching a premeditated nuclear attack on the Soviet Union before its nuclear capability matured sufficiently to make such an attack infeasible.

Nuclear weapons appeared to offer the United States a relatively cheap (i.e., "more bang for the buck") counter to the Soviet preponderance of conventional power threatening Western Europe. Also, as long as it enjoyed an effective nuclear monopoly, the United States apparently believed that threatening the Soviet Union with "massive (nuclear) retaliation" would be a credible enough threat to deter Soviet leaders from invading Western Europe or threatening U.S. interests in any other significant way.[4]

---

[3]At the time of the Gulf War, one of my colleagues wryly observed that comparisons of Saddam Hussein to Hitler were off target—Hussein was actually more like Stalin. That was hardly reassuring.

[4]Even after the United States lost its nuclear monopoly, Eisenhower clung to his policy of massive retaliation even in the face of heavy criticism. Eisenhower apparently believed that any major war would quickly escalate to a general nuclear war, so

As the capability of Soviet nuclear weapons and delivery systems increased, making the prospects for a successful disarming first strike against them less and less likely, the United States shifted the emphasis of its nuclear strategy as the 1950s progressed to maintaining the capability to retaliate massively to any Soviet nuclear attack. Accordingly, it had to worry about the vulnerability of its own nuclear forces to a nuclear surprise attack. That became a transcendent concern of U.S. nuclear planning, which continues to this day.

The theory was that the threat of nuclear retaliation would raise the price of a Soviet attack on the United States or its allies to the point where the costs would exceed any conceivable gain, thereby hopefully deterring the Soviets from making the attack. Although this notion of deterrence was fraught with ambiguities, not the least of which was how far effective deterrence extended (e.g., Would the United States use nuclear weapons in response to some relatively minor provocation?), a strategy of deterrence by "assured destruction" of a major portion of hard-won Soviet economic and military capabilities seemed to be the best option available to the United States as U.S. and Soviet nuclear arsenals matured.

Still, no one had any illusions: Nobody could be certain what deterred whom from doing what to whom. However, all the alternatives appeared to be worse: wasteful, dangerous (i.e., likely to increase the chances of nuclear war), or unlikely to work. For example, Secretary of Defense Robert McNamara briefly flirted with, but ultimately rejected, the idea of structuring U.S. forces to threaten Soviet nuclear forces. He felt it failed all three criteria. The debate on that subject has been ceaseless ever since, and U.S. forces evolved in that direction anyway. The United States also debated torturously for decades about how to develop "rules of the game" that would prevent total destruction if deterrence failed and nuclear war broke out.

Settling on what sort of forces "assured destruction" required became the preoccupation of the next several decades, and the debate has never ceased.[5] The overall shape of U.S. strategic forces—

---

preparing for lower levels of conflict was a waste of resources and sent the wrong message to potential enemies.

[5]Actually, there was less to the debate than there appeared to be. Money and "turf" were the main issues in the fierce internecine battles over force structure. The public image to the contrary notwithstanding, grand strategy was rarely an issue. From the

the so-called "triad" of ICBMs, submarine-launched ballistic missiles (SLBMs), and long-range bombers—had already been cast in stone based on decisions made in the 1950s, and the remaining decisions had to do with how many of what to buy and how to modernize the forces over time. The concept of assured destruction seemed to imply limits on the useful number of strategic weapons that the United States might need and what characteristics strategic forces should have (as well as what characteristics were either superfluous or downright dangerous). The implicit possibilities for restraint opened the door for formal arms control, which has since become a relatively permanent part of the nuclear policy scene. The formal arms control process, beginning with the Strategic Arms Limitation Talks (SALT) and followed by the Strategic Arms Reduction Treaty (START) and other agreements on nonstrategic nuclear weapons, was supposed to help stabilize the strategic balance by codifying the "rules of the game"—mutual assured destruction—that had evolved, slow the arms race, save money, and reduce the risks of war. However, in spite of this "restraint," the scorecard looked like this just prior to the end of the Cold War:

- The United States and the Soviet Union, between them, had accumulated tens of thousands of strategic and tactical nuclear weapons.

- The structure of U.S. strategic forces established in the 1950s (i.e., the triad) remained unchanged and became enshrined in strategic dogma.

- The quality of strategic offensive weapons (e.g., missile accuracy) continued to improve to the point where virtually any target could be destroyed.

- The dominance of offensive forces over defenses was codified in law by the ABM Treaty of 1972, which placed severe constraints on ballistic missile defense (BMD) deployments, thereby rendering effective nationwide BMD impossible. Moreover, the dominance of the offense continued to be underwritten by the physics of the problem and the state of technology.

---

beginning of the nuclear era, U.S. targeting plans involved attacks against a comprehensive set of Soviet targets—strategic forces, other military targets, critical industrial facilities, leadership. The plans changed relatively little over the years in spite of the ins and outs of the strategic debate.

- U.S. strategic nuclear war plans were integrated into a single plan, the Single Integrated Operational Plan (SIOP), which included several different targeting options. The first SIOP was created in 1960 to coordinate the disparate nuclear forces that could be brought to bear on the Soviet Union and its allies, and the SIOP has been the centerpiece of U.S. strategic nuclear planning ever since.

The United States also supplemented its strategic nuclear forces by thousands of tactical nuclear weapons—artillery shells, gravity bombs, short- and intermediate-range missiles, and some specialized nuclear weapons (e.g., atomic demolition munitions). The *function* of these tactical weapons was to defeat invading Soviet forces; however, their *purpose* was somewhat different. After the Soviet Union developed a strategic nuclear capability of its own, U.S. threats to launch a nuclear attack on the Soviet Union in response to a Soviet invasion of Western Europe rang a little hollow, as such statesmen as Charles de Gaulle and such analysts as Henry Kissinger never failed to point out.

Accordingly, NATO's most fundamental problem throughout the Cold War was to find a way to finesse the logical contradiction inherent in its deterrent strategy: how to make it as difficult as possible for the United States to do the "rational" thing and "sit out" a war in Europe or limit the fighting to Europe itself, thereby reducing the risk of Soviet nuclear retaliation against the United States itself. During the 1950s, the United States introduced tactical nuclear weapons into Europe to help weave that tapestry by providing a means to blunt (but not defeat) a Soviet-led Warsaw Pact attack and raise the stakes in the process, thus making it unlikely that the United States could avoid escalating the conflict at some point to a general strategic exchange. In later years, despite unavoidable tension between the United States and its NATO allies on the subject, NATO did modernize its conventional forces to increase their capability to halt a Warsaw Pact advance, thereby presumably raising the nuclear threshold, at least from *NATO's* point of view.[6] However, the nuclear

---

[6]Arguably, however, such actions might have even *lowered* the nuclear threshold for the *Soviet Union*. If it were bound and determined to attack NATO and found itself faced with a more formidable conventional threat, it might well have felt obliged to introduce nuclear weapons itself sooner rather than later.

part of the equation remained critical until the last days of the Cold War, when the United States and the Soviet Union agreed to the Intermediate-Range Nuclear Forces (INF) Treaty, which removed all intermediate-range nuclear missiles from Europe, and subsequent reciprocal unilateral actions by the United States and Russia that removed all tactical nuclear weapons from Europe except for a few hundred nuclear U.S. gravity bombs.

Despite the obvious drawbacks of a strategy that contemplated actions that could lead to the end of the world, the United States and the Soviet Union behaved prudently—with some notable exceptions—and managed risks reasonably well during most of the Cold War years. The fact that the principal antagonists were on opposite sides of the world was a fortuitous historical accident that helped reduce the friction between them and allowed the space and time necessary to provide adequate warning of an imminent attack, a necessary ingredient of the retaliatory strategies that both sides adopted. Similarly, the years that it took to build up their respective nuclear arsenals and arrays of long-range delivery systems gave both sides some "breathing space" to learn how to play the game and survive in the nuclear age.

Still, the Cold War nuclear world was inherently dangerous, and both sides occasionally behaved in ways that exacerbated the danger. Because both sides worried about being able to retaliate after a massive surprise would-be disarming attack, they maintained significant portions of their strategic forces on high levels of alert ready to respond in minutes if an attack should occur. That put immense pressure on tactical warning systems to correctly identify attacks and on command and control systems and decisionmakers to be able to respond in minutes under the most intense pressure imaginable. The dangers this posed, which still exist, have long been recognized, and incidents that could have led to accidental war occurred all too frequently during the Cold War.

Most important, of course, were the confrontations that could have led to war by miscalculation, misunderstanding, or bad judgment. The Cuban Missile Crisis is the most famous example. However, there were other near collisions, such as the incident in 1983 in which a particularly paranoid portion of the Soviet leadership had apparently convinced itself that a routine NATO exercise, known as ABLE ARCHER 83, was actually a cover for preparations for a U.S.

first strike (Pry, 1999) and was prepared to launch a preemptive nuclear attack if it concluded that a U.S. attack was imminent.

The Sea Change—The End of the Cold War

Then, everything changed. The Cold War ended. The United States apparently had achieved all its really major political objectives during the Cold War without sacrificing any vital interests. In a stunningly short time, the Soviet Union and the Warsaw Pact were no more, the Berlin Wall came down, and U.S. and Russian leaders were outdoing each other in dismantling their nuclear arsenals. The ideological conflict seemed to be resolved in the United States' favor. In short order,

- The United States and Russia agreed to and implemented the START I arms control agreement, which reduced strategic force levels to 6,000 "accountable" warheads on each side.

- Both sides agreed to further arms reductions by signing the START II treaty and eventually ratifying it, albeit with differing attached understandings. The Treaty

  — reduces strategic force levels to 3,500 warheads

  — eliminates Russian "heavy" ICBMs and U.S. Peacekeepers

  — eliminates MIRVed ICBMs.

- The United States unilaterally took its bomber force off day-to-day alert and reduced the alert status of its command and control aircraft to unprecedented levels.

- Both presidents took unilateral steps to reduce nonstrategic nuclear forces.

- The United States withdrew all its tactical nuclear weapons from overseas bases except for a few hundred nuclear gravity bombs for delivery by tactical aircraft that remain stored in Europe.

- The U.S. Army essentially got out of the nuclear business, eliminating all its nuclear weapons.

- The U.S. Navy removed all its nuclear weapons, other than SLBMs, from ships at sea and eliminated all its tactical nuclear weapons, except for a force of nuclear-armed cruise missiles that it keeps in secure storage on land.

Perhaps most importantly, U.S. and Russian leaders achieved a degree of political rapprochement that suggested there were no longer any quarrels between the United States and Russia that would justify fighting a nuclear war. The two countries even managed to cooperate on serious ventures, such as the Gulf War.

The reason for dwelling on this history is that U.S. nuclear strategy for the last decade or so has remained largely the captive of the past. The legacy of the past is overwhelming and has so far inhibited a serious reevaluation of U.S. nuclear strategy. Veteran strategic analyst Leon Sloss, for example, has argued that the basic structure of U.S. strategic forces—i.e., the triad—has proved so resilient that it is unlikely to be changed in the foreseeable future, any analytical arguments to the contrary notwithstanding.

Empirically, Sloss certainly seems to be right so far. Breaking out of the old molds seems extraordinarily difficult. For example, the 1994 NPR the late Les Aspin initiated during his brief tenure as Secretary of Defense was supposed to provide a fundamental review of nuclear policy. However, it failed, as Aspin lamented in private shortly before his death. The 1994 NPR basically reaffirmed the *status quo ante*: "business as usual, only smaller." It was a product of all the usual bureaucratic processes and pressures. Perhaps because there was no real money to be saved by radically restructuring the strategic forces and the strategic environment was sufficiently benign, there was no particular pressure to do anything different.

The other major factor impeding a serious reexamination of basic U.S. strategic nuclear policy has been simple neglect. There is a widespread view that nuclear issues simply do not matter much anymore. That means discussions of nuclear questions receive less attention than other, higher-priority matters. Military officers are likely to be less willing to tie their careers to nuclear weapons for fear of heading into a professional cul-de-sac. Nuclear weapon-related programs receive less attention and fewer resources from the government, the defense contractors, or the national nuclear weapon laboratories, which are trying to decide how best to invest their research and development dollars. Competing programs nip away at the modest resources that remain for nuclear matters.

Unfortunately, this understandable neglect of nuclear matters may someday have serious consequences. Since the heady days of the post–Cold War "honeymoon," international politics has gotten tougher. The September 11, 2001, attacks on the World Trade Center and the Pentagon demonstrated in the most dramatic way that the United States could be damaged seriously, even by a handful of terrorists. These events seemed to reinforce President Bush's oft-stated concern with what had traditionally been considered minor players on the world scene—i.e., "rogue" states and terrorists—as opposed to major powers. Moreover, the sobering reality of the amount of damage that terrorists could cause using aircraft as weapons led to inevitable speculation: How much greater damage could terrorists or rogue states cause if they could attack the United States with some kind of weapon of mass destruction? That concern and how to deal with it were at the heart of President Bush's May 2001 speech.

President Bush has stated repeatedly that Russia is no longer our enemy. However, U.S.-Russian relations have become more problematic over the last decade, although Russian President Putin's support of the U.S. military campaign in Afghanistan in response to the September 11 attacks has certainly won him friends in the United States. Russia's response to the U.S. withdrawal from the ABM Treaty was predictably negative, but the criticism was relatively muted. The Russians should welcome some of the elements of the new NPR, especially the planned substantial unilateral reductions in deployed U.S. nuclear forces and the reduced emphasis on targeting Russia. However, some Russian commentators have noted disapprovingly that Russia is still one of seven countries that the NPR lists as potential targets for U.S. nuclear weapons (Richter, 2002; Arkin, 2002) and take that as evidence of American ill will (Holland, 2002). Apparently, the balance lies in the eye of the beholder, and the U.S.-Russian relationship remains a work in progress.

Meanwhile, India and Pakistan have each conducted a series of nuclear tests and seem committed to developing usable nuclear arsenals even as they face off yet again over disputed territory in Kashmir. Not only does this pose the risk of a confrontation escalating into a regional nuclear war with possible consequences for others, but also the war in neighboring Afghanistan has raised questions about the security of Pakistan's nuclear weapons. Early in the Afghanistan campaign, there were fears that civil war could break out in Pakistan, which might lead to Muslim fundamentalists allied with Osama bin Laden gaining control of some of Pakistan's nuclear weapons. For the United States and others, that would be a nightmare. While assertive action by the Pakistani government appears to have prevented that disaster this time around, there could always be a next time.

The Chinese nuclear program appears quite vigorous, and China's political agenda remains uncertain. Moreover, Chinese-U.S. relations have suffered a series of blows, including the collision of the U.S. intelligence-gathering aircraft and the Chinese fighter, with the subsequent loss of the Chinese fighter pilot, the brief detention of the U.S. crew, and the longer retention of the U.S. plane; the continuing—and possibly escalating—tension over Taiwan; the detention of U.S. citizens visiting China as "spies"; and the Bush administration's decision to pursue BMD. Both governments have subsequently taken steps to reduce the tension and try to normalize relations, but it is unclear how deep the divisions are.

Meanwhile, alternatives have been proposed in the United States for very different approaches to the nuclear problem—everything from radical arms reductions to active defenses to more vigorous nuclear weapon development. Charting a course for the future requires a breadth of thinking that has been in short supply in recent years. The new NPR is a start. A "back-to-basics" review of nuclear weapons and their possible applications could help frame the problem.

WHY NUKES?

Nuclear weapons are generally perceived by those who possess them and those who would like to as the ultimate guarantors of a nation's security. The almost unlimited destructive power of thermonuclear weapons makes it possible to put massive explosive power in very small packages. A single nuclear detonation of even modest size—the bomb that destroyed Hiroshima was about the size of a nuclear artillery shell—can destroy virtually any individual target or lay waste to large areas (e.g., destroy a city). That capability changed the nature of war dramatically. It appeared to make defense, in the traditional sense, virtually impossible because of the damage that even a single nuclear weapon that leaked through defenses could cause. Also, when coupled with long-range delivery systems (particularly long-range bombers and ballistic missiles), nuclear weapons allow those possessing them to destroy an enemy's homeland without necessarily having to defeat its military forces first. Moreover, the small size of modern nuclear weapons makes possible a variety of delivery means in addition to the more traditional missiles and aircraft (e.g., truck or "suitcase" bombs). Thus, nuclear weapons, if used effectively, could prevent an enemy's military from achieving the most fundamental objective of any military establishment: protecting its homeland. That stood traditional concepts of war on their heads.

Even in strictly military terms, nuclear weapons are simply more effective than other weapons in destroying targets. The following are some classes of targets against which nuclear weapons are particularly well suited:

* massed formations of troops, particularly armor
* large military complexes (e.g., airfields, ports)
* hardened military installations (e.g., missile silos, underground command centers)
* inherently hard natural or man-made structures (e.g., concrete bridges or dams, cave or tunnel entrances)
* large warships
* arriving ballistic missile warheads
* satellite constellations
* some kinds of communications and electronic systems
* industrial capacity and cities.

Nuclear weapons offer attractive strategic advantages to those who own them:

* a capacity to coerce enemies by threat or actual use of nuclear weapons
* an instrument to deter adversaries from a range of actions by threat of nuclear use
* a cheap means of offsetting an imbalance of conventional military power
* an effective means of fighting a large-scale war
* prestige and a "place at the table" among other "great powers."

Of all the types of so-called "weapons of mass destruction," nuclear weapons are clearly the best to have. As terror weapons, they are unmatched. As military weapons, they are more effective and more difficult to protect against than chemical, biological, or advanced conventional weapons.

AND WHY NOT

On the other hand, nuclear weapons have significant disadvantages and inherent risks as well, stemming mainly from the same characteristics responsible for their unique advantages. Some of the primary risks are the following:

- excessive damage
- incidents, accidents, mistakes, and miscalculations
- unauthorized use
- theft
- operational difficulties
- "pariah" status and the taboo against nuclear use
- increased proliferation
- environmental hazards and infrastructure problems.

The fearsome destructiveness of nuclear weapons has been a source of major concern since the beginning of the nuclear age. The fundamental concern has been that the damage resulting from actual use of nuclear weapons would be far out of proportion to any legitimate political or military ends. For conflicts involving major nuclear powers, the danger has always been perceived as particularly acute due to the sheer scale of the potential effects (e.g., large-scale fallout, climatic effects) of an unlimited nuclear exchange, should one ever occur. Thus, risk of escalation to large-scale nuclear use has always been a major issue in superpower confrontation.

Even limited nuclear use could fail the proportionality test inherent in the notion of "just wars" and reinforce the long-standing moral argument against nuclear weapons. Even the precision conventional bombing of Yugoslavia during the Kosovo campaign of 1999 did enough damage to raise arguments about the morality and effectiveness of coercive strategic bombing. That could render nuclear use almost out of the question in any but the most extreme circumstances.

Aside from their horrific impact on civilians, collateral effects of nuclear weapons can complicate military operations and cause a variety of headaches for field commanders. In addition to the obvious problems of operating in a radiation environment, there are more subtle difficulties as well. For example, nuclear detonations can black out some types of radar and communications systems, in some cases for extended periods of time.

On a more mundane level, the problems associated with special handling of nuclear weapons, the need to obtain release authority to actually use them, and the competition for scarce support resources might readily convince military commanders that nuclear weapons are more trouble than they are worth, unless the need is really compelling. An interesting aspect of the current policy debate about nuclear weapons is the number of senior former military officers who have become disenchanted with nuclear weapons and are actively seeking ways to eliminate them or drastically reduce their numbers.

Finally, the flip side of the argument that nuclear powers acquire a special status is that they might also be regarded as pariahs. This has always been sort of a delicate balancing act for the established nuclear powers, which are obliged by the Nuclear Non-Proliferation Treaty (NPT) to move toward nuclear disarmament. New nuclear powers—the few that have emerged—have not obviously improved their security by demonstrating their nuclear capability. India and Pakistan have both suffered economic and political sanctions as a result of their nuclear tests, and both are probably less secure than they were before.

WHERE NUCLEAR WEAPONS MIGHT FIT

Given this complicated array of pluses and minuses, it is not entirely clear that the United States derives substantial benefits from its nuclear arsenal, particularly in its current configuration. Evaluating whether it does requires matching up the U.S. nuclear arsenal with its security needs. While the United States in general enjoys unparalleled security and a large superiority in both nuclear and conventional forces, all is not entirely well. More importantly, the world scene remains volatile, so more "nonlinear"—i.e., unpredictable—change is likely.

As other contributions to this volume make clear (see, for example, Chapter Two), the evolving world will likely have a more diverse set of challenges and players than in the past. The key point is that "one size does not fit all": Different kinds of opponents and different challenges will require different strategies. To cope with the new context, the United States will need a variety of tools, of which nuclear weapons are only one. At the same time, U.S. nuclear force could conceivably be put to a wide range of possible uses, from traditional deterrence to warfighting uses, as elaborated below. However, as Figure 7.1 suggests, there are very few roles for which nuclear weapons have really decisive advantages over modern conventional weapons, and some of those may be of dubious value. A major strength of the new NPR is that it explicitly recognizes the capabilities of advanced conventional weapons and the diminishing role of nuclear weapons in U.S. policy. The United States has considerable latitude in choosing what it would like its nuclear forces to do in the contemporary world and should have the luxury of choosing its quarrels carefully in the future. However, it must also consider that the tradition of nonuse of nuclear weapons probably serves U.S. interests more than those of any other country.

Terror Weapons for Traditional Deterrence

Nuclear weapons are unmatched as terror weapons and, therefore, have the potential to be the most effective possible weapons to implement a policy of deterrence by threat of punishment. This is the most enduring role for nuclear weapons and the one for which they are most uniquely suited. The only issue is whether the United States wants to continue to have this kind of capability, and whether it needs to inflict the levels of damage that nuclear weapons can cause. Moreover, even without using nuclear weapons, the United States demonstrated in Iraq and Yugoslavia an ability to destroy virtually any kind of target with modern conventional weapons. Thus, for adversaries that are amenable to deterrent or coercive threats, the United States probably has the means to make such threats without having to pay the political cost of crossing the nuclear threshold.

As noted earlier, nobody can be sure what deters effectively. In practical terms, that could be even more complex than usual in the future, but large nuclear forces would probably not be required. For

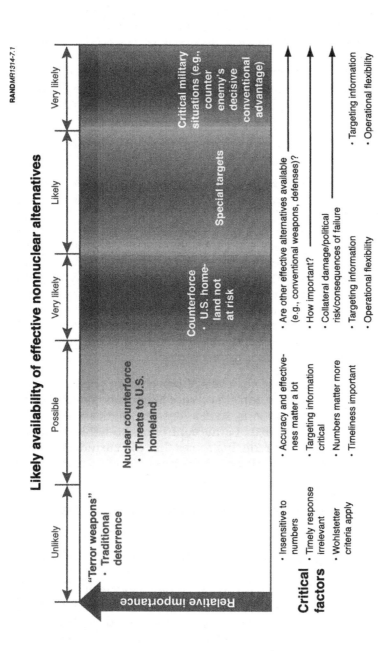

Figure 7.1—Why the United States Might Want Nuclear Weapons in the Contemporary World

example, notwithstanding the Bush administration's policy statement that Russia is not an enemy, if the United States still believed it necessary to target Russia as a deterrent, what would it target?

- The Russian economy seems hardly worth targeting with nuclear weapons, considering what bad shape it is in already. (Perhaps a handful of military-oriented industrial facilities might warrant attack with nuclear weapons.)

- Similarly, Russian conventional forces have deteriorated to the point that they hardly warrant nuclear attack.

- Leadership attacks have always been problematic—good idea or bad, straightforward or very difficult, depending on the details of the situation.

Of the traditional target categories, that leaves only nuclear counterforce targets. Threatening those is a warfighting issue, not one of deterrence by threat of punishment. Thus, not only might small nuclear forces be adequate as a deterrent, but also even selecting suitable targets for those could be difficult. A senior Russian defense analyst remarked a few years ago that the United States and Russia need to find a different way to relate to each other than through a nuclear suicide pact. He is probably right.

Indeed, "requirements" for levels of punishment needed to deter are largely arbitrary. Thus, force levels are likely to be determined more by practical considerations, such as cost, or political considerations, such as the constraints that arms control has sometimes imposed or the opportunities for reducing forces that the current Russian fiscal problems could allow. However, nuclear forces and their associated command and control system do need to be survivable against competent attacks. They also must be operated safely and securely. Interestingly, timely use of nuclear weapons is *not* required. Response has to be certain and properly directed, but not swift.

Counterforce

The other role for which nuclear weapons are especially well suited is nuclear counterforce, particularly counterforce attacks against enemy weapons that can reach the United States. With the improvements in weapon accuracy that have occurred in the last couple of decades, debates about the potential effectiveness of nuclear weapons in destroying hardened fixed targets, such as missile silos, have long since been resolved. Mobile targets remain elusive, but if adequate surveillance capability to locate them reliably can be achieved, destroying them with nuclear or conventional weapons should be relative easy. For the time being, however, nuclear weapons remain the most effective weapons for counterforce operations.

An irony of the end of the Cold War is that the sort of counterforce strategies that the United States planned for during the Cold War—and which had no chance of being effective once the Soviets developed hardened silos, mobile ICBMs, and missile-launching submarines—might actually work in the current world, particularly against fledgling nuclear powers that have not yet learned how to play the game (e.g., have not developed the technical and operational expertise necessary to employ mobile systems effectively). Deciding how robust a nuclear counterforce capability the United States wants to maintain will be a major determinant of how far the United States is willing to reduce its nuclear forces and how it chooses to operate them.

Special Targets

In recent years, there has been even more interest than in the past in developing effective ways to attack special kinds of targets that are very difficult to destroy. Underground facilities, especially deeply buried targets, have received considerable attention in this regard. Such facilities could include command centers, manufacturing plants or storage sites for special weapons, or other types of high-value installations.

Nuclear weapons may indeed be necessary to destroy really deeply buried installations. However, even nuclear weapons may not be adequate if the details of the underground structure are not adequately known. Moreover, literally destroying the structure may not be necessary to defeat its function. Thus, there may be other, less drastic ways to solve these problems, particularly since nuclear attacks on underground targets will inevitably cause substantial fallout (Nelson, 2001; Drell, Jeanloz, and Peurifoy, 2002).

Critical Military Situations

Finally, there is a whole set of particular military situations that could arise in which nuclear use might be an option if the situation were important enough and other options appeared inadequate. During the Cold War, unfavorable conventional balances in Europe and, to a lesser degree, in Korea drove the United States to develop tactical options for nuclear use to offset its conventional deficiencies. Since then, conventional weapons have improved considerably, particularly U.S. conventional weapons. There are fewer potential situations in which the United States would have a compelling stake in the outcome of a conflict and would be at a serious conventional disadvantage. Still, this remains one of the few plausible possibilities for the United States needing nuclear weapons in the future, and the NPR reportedly recognizes it explicitly (Gordon, 2002).

Still, designing a future U.S. nuclear policy requires addressing the questions directly:

• Are any contemporary situations important enough and difficult enough to warrant the threat or actual use of U.S. nuclear weapons?

• If so, what sort of nuclear forces and operational procedures would be necessary or appropriate for dealing with these contingencies? Specifically, are different kinds of nuclear weapon systems required, and is current nuclear operational practice appropriate for future tactical situations that might emerge?

• How much more effective than conventional weapons would nuclear weapons be, and are the differences worth the cost (e.g., collateral damage, political costs, long-term impact on "rules of the game")?

The answers to some of these questions turn on analytical issues, specifically the absolute and relative effectiveness of modern conventional and nuclear weapons. Others, such as the political repercussions of actual use of nuclear weapons, are more metaphysical.

A SPECTRUM OF NUCLEAR OPTIONS

The relative importance of developing actual warfighting nuclear options is the critical element in deciding on the future course of U.S.

strategy. The broad problem of nuclear strategy is bifurcated: Nuclear weapons could be brandished to deter others by threatening punishment, or they could actually be used as warfighting instruments. The first is familiar and straightforward, and the current administration continues to emphasize deterrence as declaratory policy (Savage, 2002). The second includes a richer set of possibilities and has different implications for forces and operational procedures. All these capabilities are achievable to some degree, and they would lead to significant variations in future U.S. nuclear posture. Thus, choices abound.

Even putting aside the domestic political issues inherent in nuclear weapon-related issues, choosing among the options will depend to a significant degree on nitty-gritty questions, such as cost and technical effectiveness of various approaches. Below, I will examine some of the major strategic policy options available to the United States and some key issues that will affect the choices.

The choices available to the United States run the gamut from renouncing nuclear weapons entirely to much more aggressive nuclear strategies than the United States has entertained in the recent past. A variety of alternatives to current U.S. policy are already on the table. Below are five generic approaches that cover the spectrum of possibilities. Options also exist to create combination strategies:

- abolition of U.S. nuclear weapons, with or without formal arms control
- aggressive reductions and "dealerting"
- "business as usual, only smaller"
- more aggressive nuclear posture
- nuclear emphasis.

Abolition

Doing away with U.S. nuclear weapons entirely must logically be a part of any complete set of strategic nuclear options, in spite of the fact that President Bush—following the course set by previous administrations—has made it clear that nuclear weapons remain an important part of U.S. national security strategy (Bush, 2001). Even most advocates of deep reductions in nuclear forces stop short of calling for abolition of nuclear weapons, at least for the foreseeable future (see Feiveson, 1999, and Turner, 1997). Still, there is a case to be made for abolition, and strictly speaking, the NPT commits the United States and other nuclear-armed signatories to the treaty to divest themselves of their nuclear weapons eventually. Moreover, as advocates correctly point out, the logic of totally eliminating nuclear weapons is different from merely reducing them to low levels.[7] Thus, if elimination is the long-term objective, the argument has to be made that way.

If the United States were to choose to divest itself of its nuclear weapons, it would presumably be for some combination of the following reasons:

• No military or political threat to the United States is serious enough to require a threat of nuclear retaliation to deter.

• Alternatives to nuclear weapons are adequate to solve any military problem.

• The danger, trouble, expense, and political baggage associated with maintaining nuclear weapons exceed whatever residual value they might have.

• Nuclear weapons are not "usable" politically or militarily.

• Giving up its nuclear weapons would do more to restrain nuclear proliferation than maintaining a dominant nuclear capability.

The first two points are key: a lack of a compelling need for nuclear weapons and the availability of adequate alternatives. The dramatic improvements in the accuracy and lethality of conventional weapons clearly make them attractive alternatives to nuclear weapons for many applications. However, it is not as yet clear that they can really achieve these capabilities at an affordable cost. In particular, improved information collection and processing technologies "enable" most of these weapon concepts, and not all the requisite capabilities are in hand. Still, the performance of modern precision-guided weapons in the Gulf War, Kosovo, and Afghanistan is certainly encouraging.

---

[7]See, for example, McNamara and Blight (2001) and Butler (1998).

One of the key issues involving abolition of nuclear weapons is whether the United States would eliminate its nuclear weapons unilaterally or as part of a more sweeping agreement among several—or all—nations to abolish all nuclear weapons. If there were actually a way to abolish nuclear weapons broadly (and leaving aside for the moment the inevitable questions about verification, hidden weapons, and nations' relative ability to regenerate nuclear capability), the United States would clearly be the major beneficiary. The United States' economic and conventional military power would leave it in a position of greater relative strength in a world without nuclear weapons. In fact, since the United States would benefit most from worldwide elimination of nuclear weapons, perhaps it should be leading the charge for abolition. The problem, of course, is that the political processes necessary to produce such an agreement boggle the mind. The bar would have to be set so high that the question becomes academic. If the international political climate became that benign, nobody would need nuclear weapons anyway. A more interesting case is the one in which the United States considers the possibility of eliminating its nuclear weapons unilaterally.

Finally, there is the possibility that the United States may not be able to retain its nuclear infrastructure and design and operational expertise indefinitely, no matter what policymakers would prefer (Defense Science Board, 1998; Buchan, 1994, p. 77). If that proves to be true, the issue for the United States will not be whether to eliminate its nuclear weapons. Rather, the only question will be when and how. I will return to this point later.

Aggressive Reductions and "Dealerting"

A less aggressive, but still quite radical, set of proposals has been put forward to reduce the size of U.S. nuclear forces far beyond what even START III currently envisions (e.g., to a level of a few hundred warheads) and to reduce their alert levels dramatically so that nuclear weapons could not be used quickly.[8] The possibility of going to smaller force levels has been on the table, at least implicitly, since the beginning of the nuclear age. Admiral Arleigh Burke made a very articulate argument in the 1950s, when the key decisions that would

---

[8]Feiveson (1999) presents the most comprehensive treatment of this option. Individual elements of the proposal have been around for some time.

shape U.S. strategic force structure for the next 50 years were being made, that the United States ought to limit its nuclear force to a modest number of missiles deployed on submarines. Such a "finite deterrent" force, he argued, would provide the United States with the capability to respond to a nuclear attack by launching a retaliatory attack on an enemy's cities, and that was all that U.S. nuclear forces needed to be able to do.

Burke lost that battle, but the argument has been around in one form or another ever since. Recently, a prominent Russian defense expert said that was an option that Russia might pursue in the future. The United States could move in that direction as well, regardless of any formal arms control arrangements. Several contemporary U.S. authors have made essentially that argument (see Bundy et al., 1993; Blair, 1995; and Feiveson, 1999). The contemporary version is that, in the current world, no reasonable use of nuclear weapons would require a large number of warheads.

The proposal to "dealert" nuclear forces—that is, to reduce the launch readiness of some or all U.S. nuclear forces during normal day-to-day operations—is more recent, although its origins go back almost 20 years (Blair, 1985, pp. 288–295). It seeks to solve a different problem: war by accident or miscalculation. The premise is that, in the contemporary world in particular, the greatest nuclear danger comes not from a purposeful attack but from an accident or mistake on someone's part. Of particular current concern are the potential vulnerabilities of the Russian nuclear forces and the deterioration of their tactical warning systems. The danger is that, the end of the Cold War notwithstanding, Russia still worries about the vulnerability of its strategic forces to a preemptive attack. Unable to afford to keep its nuclear submarines (SSBNs) at sea and perhaps even its mobile ICBMs out of garrison, Russia might opt—as suggested earlier—to rely even more heavily on being able to launch its vulnerable missiles on receipt of tactical warning of an attack in progress. That in itself is bad enough, the theory goes, especially because of the deteriorating state of Russian tactical warning systems, since inaccurate warning increases the risk of an error that could lead to an accidental nuclear war.[9] Even worse, lacking faith in their tactical warning systems, the

---

[9]For examples of such errors, see Blair (1995), p. 51; and Blair et al. (1997).

Russians might feel desperate enough to launch a preemptive attack if they suspected they were about to be attacked.

Dealerting is intended to solve these problems by making it more difficult for the United States to launch a nuclear attack quickly (see, for example, Turner, 1997; Feiveson, 1999; and Blair, 1995). Details of individual proposals vary. Most include separating warheads from delivery vehicles or key components from missiles so that some time would be required to prepare to launch a nuclear attack (Blair et al., 1997). The objectives would be to avoid a premature launch of U.S. (and possibly Russian) nuclear forces prompted by a false alarm or some other mistake, to allow time to think about the wisdom and nature of a nuclear response, and to reassure the Russians and other nuclear powers that the United States does not intend—and cannot execute—a large-scale surprise nuclear attack. The presumption is that other powers—particularly Russia—would reciprocate by reducing the day-to-day alert posture of their forces, thus eliminating "hair trigger" response options of their own.

The dealerting proposals raise several fundamental strategic issues:

- Is the premise that accidental war is now a greater danger than a deliberate surprise attack valid?

- How likely would the Russians be to respond to U.S. dealerting initiatives by reducing the day-to-day readiness of their own vulnerable ballistic missiles?

- Will delaying a response weaken the deterrent effect of a threat of nuclear retaliation?

- What effect would the reduction in day-to-day launch readiness have on other potential uses of U.S. nuclear forces?

- Are specific dealerting proposals practical, and do they cause more problems than they solve?

The first two points are critical to the case for dealerting and amount to a judgment call. War resulting from accidents, misinterpretation of events, or unauthorized use of nuclear weapons has been a matter of intense concern since the earliest days of nuclear weapons. There has always been a tension between decreasing U.S. vulnerability to surprise attacks and increasing the risks of accidental war, and the balance has shifted over the years to reflect altered perceptions of the threat to the United States and the costs and benefits of the remedies available at any particular time. For example, for a while during the 1960s, the United States kept part of its bomber force on airborne alert to reduce its vulnerability to a surprise attack. It ended that practice, however, deciding that there were adequate solutions to the potential bomber vulnerability problem that were less dangerous and expensive than airborne alert and that the growing U.S. ICBM and SLBM forces made bomber vulnerability less critical. Even during the most intense periods of the Cold War, the United States recognized the need for—and made—trades between the vulnerability of its forces to a first strike and the risks of unintended nuclear war.

As the Cold War was ending, the United States took its bomber force off of day-to-day alert as a part of a series of reciprocal initiatives that Russia and the United States took to reduce the size and alert levels of their nuclear forces. Presumably, U.S. leaders understood that they were increasing the overall vulnerability of U.S. strategic forces to a surprise attack when they took these actions. They simply made the *political judgment* that the risks of a surprise attack had decreased sufficiently that the costs and stress of maintaining bombers on strip alert were no longer justified. Moreover, the climate at the time encouraged both sides to take initiatives to reduce their capability and readiness in the nuclear arena. Thus, there is precedent for taking these kinds of actions and expecting some kind of reciprocity in a relatively benign political environment. The real issues are how far to go and under what conditions, and how to avoid increasing the incentives to strike first.

The issue of Russian reactions is crucial to the dealerting argument. The United States might argue—and has in fact (Warner, 1999)—that it has adequately dealt with its own command and control problems to minimize risks of accidental or unauthorized nuclear launches. If so, the only rationale for further U.S. dealerting would be to persuade the Russians that their strategic forces were not in enough danger from the United States to justify such dangerous strategies as launch-on-warning or preemptive attack. If the Russians are actually worried about U.S. motives, they might refuse to respond to anything other than very transparent dealerting measures. (They also would have political incentives to make the price to the United States as high as possible.)

The issue of the effect of a delay in a U.S. nuclear response is clearer. There is nothing about a strategy of deterrence based on nuclear retaliation that requires a prompt response. There never has been. The only rationale for a quick response during the Cold War was the fragility of U.S. forces and command and control systems. The choice might have been between a quick U.S. response and no response at all. Nothing about the Soviet target base ever required a quick response.[10] That is even more true in the current world. Indeed, a prerequisite for credible contemporary deterrence enforced by a threat of nuclear retaliation is a certainty about what happened and whom to blame. That puts a premium on being able to delay a response. Thus, a "dealerted" U.S. nuclear force, assuming it can be made survivable, should still be capable of enforcing a strategy of deterrence based on a threat of devastating nuclear retaliation.

Evaluating specific dealerting proposals requires more-focused analysis. Details could matter a lot. For example, alerts do a number of different things. One of them is to increase—usually substantially—the fraction of nuclear forces that could survive a first strike, which tends to decrease rather than increase the sensitivity and stability of nuclear force interactions during a crisis. Indeed, alerts have pluses and minuses (see Buchan, 1992, and Lodal, 2001). Dealerting focuses only on the minuses, and specific schemes may not have the

---

[10]"Prompt, second-strike counterforce"—a cliché from the Cold War days—never made any sense even when the U.S. and Soviet arsenals were so large and capable. It was, at best, a "politically correct" cover for acquiring strategic forces—mainly ICBMs, but also SLBMs—that were accurate enough to destroy hardened missile silos and could, therefore, be potentially effective first-strike weapons. Having a second-strike counterforce capability could only improve the outcome of a nuclear exchange from the U.S. point of view if all the following conditions applied:

- The attacker launches a deliberately limited initial nuclear strike (a distinctly Western notion), retaining a substantial vulnerable nuclear force in reserve.
- Having launched such an attack (presumably not an entirely successful one if the United States has remaining hard-target killing missiles with which to respond), the attacker would be incompetent enough to allow his remaining vulnerable missiles to sit in their silos while the United States destroys them with a counterattack.
- The difference in damage to its society between the initial nuclear attack and a possible follow-up attack has to be substantial enough to matter to U.S. policymakers.

Thus, prompt second-strike counterforce only has benefit in the narrowest of circumstances against an incompetent opponent schooled in Western nuclear lore. In the meantime, the necessary forces look—and are—indistinguishable from forces designed to execute a preemptive disarming attack.

intended effects. Obviously, survivability of the forces is an issue, as is having some reasonable plan for generating the forces again under difficult conditions, should the need arise. Also, supporters of dealerting often dismiss too cavalierly the problem of creating instabilities in crises by generating forces (Feiveson, 1999, pp. 121–122).[11] The very act of force generation sends a signal, and political leaders may well be either too hesitant or too eager to generate forces as a result.

It is premature to write a bottom line on dealerting and deep cuts. The proponents may not have worked out all the details adequately, but opponents have been even less convincing with their critiques. If all that the United States expects its nuclear forces to do is deter through threat of retaliation, some much-smaller force, perhaps operated differently, probably would suffice.

"Business as Usual, Only Smaller"

"Business as usual, only smaller" best characterized the official U.S. nuclear posture and strategy during the Clinton administration, as easily inferred in the aftermath of the 1994 NPR. Although the NPR was not made public initially, the general thrust was easy to discern from all the visible things that did not change.[12] Subsequently, in an exquisite vivisection of the NPR, Nolan showed how it inevitably reinforced the status quo (Nolan, 1999a; see also 1999b). The basic U.S. force structure has remained largely unchanged, although it would have shrunk substantially if START II had eventually been implemented and if START III cuts were achieved. With the end of the ABM Treaty, the START process is presumably dead. However, the force-level goals specified in the Bush NPR are essentially the same as the proposed START III goals, so the resulting force is likely to be similar.

The current force structure and its operational procedures are more appropriate for the Cold War than the contemporary world scene. (About the best argument in their favor is the old saw, "If it ain't

---

[11]Kahn (1960), pp. 268–269, discussed the dangers of a "mobilization race" and used the analogy of World War I to illustrate the dangers that competitive force generation could cause in the nuclear age.

[12]Subsequently, an unclassified summary was made available (DoD, 1994).

broke, don't fix it.") This kind of force is larger than it needs to be if deterrence by threat of nuclear retaliation is the sole objective of U.S. nuclear strategy. Even a mildly expanded target base that included selected targets in emerging nuclear powers, as well as chemical and biological weapon facilities in a larger set of countries, would not necessarily require the sort of force that the United States planned to maintain during the last administration. *What that force appears best suited to provide beyond the needs of traditional deterrence is a pre-emptive counterforce capability against Russia and China.* Others have made similar observations (see Lodal, as quoted in Dudney and Grier, 2002, p. 31).

A More-Aggressive Nuclear Posture

The United States could try to exploit its currently dominant nuclear position more aggressively. Beyond deterrence, large-scale counter-force, and selected use against other countries' chemical, biological, and nuclear weapon-related facilities, the United States might choose to use its nuclear weapons more aggressively to solve any problems that were both important enough to use nuclear weapons and difficult to solve any other way.

Such a strategy need not require a large nuclear force. It would, how-ever, require much greater targeting flexibility and support (e.g., surveillance, command and control) than has traditionally been associated with nuclear forces. Interestingly, the new NPR calls for such flexibility.

For truly tactical applications, relatively short-range air-delivered nuclear weapons have some advantages (e.g., a short time of flight against movable targets). Alternatively, longer-range weapons could be equipped with in-flight updates, a possibility that has been con-sidered for decades. Other than that, current weapons are probably adequate for a wide range of possible applications.

Nonetheless, analysis suggests that nuclear weapons are probably not necessary for most foreseeable routine military tasks *if the United States procures adequate advanced conventional weapons.* Still, the future is inherently unpredictable. A force adequate to both deter by threat of punishment and deal with any emerging situation would not necessarily have to be big, but it would have to be flexible. It would also call for a coherent nuclear strategy, both actual and declaratory.

Nuclear Emphasis

Finally, there is the possibility of a much more radical nuclear force, one that is the main focus of U.S. military operations. This would be a significant departure from anything the United States currently plans. In essence, it would do something similar to what Russia says it wants to do: rely heavily on nuclear forces—strategic forces supplemented by a larger arsenal of smaller nuclear weapons designed for battlefield use—to be the mainstays of U.S. combat capability.

Such an approach, while possible in principle, seems hard to justify based on the state of the world as it seems to be evolving and the effectiveness of modern conventional weapons. At the very least, it would almost certainly end any hope of limiting nuclear proliferation.

Only under some combination of the following conditions would a real emphasis on nuclear weapons in future U.S. nuclear defense policy seem worthy of serious consideration:

- A U.S. financial crisis comparable to what Russia is facing now that would force the United States to again turn to nuclear weapons seeking "more bang for the buck."

- The evolution of a much more violent world in which many other countries had the capacity for large-scale violence.

- A U.S. choice to involve itself in military situations that are too difficult to handle with either conventional weapons or rigid SIOP-like nuclear "strategic" targeting plans.[13]

---

[13]Unfortunately, this possibility is not quite as remote as it sounds. For several years, the United States has seemed to be turning its attention to China as a potential enemy. (See Pfaff, 2001, for example, on the Bush administration's apparent attitudes toward China.) Some of the Chinese scenarios—the possibility of a Chinese nuclear attack on the United States, a Chinese invasion of or attack on Taiwan, Chinese involvement in one of the variants of a new Korean War—are familiar. However, in recent years other scenarios—e.g., an invasion of neighboring countries, such as Russian Siberia, by a more mobile future Chinese army—have been widely discussed. Massing enough firepower from intercontinental distances to halt such an invasion could require the

The Bush NPR appears to reject this option, since it calls for large-scale nuclear force reductions.

ISSUES AFFECTING U.S. CHOICES OF A FUTURE NUCLEAR STRATEGY

A number of technical and political issues must be resolved as part of the process of selecting a future U.S. nuclear strategy. Some have Cold War antecedents, but the new world environment can introduce new wrinkles.

Political Sustainability

First, there is the question of whether any of these strategies (except abolition of nuclear weapons) will continue to be politically sustainable. The punishment inflicted on Yugoslavia by precision conventional bombing during the Kosovo campaign to coerce Serbian acquiescence to NATO's conditions for peace undoubtedly caused some Americans to have moral qualms about inflicting even that degree of pain on civilian populations. Even the sustained economic sanctions on Iraq appear to have done much more damage to Iraqi civilians than to Saddam Hussein's government. With that sort of recent experience, will the American public continue to support a policy that threatens vastly greater damage?

Maintaining a Robust Nuclear Deterrent

One of the traditional concerns of U.S. nuclear strategy that continues to this day is how to maintain a robust deterrent capability. During the Cold War, that process was relatively straightforward, but it is much less clear in the contemporary world what a "robust deterrent" even means.

In the past, robustness measures were taken mainly to preserve the effectiveness of U.S. nuclear forces in the event of changes in the threat to U.S. forces—e.g., a technical breakthrough in antisubmarine warfare—or an unexpected failure in part of the force—e.g., a systemic reentry vehicle problem. Such problems were common

---

United States to use nuclear weapons, although it requires a Clancy-esque imagination to understand why the United States should choose to involve itself in such a quarrel.

during the Cold War and are still possible. Responses included developing both technical and operational countermeasures. This sort of activity was an integral part of the Soviet-U.S. arms competition during the Cold War and sometimes helped fuel the arms race. However, in a more complex—albeit less threatening—world, the standard reactions might be irrelevant or even wrong. If all the United States expects its nuclear forces to do is inflict damage to punish or coerce enemies, then the traditional approach is probably adequate. In fact, absent a threat to the survivability of U.S. nuclear forces and command and control systems comparable to that mounted by the Soviet Union during the Cold War, guaranteeing the robustness of the U.S. nuclear forces against external factors should not be very demanding.

On the other hand, if the United States has different primary objectives—countering nuclear proliferation, for example—the demands on its nuclear forces could be different, and the robustness criteria could change accordingly. For example, in an unstable post–Cold War world in which coping with proliferators was a principal concern, the United States might prefer nuclear forces with a robust capability for preemptive attacks against nuclear forces or facilities in emerging nuclear countries. Of course, that is precisely the kind of nuclear capability that would have been considered destabilizing according to traditional Cold War criteria.

What this means as a practical matter is that the United States needs to address exactly what it expects its nuclear forces to accomplish in the future and what exactly that means for the characteristics of the forces and the way they are operated. On the other hand, depending on the alternatives available, nuclear forces may not be either required or adequate to deal with a wide range of contemporary problems. Improved conventional weapons, special operations forces, or BMD may be more appropriate responses to many threats.

Preparing for Operational Use of Nuclear Weapons

As I have already pointed out, the most fundamental choice that the United States has to make in shaping its future nuclear strategy is whether it is willing to limit the role of its nuclear forces to deterrence or coercion by threat of punishment, perhaps supplemented by a preemptive nuclear counterforce capability against fixed enemy nuclear forces, or whether it wants to maintain any capability to use nuclear weapons flexibly in actual combat if the situation demands. The question regarding battlefield use is simple: Does the United States want to have the capability to use nuclear weapons to solve tactical problems if the stakes are high enough and other means are inadequate? In the past, the United States explicitly developed a nuclear posture with that in mind. Since the end of the Cold War, the rationale for the old kind of U.S. tactical capabilities has largely disappeared, and the capabilities have atrophied accordingly. If the United States wants to resurrect any capability for "tactical" use of nuclear weapons in the future, there are some problems that it is going to have to address.

Numerous studies since the end of the Cold War (see, for example, Feiveson, 1999; Buchan, 1994; and National Academy of Sciences, 1997) have called for the traditional, rigid SIOP process to be replaced by a much more flexible approach analogous to what is currently used to develop conventional targeting plans, and the new NPR appears to endorse that kind of approach. Absent the set-piece nuclear scenarios of the Cold War, the details of situations dire enough to warrant U.S. nuclear use are so unpredictable that preplanning is likely to be useless. Feiveson (1999, p. 56) said it best:

> The circumstances in which the United States might seriously consider the use of nuclear weapons are so uncertain and unforeseeable that it makes little sense to focus on a handful of preplanned options.

Even predictable kinds of warfighting plans (e.g., interdiction, attacking critical time-urgent targets) require flexible planning and execution capability. Conventional forces increasingly tend to operate that way. Nuclear forces should as well.

U.S. Strategic Command (STRATCOM) has increased the flexibility of its planning process considerably compared to the Cold War days. In fact, it is unclear why fully flexible targeting could not have been implemented years ago.

Improving the mechanics of the planning process is necessary, but not sufficient for developing real operational capability to employ nuclear forces flexibly and "tactically." At least the following are also required:

- targeting support roughly comparable to that needed for conventional forces
- suitable command and control
- nuclear planning expertise at the theater level
- adequate training for nuclear operators
- adjustments to the weapon systems, if necessary.

The first point merely reflects the need to be able to find, identify, and locate suitable targets, particularly mobile targets, such as invasion forces or mobile missiles, well enough and in a timely enough way to target nuclear weapons against them. That, of course, is also a prerequisite for targeting conventional weapons. The second means adding sufficient command and control capability to bring the weapons to bear effectively against the targets. This was one of the shortcomings that would have seriously complicated, for example, any U.S. attempt to use nuclear weapons to counter an Iraqi invasion of Saudi Arabia during the early stages of the Gulf War, had such an event occurred.

Nuclear expertise on theater planning staffs is both critical and currently in short supply. Unlike the Cold War days, when nuclear weapons were an integral part of U.S. war plans and theater planners were well-versed in nuclear matters, questions about nuclear use apparently do not arise very often today. As a result, nuclear expertise at the theater level has atrophied.[14]

A related issue is training. While much of the training for nuclear use—particularly simulated missile launching—is routine and relatively independent of context, some is not. For example, the United States still retains a modest number of nuclear bombs in Europe for use by tactical aircraft. Tactical delivery of nuclear bombs requires practice if pilots are to retain their proficiency.

A more important and difficult aspect of the training problem is integrating nuclear weapons into an overall campaign. A necessary condition for doing that is including nuclear use in exercises.[15] That

---

[14]This observation is based primarily on conversations with U.S. STRATCOM staff.

[15]I have been raising the issue of involving nuclear weapons more actively in exercises for some years now. See, for example, Buchan (1994), pp. 42–43.

is the closest thing to actual combat for gaining experience with both the planning and operational aspects of nuclear use in a larger context. Exercises are probably also the best "laboratory experiments" to identify what operational concepts for the use of nuclear weapons actually make sense, if any.

Characteristics of Nuclear Weapon Systems

The characteristics of the particular nuclear weapon systems that the United States maintains in its inventory could be at issue as well. If the United States were to opt for heavy reliance on nuclear weapons for tactical use, as the Russians say they might do now, it would almost certainly lead to an arsenal of more numerous, smaller nuclear warheads. The United States could still choose that path— even though the new NPR apparently rejects it—as long as it had sufficient critical nuclear materials (i.e., plutonium, tritium, enriched uranium), its nuclear weapon design skills, and the ability to develop new warheads (assuming that existing designs would not suffice).[16] However, that would run strongly counter to current political trends in the United States and would almost certainly be viewed as inflammatory on the international scene.

More fundamentally, the need is not really there, or at least it should not be if the United States plans its conventional forces properly (Buchan, 1994, p. 12).[17] Even in the conventional world, our past studies have shown that large numbers of relatively small (e.g., 500 pound), accurate weapons were more effective than fewer, larger (e.g., 2,000 pound) precision-guided weapons. Thus, making a case for new tactical nuclear weapons, even so-called "mininukes," would be very difficult unless both the external world and U.S. fortunes

---

[16]Developing new nuclear warheads has traditionally required testing, which the Comprehensive Test Ban Treaty prohibits. Although the U.S. Senate failed to ratify the treaty, the United States is still obliged by international law to comply with its provisions. Moreover, the Bush administration has recently been told that the president cannot simply withdraw the treaty once it has been submitted to the Senate for ratification (Shanker and Sanger, 2001). Thus, there is no graceful way for the United States to resume nuclear testing. Relying on computer simulation in lieu of testing is a theoretical possibility but has never had sufficient credibility with weapon designers.

[17]It is not reassuring, however, that the Air Force in particular has a track record of underinvesting in weapons as opposed to platforms, in spite of the mass of studies that have emphasized the relative importance of weapons.

change drastically.[18] So far, the Bush administration appears only to be considering developing a new generation of smaller nuclear weapons for earth penetration (Brumfiel, 2002). Other possibilities include producing a new generation of fission weapons that could be developed and maintained indefinitely without testing.

Beyond the warheads themselves, there are some issues regarding the delivery systems. For most applications, any of the current U.S. systems is likely to be adequate. The specific exception is attacking mobile targets—e.g., invasion forces or mobile missiles. Even assuming good initial targeting, weapon flight time could be an issue, particularly for either long-range ballistic missiles or cruise missiles. For this reason, relatively short-range, aircraft-delivered weapons are likely to be the weapons of choice for attacking such targets. However, gravity bombs are appropriate weapons only in a relatively benign air defense environment. Short-range attack missile–class weapons would be particularly appropriate, but the United States no longer has these weapons in the inventory and does not, as far as we at RAND know, plan to develop a replacement. Alternatively, long-range missiles could be modified to receive in-flight targeting updates to allow targeting flexibility. The United States has considered such technical options for decades but has never felt the need to pursue them. In sum, there are several ways to solve this problem, and solving it is probably important *if* the United States wants to be able to threaten mobile targets with nuclear weapons as part of its overall security strategy.

Exploiting Asymmetries

A general thread that runs through the analysis of future U.S. security needs is the importance of asymmetries. Coping with future nuclear threats may require more than just deterrence, and deterrence might need more than just nuclear threats. For example, as we noted earlier, a nuclear strike might not be an appropriate response *even to a nuclear attack*. Depending on who is responsible, a rifle bullet might be a more appropriate response and a more effective deterrent. Similarly, U.S. nuclear weapons might be a suitable response or deterrent to nonnuclear threats that were important enough, and not

---

[18]I have made this point in the past (see Buchan, 1994, p. 65).

just to other so-called "weapons of mass destruction." The point is that U.S. security strategy in general needs to be richer and more nuanced to deal with contemporary security problems. U.S. nuclear strategy needs to be flexible enough to recognize and exploit these asymmetries. The new NPR apparently acknowledges this sort of asymmetry, and that is a welcome change.

Nuclear Proliferation

Another contentious issue affecting future U.S. nuclear strategy is the effect of U.S. policy on nuclear proliferation. Central to the arguments of many advocates of dramatic nuclear arms reductions and dealerting is the presumption that other nations will be more inclined to acquire nuclear weapons themselves unless the United States and other established nuclear powers drastically reduce or eliminate theirs. That presumption is codified in the NPT.

While that assumption may be true, the issue is not so clear cut. It could also be that maintaining high-quality U.S. nuclear forces might be a more-effective deterrent to some states considering joining the nuclear club. In essence, the message to the states would be, "We may not be able to stop you from acquiring nuclear weapons, but if you do, you put yourselves at much greater risk if our interests ever collide." Thus, U.S. nuclear capabilities could either encourage or dissuade others from developing nuclear weapons. Individual cases could vary considerably.

Most likely, however, regional powers will make decisions about acquiring nuclear weapons with little regard to what the United States or others think or do themselves. Even if potential proliferators might be influenced by U.S. attitudes, it is problematic at best whether the various players will be able to read each other accurately. For example, as Perkovich points out, India and the United States read each other's intentions quite differently prior to the 1998 Indian nuclear tests (Perkovich, 1999, p. 421). The United States thought it was responding adequately to India's concerns about being largely ignored in U.S. policy. By contrast, India thought the United States was placing more weight on its relations with China at India's expense.

Finally, although the United States has made clear its opposition to further proliferation of nuclear weapons, long-range delivery systems, or other particularly threatening weapons, there are limits to how much the United States should be willing to pay to prevent proliferation. Otherwise, potential proliferators have too much bargaining leverage and, therefore, incentives to develop nuclear or other advanced weapons. That is a dangerous game. Ukraine, for example, played its hand very skillfully on the issue of giving up its nuclear weapons when the Soviet Union dissolved. It managed to get some political and economic benefits in exchange for giving up its nuclear weapons without either antagonizing Russia excessively or alienating the United States. North Korea has "pushed the envelope" in trying to extract the maximum political mileage from its latent nuclear capability. So far, it has worked, but the future remains uncertain and potentially very dangerous for North Korea and for others.

Individual countries' decisions about whether or not to develop nuclear weapons are probably going to be influenced only at the margin by U.S. nuclear policy. Even then, it is not at all clear whether U.S. restraint or enhanced capability is more likely to discourage proliferation. The most effective approach is probably a balanced combination of the two: minimizing the importance of U.S. nuclear capabilities in most international dealings, as a "carrot," but making it clear that the United States has nuclear and conventional weapon systems capable of dealing with an emerging nuclear power more forcefully should the need arise, as either a spoken or unspoken "stick." That is basically the policy that the United States has followed for some time.

Is "Withering Away" of U.S. Nuclear Capability Inevitable?

Earlier, I raised the possibility that U.S. nuclear capability might wither away over time. If that were to occur, the United States would be facing elimination of its nuclear arsenal sometime in the future whether it liked it or not, and U.S. security strategy would have to reflect that reality. This is a very serious issue and is increasingly recognized as such (see, for example, Defense Science Board, 1998, and Buchan, 1994). In fact, withering away of U.S. nuclear capability could, arguably, be the inevitable result of the combination of historical momentum and benign neglect that have dominated U.S. nuclear policy for the last decade or so. It could take some time, but, ironically, eventual elimination of U.S. nuclear capability might be much more certain if the United States follows its current course than if it pursues formal negotiations with that aim.

Problems occur at several levels. First is the country's nuclear infrastructure. This issue has received a considerable amount of attention, primarily from the Department of Energy, which has the responsibility for maintaining the U.S. nuclear stockpile. The United States has shut down its production reactors that create plutonium. It also has drastically reduced its capability to produce key nuclear weapon components (e.g., "pits" for bombs). It relies heavily on its stockpile of existing warheads for both warhead components and nuclear materials. At the moment, this problem appears to be under control, but there is not much slack.

More fundamental, and much harder to solve, are problems of retaining expertise in critical areas. Nuclear weapon design expertise is going to be very difficult to maintain at a time when the United States plans no new nuclear warheads. To be sure, there is work for nuclear weapon designers to do maintaining the current nuclear stockpile and implementing the science-based stockpile stewardship program. The problem is that, given the relatively low priority that nuclear weapon–related issues appear to have now in the United States and the low likelihood that a weapon designer would actually get a chance to design new nuclear warheads, there are few career incentives for the "best and brightest" to get into this business. Thus, over time, U.S. nuclear weapon design capabilities will erode. Even if the science can be preserved, the art and engineering of warhead design are likely to be lost. The Bush administration's apparent plan to maintain small teams of weapon designers to preserve these skills hardly seems sufficient to deal with the long-term problem.

The same is true in other weapon system–related areas as well, although in some cases to a somewhat lesser extent. For example, absent plans to build new ICBMs or SLBMs, there is little incentive for the few aerospace companies that still have the design and manufacturing skills to build big missiles to maintain the capabilities. Similarly, there is little incentive for bright young engineers to choose this career path. Actually, since the United States still needs space boosters, maintaining that capability should alleviate the missile design problem to some degree. Still, missile system integration for nuclear weapon applications remains an issue, as do specific subsystem technology areas (e.g., reentry vehicle technology, where the United States has traditionally led the world).

At least as important as the technical skills are those of the military operators. Given current service priorities, nuclear weapon skills and experience are likely to lose the luster that they once had. Traditionally, both in the Air Force and the Navy, nuclear service has been considered an elite assignment and was sought after accordingly. With the current lack of interest in nuclear issues, it will be difficult to persuade talented officers and enlisted personnel to enter nuclear career fields.

These problems will be extraordinarily hard to solve because solutions will require influencing the decisions of large numbers of disparate individuals, as well as various organizations, large and small, public and private. During the Cold War, it was easy to persuade individuals and organizations of the importance of dealing with nuclear weapons. Now, that will be much more difficult. Withering away by design is one thing and could be more effective than even the most radical arms control at "denuclearizing" the United States. Withering away by default could be quite dangerous.

SO, WHERE DO WE GO FROM HERE?

Nuclear weapons remain the ultimate guarantor of U.S. national security even as their day-to-day role diminishes, and the Bush administration has reaffirmed their importance in the 2002 NPR. Because of the massive destruction that even a single nuclear detonation could cause, nuclear weapons trump all other types of weapons either as a deterrent—a threat of punishment—or as a military instrument to be used if the situation were serious enough to warrant such drastic action. Even when not actually used or overtly brandished, their mere existence in the U.S. arsenal provides a certain amount of implicit leverage in any serious crisis, even if the likely short- and long-term political consequences appear to rule out actual use of nuclear weapons as a "sensible" option. They form a nuclear "umbrella" over all other U.S. military forces and instruments of policy.

However, nuclear weapons have significant disadvantages as well, most of which result from the same characteristics that make them potentially attractive:

- Their sheer destructiveness means that actual use of nuclear weapons, particularly on a large scale, is likely to produce damage out of all proportion to any reasonable military or political objectives and is likely to set an unfortunate historical precedent. As a result, a *tradition of nonuse has evolved, a tradition that particularly serves the interest of the United States*, so the United States has very strong incentives not to use nuclear weapons unless the world changes drastically.

- Actual battlefield use of U.S. nuclear weapons can cause headaches for field commanders—e.g., radiation, blackout, fallout, problems obtaining release authority, planning problems. Such problems associated with actual employment of nuclear weapons may make their use more trouble than it is worth unless the need is overwhelming.

- Because the consequences are so great, the need for safeguards to avoid accidents, incidents, unauthorized use, mistakes, or theft of nuclear weapons is overwhelming. *The weight given to this factor in the equation will have a major impact on the future nuclear strategy that the United States selects and how it chooses to implement that strategy.* It is one of the two or three factors at the heart of the current dispute over future U.S. nuclear policy.

As a mature and experienced nuclear power—especially one that also dominates the conventional military and economic arenas—the United States has a wide range of options in crafting a nuclear strategy for the future. Choosing an appropriate role for U.S. nuclear weapons will require balancing some potentially competing objectives:

- creating an overall force structure, including nuclear and conventional elements, that allows the United States to impose its will on others when it really matters

- making nuclear weapons less important rather than more important in world affairs to reduce the incentives for others to acquire or to use them

- avoiding operational practices that might appear overly provocative to other nuclear powers and thus might prompt unfortunate responses

- operating nuclear weapons in such a way that risks of accidents, unauthorized use, and theft are minimized.

The most obvious transcendent role for U.S. nuclear weapons in the current world is to continue to provide a deterrent force capable of threatening any nation (or nonnation that controls territory or valuable facilities) with massive destruction. That is what nuclear weapons are particularly well suited to do, even granting the uncertainties and ambiguities inherent in any concept of deterrence.

The only real threat to the United States' existence as a functioning society remains Russia's nuclear arsenal, even if it shrinks to much lower levels, as projections suggest. Although U.S.-Russian relations have had their ups and downs, it is very unlikely that the Cold War nuclear standoff between the United States and Russia would return with the same force as in the old days. If it did, or if other similar threats emerged, the familiar solution of deterrence by threat of nuclear retaliation, with all its theoretical flaws, is still probably the best option for the foreseeable future. In the contemporary world that probably requires

- survivable forces and command and control, as in the past

- a force of almost any reasonable size (e.g., hundreds of deliverable warheads)[19]

- an adequate mix of forces to hedge against technical or operational failures: *The key Air Force systems to ensure variety are air-breathing weapons* (i.e., bombers and cruise missiles).

An important point is that there is no need for a prompt attack. Indeed, prompt responses could be dangerous under some conditions. That means that even small, dealerted forces could, in princi-

---

[19]Damage requirements were always largely arbitrary. In the contemporary world, there is an even less compelling need for a large force. For example, if the United States were to target Russia, what would it target? The economy and the conventional military hardly seem worth attacking with nuclear weapons. Attacking leadership is problematical. That leaves only strategic forces, and targeting them is a separate strategic issue. It would be a supreme irony of the contemporary world if strategic forces were the only suitable Russian targets for U.S. nuclear weapons *now*, when such attacks would have been ineffective and possibly counterproductive during the Cold War and should be irrelevant in the post–Cold War world.

ple, have considerable deterrent power if the problems of survivability and force generation were solved adequately.

These are familiar problems from the old Cold War days with some modifications to accommodate the changes in the relationship between the United States and Russia. The contemporary world also has some new wrinkles in addition to the usual elements:

- Identifying attackers may be harder with more players and diverse available delivery options.

- A broader range of options than just nuclear weapons may be needed to deter or deal with some kinds of threats (e.g., terrorists who cannot be threatened directly by U.S. nuclear weapons).

- No threat of punishment may be sufficient to deter some nuclear threats to the United States (e.g., nations with nuclear weapons and nothing left to lose). *An established nuclear power coming unglued and lashing out is the worst possible threat to U.S. security for the foreseeable future*, much worse than so-called "rogue nations." Something other than deterrence will be necessary to deal with them.

An even more challenging issue is the degree to which the United States wants to include actual warfighting use of nuclear weapons in its overall strategy. The first possibility is nuclear counterforce. Ironically, nuclear counterforce, which probably would not have worked during the Cold War, might be feasible in the current world, particularly against new nuclear powers that have not developed high-quality mobile systems and survivable command and control. A counterforce emphasis would provide a more quantitative basis for sizing forces than "simple" deterrence. It would put more of a premium on timely delivery, although that need not mean that U.S. forces must be kept on a high level of day-to-day alert. Also, to the degree that U.S. nuclear strategy included counterforce as a hedge against nuclear proliferation, it could be viewed as part of the "robustness" criteria normally associated with keeping a deterrent force effective.

The current U.S. counterforce advantage is probably fleeting. Counters are well-known. They just require resources, time, and experience to implement. Thus, there is an issue about how much contemporary U.S. nuclear strategy ought to emphasize counterforce. To some degree, the strategic issue is almost moot, since any nuclear force the United States maintains is likely to have considerable inherent counterforce capability if it operates more or less the way U.S. strategic forces operate currently. Interestingly, *only a large-scale commitment to a counterforce-heavy strategic doctrine focused on a major nuclear power (e.g., Russia) is likely to require the "business as usual, only smaller" type of force structure recommended by the 1994 NPR and apparently reinforced by the 2002 NPR.* That point will not be lost on others who might infer U.S. intentions from the force structure that they observe and might then react badly to what they could view as a serious U.S. threat.

Using nuclear weapons against a broader set of military targets is a policy option as well. It is actually a more interesting possibility because it follows a broader policy logic: *One of the reasons the United States maintains nuclear weapons is to deal with any situation that should emerge that threatens vital U.S. interests and cannot be dealt with adequately in any other manner.* The real issue is the effectiveness of conventional weapons. If the United States invests adequately in advanced conventional weapons, there should be no need for nuclear weapons to be used "tactically," except possibly for attacks on deeply buried targets.[20] Thus, decisions on future U.S. nuclear strategy depend critically on issues not associated directly with nuclear weapons (e.g., conventional weapons, BMD).

If the United States wanted to maintain the option to use nuclear weapons tactically if a really desperate need arose, the problems are not generally with the weapons themselves but in planning and operational flexibility. Indeed, there is a strong *a priori* case for developing this kind of operational flexibility for U.S. nuclear forces in place of the traditional SIOP-like process, precisely because the circumstances under which U.S. nuclear weapons might actually have to be used in the future are so hard to predict that they cannot be planned for in advance.

---

[20]There is a certain irony in this current interest in nuclear earth penetrators to attack underground targets. Even if the United States were to pursue this option more vigorously (i.e., beyond the nuclear earth-penetrating gravity bombs that the B-2 bomber can currently carry), placing such importance on this limited role for nuclear weapons clearly marginalizes their overall importance in U.S. defense strategy. It simply lacks the *grandeur* that has traditionally been associated with nuclear weapons.

Achieving such nuclear operational flexibility would require radical changes in current U.S. nuclear operational practice. It would require at the very least

- suitable planning systems (e.g., near-real-time target planning)
- training
- including nuclear weapons in exercises
- nuclear expertise on theater planning staffs
- suitable command and control
- intelligence support comparable to that for conventional forces.

In the long term, there are other practical problems to solve if the United States is to remain a viable nuclear power. "Withering away" of U.S. nuclear operational expertise, support infrastructure, and weapon design capabilities may be unavoidable, given current career incentives, fiscal constraints, political realities, and service priorities. Thus, U.S. nuclear capability may diminish over time whether it likes it or not.

In considering overall contemporary U.S. strategic options, one striking possibility is that a new strategy could simultaneously be both more "dovish" and more "hawkish." That might involve a much smaller nuclear force intended to deter egregious behavior with threats of retaliation but operated flexibly enough so that the weapons could actually be used if a serious enough need arose against whatever particular set of targets turned out to be important. That sort of nuclear strategy would lend itself to a succinct description along the following lines:

> The United States views nuclear weapons as the ultimate guarantor of its security. They provide a means of deterring an enemy from damaging vital U.S. interests by threatening to punish him with massive damage. In particular situations, the United States might use nuclear weapons directly to resolve a crisis if U.S. vital interests were at stake and other means appeared inadequate.

Such a nuclear strategy would also have to be supplemented by a broader spectrum of other options to deal with contemporary problems that nuclear threats or use alone could not handle.

272 Strategic Appraisal: United States Air and Space Power in the 21st Century

There is room for debate about some of the details of the nuclear force structure that the United States should retain in the relatively near term, but there is a remarkably broad consensus within the nuclear intelligentsia—if not the nuclear bureaucracy—on the general direction in which things ought to go. First, the idea of deep cuts in U.S. nuclear forces has almost universal support, apparently including President Bush (Bush, 2001). Outside the nuclear bureaucracy, cuts substantially below the proposed NPR goals (i.e., 1,700 to 2,200 warheads) appear both feasible and desirable. Schemes vary, but force levels from a few hundred to 1,000–1,500 warheads appear more than adequate to provide whatever deterrent value nuclear weapons can deliver, potential counterforce capability against vulnerable nuclear forces of emerging nuclear powers, plus any bizarre battlefield situations that are likely to come up. In fact, at these force levels, the *exact numbers probably do not matter all that much*, since "requirements" are largely arbitrary. Rather, more-practical considerations may be the driving factors. For example, the core of any future U.S. force will be the SSBN fleet. Modern SLBMs are accurate, survivable, and—if deployed properly—can cover all potential targets of interest. They can provide all the capability that long-range ballistic missiles are capable of providing without the vulnerability problems that have plagued ICBMs for decades and have led to pressures to launch them quickly rather than lose them.

Thus, a place to start in structuring the future U.S. nuclear force is to determine the minimum number of submarines that is practical to operate out of a single port and double that number to allow SSBNs to operate in both the Pacific and Atlantic. Since each Trident SSBN carries 24 missiles, that fixes the total number of missiles available.[21] The number of warheads per missile can then be determined based on whatever total number of warheads the United States decides it wants to maintain. Actually, some degree of "deMIRVing," while

---

[21]Lodal (2001) proposes an interesting and strategically attractive variation of the usual scheme. He suggests keeping all SLBMs continuously at sea, transferring the missiles from SSBNs returning from patrol to outgoing SSBNs as the boats rotate between sea duty and port. Assuming that the United States continues its current practice of keeping half of its SSBN force at sea at any one time, that would mean only half of the total fleet would be armed at any one time, but that half would be invulnerable. That should eliminate any incentive for a preemptive attack against SSBNs in port on day-to-day alert.

adding to the cost of a force with a fixed number of warheads, has some strategic advantages. Reducing the number of warheads on a missile increases both its range and "footprint"—the total area within which warheads can be targeted—and provides more flexibility to add penetration aids to counter any BMD that enemies might deploy in the future. The arithmetic of the trades is simple.

If the SLBM warheads are to be supplemented by other weapons, air-launched cruise missiles or higher-performance, shorter-range missiles (which the United States no longer has in its inventory) offer particular advantages.[22] They stress defenses in different ways than long-range ballistic missiles and can be used in small numbers with more ease. Fast, short-range missiles may be the most appropriate weapons for tactical applications if the United States wants to keep the options open. Since the United States already owns suitable delivery aircraft and needs them for conventional applications in any case, the added cost should be minimal.

The new NPR could provide a vehicle for beginning to institute concrete changes in U.S. nuclear posture. Some of the changes—e.g., deep cuts in nuclear force levels, increased flexibility of nuclear forces, recognition of the importance of conventional forces—are long overdue. Other elements, such as the role of BMD, still must demonstrate their value. In other areas—e.g., restructuring nuclear forces—the NPR probably does not go far enough. Implementation will be the key. So far, the Bush administration appears serious about revamping U.S. nuclear policy. Creating change will require the continued attention of high-level U.S. political and military leadership.

---

[22]Nuclear gravity bombs, while much cheaper and currently more numerous, are much less attractive for most likely applications. For example, the United States maintains several hundred nuclear gravity bombs in Europe. These weapons serve a political function: underwriting the United States' nuclear commitment to NATO. However, it is hard to imagine what useful military function they could perform. Tactical fighter-bombers carrying the bombs would have very limited range. They are also vulnerable to air defenses. Thus, these nuclear weapons would be militarily useful only in situations where *both* of the following conditions applied:

- A military crisis arose within fighter range of NATO bases that conventional weapons could not handle adequately.
- The enemy that could overcome U.S.-NATO conventional forces did not have air defenses capable of defeating U.S. tactical aircraft, particularly nonstealthy fighter-bombers.

BIBLIOGRAPHY

Allison, Graham, and Philip Zelikow, *Essence of Decision: Explaining the Cuban Missile Crisis,* 2nd ed., New York: Addison Wesley Longman, 1999.

"America and the Bomb: Fission and Confusion," *The Economist,* March 16, 2002, p. 15.

Arkin, William M., "Nuclear Warfare: Secret Plan Outlines the Unthinkable," *Los Angeles Times,* March 10, 2002, pp. M1, M6.

Arkin, W., R. Norris, and J. Handler, *Taking Stock: Worldwide Nuclear Deployments 1998,* New York: Natural Resources Defense Council, March 1998, Appendix A and Table 2. Also online at http://www. nrdc.org/publications/ (as of December 14, 2001).

Associated Press, "France Sets off 5th Nuclear Test in Pacific," *New York Times,* December 28, 1995, p. A13.

Barry, John, and Evan Thomas, "Dropping the Bomb," *Newsweek,* June 25, 2001, pp. 28–30.

Bering, Helle, "War by Other Means," *Washington Times,* September 1, 1999.

Blair, Bruce G., *Strategic Command and Control: Redefining the Nuclear Threat,* Washington, D.C.: The Brookings Institution, 1985.

_____, *The Logic of Accidental Nuclear War,* Washington, D.C.: The Brookings Institution, 1993.

_____, *Global Zero Alert for Nuclear Forces,* Washington, D.C.: The Brookings Institution, 1995.

Blair, Bruce G., et al., *Toward True Security: A U.S. Nuclear Posture for the Next Decade,* Washington, D.C.: Federation of American Scientists, Natural Resources Defense Council, Union of Concerned Scientists, June 2001.

Blair, Bruce G., Harold A. Feiveson, and Frank N. von Hipple, "Taking Nuclear Weapons off Hair-Trigger Alert," *Scientific American,* November 1997, pp. 74–81.

Bracken, Paul, *The Command and Control of Nuclear Forces*, New Haven, Conn.: Yale University Press, 1983.

Brodie, Bernard, ed., *The Absolute Weapon: Atomic Power and World Order*, New York: Harcourt, Brace and Co., 1946.

_____, *Strategy in the Missile Age*, Princeton, N.J.: Princeton University Press, 1959.

_____, *Escalation and the Nuclear Option*, Princeton, N.J.: Princeton University Press, 1966.

Brumfiel, Geoff, "Nuclear-Weapons Design Plan Raises Fresh Proliferation Fears," *Nature*, Vol. 415, February 28, 2002, pp. 945–946.

Buchan, Glenn C., "De-Escalatory Confidence-Building Measures and U.S. Nuclear Operations," in Joseph L. Nation, ed., *The De-escalation of Nuclear Crises*, New York: St. Martin's Press, 1992.

_____, *U.S. Nuclear Strategy for the Post–Cold War Era*, Santa Monica, Calif.: RAND, MR-420-RC, February 1994.

_____, *One-and-a-Half Cheers for the Revolution in Military Affairs*, Santa Monica, Calif.: RAND, P-8015-AF, October 1997.

_____, "Nuclear Strategy," in Frank Carlucci, Robert Hunter, and Zalmay Khalilzad, eds., *Taking Charge: A Bipartisan Report to the President-Elect on Foreign and National Security, Discussion Papers*, Santa Monica, Calif.: RAND, MR-1306/1-RC, 2001.

Buchan, Glenn, Dave Frelinger, and Tom Herbert, *Use of Long-Range Bombers to Counter Armored Invasions*, Santa Monica, Calif.: RAND, WP-103, March 1993.

Buchan, Glenn C., David Matonick, Calvin Shipbaugh, and Richard Mesic, *Future Role of U.S. Nuclear Forces: Implications for U.S. Strategy*, Santa Monica, Calif.: RAND, MR-1231-AF, forthcoming.

Builder, Carl H., *The Future of Nuclear Deterrence*, Santa Monica, Calif.: RAND, P-7702, 1991.

Bundy, McGeorge, *Danger and Survival: Choices About the Bomb in the First Fifty Years*, New York: Random House, 1988.

Bundy, McGeorge, William J. Crowe, Jr., and Sidney D. Drell, *Reducing the Nuclear Danger: The Road Away from the Brink*, New York: Council on Foreign Relations, 1993.

Bush, George W., Transcript: President Bush Speech on Missile Defense, Cable News Network, May 1, 2001. Online at http://www.cnn.com/2001/ALLPOLITICS/05/01/bush.missile.trans/ (as of March 22, 2002).

Butler, Lee, "A Voice of Reason," *Bulletin of the Atomic Scientists,* May–June 1998, pp. 58–61.

Costa, Keith J., "Authorizers Call for Comprehensive Review of U.S. Nuclear Posture," *Inside Missile Defense,* October 18, 2000, p. 33.

_____, "Lawmaker Introduces Legislation to Replace MAD Strategic Doctrine," *Inside Missile Defense,* November 1, 2000, pp. 22–23.

Craig, Campbell, *Destroying the Village: Eisenhower and Thermonuclear War*, New York: Columbia University Press, 1998.

Crouch, J. D., Special Briefing on the Nuclear Posture Review, news transcript from the U.S. Department of Defense, January 9, 2002.

Cushman, John H., Jr., "Rattling New Sabers," *New York Times,* March 10, 2002, p. A1.

Dao, James, "Pentagon to Ask for Retirement of MX Missiles," *New York Times,* June 28, 2001, p. A1.

Defense Science Board, *Report of the Defense Science Board Task Force on Nuclear Deterrence*, Washington, D.C.: Office of the Under Secretary of Defense for Acquisition and Technology, October 1998.

Dixon, Robyn, "Putin Vows to Rid Russia of Excess Nuclear Weapons," *Los Angeles Times,* April 1, 2000, pp. A1, A12.

DoD—*See* U.S. Department of Defense.

Douhet, Giulio, *The Command of the Air*, New York: Coward-McCann, 1942.

Drell, Sidney, Raymond Jeanloz, and Bob Peurifoy, "Bunkers, Bombs, Radiation," *Los Angeles Times,* March 17, 2002, p. M5.

Dudney, Robert S., and Peter Grier, "Bush's Nuclear Blueprint," *Air Force Magazine*, March 2002, pp. 26–31.

Efron, Sonni, "'Missile Attack' on Russia Was Just a Science Probe," *Los Angeles Times*, January 26, 1995, p. A1.

Ensor, David, "Nuclear Review Urges More Reliance on Precision Weapons," Cable News Network, January 10, 2002. Online at http://www.cnn.com/2002/US/01/10/nuclear.posture/ (as of March 22, 2002).

Feiveson, Harold A., ed., *The Nuclear Turning Point: A Blueprint for Deep Cuts and De-Alerting of Nuclear Weapons*, Washington, D.C.: The Brookings Institution, 1999.

Freedman, Lawrence, *The Evolution of Nuclear Strategy*, New York: St. Martin's Press, 1986.

Frelinger, David, et al., *Use of Heavy Bombers in Conventional Operations*, Santa Monica, Calif.: RAND, August 1994. Government publication; not releasable to the general public.

Friedman, Norman, *The Fifty Year War: Conflict and Strategy in the Cold War*, Annapolis, Md.: Naval Institute Press, 2000.

Friedman, Thomas L., "The French Ostrich," *New York Times*, October 4, 1995, p. A21.

Gertz, Bill, *Bomber Flexibility Study: A Progress Report*, Santa Monica, Calif.: RAND, DB-109-AF, 1994.

_____, "Russians Practiced Nuclear Counterattack on NATO," *Washington Times*, July 8, 1997, p. A1.

Glasstone, S., and P. Dolan, *The Effects of Nuclear Weapons*, Washington, D.C.: U.S. Department of Defense and U.S. Department of Energy, 1977.

Gordon, Michael R., "U.S. Nuclear Plan Sees New Targets and New Weapons," *New York Times*, March 10, 2002, p. A1.

Graham, Bradley, and Walter Pincus, "Nuclear Targeting Draft Shifts Focus from Russia: More Emphasis Given to China, N. Korea, Mideast," *Washington Post*, March 10, 2002, p. A27.

Herken, Gregg, *Counsels of War*, New York: Knopf, 1985.

278 Strategic Appraisal: United States Air and Space Power in the 21st Century

_____, *The Winning Weapon, The Atomic Bomb in the Cold War: 1945–1950*, Princeton, N.J.: Princeton University Press, 1988.

Hersh, Seymour M., *The Samson Option: Israel's Nuclear Arsenal and American Foreign Policy*, New York: Random House, 1991.

Holland, Gale, "Foreign Press: Nuclear Fears Abound," *Los Angeles Times*, March 17, 2002, p. M3.

Holloway, David, *Stalin and the Bomb*, New Haven, Conn.: Yale University Press, 1994.

Jervis, Robert, "Weapons Without Purpose? Nuclear Strategy in the Post–Cold War Era," *Foreign Affairs*, Vol. 80, No. 4, July–August 2001, pp. 143–148.

Kahn, Herman, *On Thermonuclear War*, New York: Princeton University Press, 1960.

_____, *On Escalation: Metaphors and Scenarios*, New York: Praeger, 1965.

_____, *Thinking About the Unthinkable*, New York: Avon, 1968.

_____, *Thinking About the Unthinkable in the 1980s*, New York: Simon & Schuster, 1984.

Kaplan, Fred, *The Wizards of Armageddon*, New York: Simon & Schuster, 1983.

Kempster, Norman, and Doyle McManus, "Huge Warhead Cuts Approved: Bush, Yeltsin Act to End the 'Nuclear Nightmare,'" *Los Angeles Times*, June 17, 1992, pp. A1, A6, and A8.

Koch, Andrew, "U.S. Rethink Could Spawn 'Mini-Nukes,'" *Jane's Defence Weekly*, May 23, 2001, pp. 22–23.

Krepon, Michael, "Moving Away from MAD," *Survival*, Vol. 43, No. 2, Summer 2001, pp. 81–95.

Lewis, John Wilson, and Xue Litai, *China Builds the Bomb*, Stanford, Calif.: Stanford University Press, 1988.

Lodal, Jan, *The Price of Dominance: The New Weapons of Mass Destruction and Their Challenge to American Leadership*, New York: Council on Foreign Relations, 2001.

Marshall, Tyler, and Jim Mann, "Goodwill Toward the U.S. Is Dwindling Globally," *Los Angeles Times*, March 26, 2000, pp. A1, A30–A31.

McKinsie, Matthew, et al., *The U.S. Nuclear War Plan: A Time for a Change*, Washington, D.C.: Natural Resources Defense Council, June 2001.

McManus, Doyle, "U.S. Casts About for Anchor in Water of Post–Cold War World," *Los Angeles Times*, March 27, 2000, pp. A6, A8.

McNamara, Robert S., "The 'No-Cities' Doctrine," in Robert J. Art and Kenneth N. Waltz, eds., *The Use of Force: Military Power and International Politics*, 3rd ed., Lanham, Md.: University Press of America, 1983.

McNamara, Robert S., and Blight, James G., *Wilson's Ghost: Reducing the Risk of Conflict, Killing, and Catastrophe in the 21st Century*, New York: PublicAffairs, 2001a.

_____, "Nuclear Arms: The Threat of Annihilation Is Still Real," *Los Angeles Times*, June 24, 2001b, p. M3.

National Academy of Sciences, Committee on International Security and Arms Control, *The Future of U.S. Nuclear Weapons Policy*, Washington, D.C.: National Academy Press, 1997.

Nelson, Robert W., "Low-Yield Earth-Penetrating Nuclear Weapons," *FAS Public Interest Report*, Vol. 54, No. 1, January/February 2001, pp. 1–5.

Nolan, Janne E., *An Elusive Consensus: Nuclear Weapons and American Security After the Cold War*, Washington, D.C.: The Brookings Institution, 1999a.

_____, "The Next Nuclear Posture Review," in Harold A. Feiveson, ed., *The Nuclear Turning Point: A Blueprint for Deep Cuts and De-Alerting of Nuclear Weapons*, Washington, D.C.: The Brookings Institution, 1999b.

Nordon, Paul, "Hardness and Survivability Requirements," in Wiley J. Larson and James R. Wertz, eds., *Space Mission Analysis and Design*, 2nd ed., Torrance, Calif.: Microcosm, Inc., 1996.

"The Nuclear Posture Review: What's New?" *The Economist*, March 16, 2002, p. 35.

Pape, Robert A., *Bombing to Win: Air Power and Coercion in War*, Ithaca, N.Y.: Cornell University Press, 1996.

Payne, Keith B., *Deterrence in the Second Nuclear Age*, Lexington, Ky.: University Press of Kentucky, 1996.

Payne, Keith B., et al., *Rationale and Requirements for U.S. Nuclear Forces and Arms Control*, Vol. 1: *Executive Support*, Fairfax, Va.: National Institute for Public Policy, 2001.

Perkovich, George, *India's Nuclear Bomb*, Berkeley, Calif.: University of California Press, 1999.

Pfaff, William, "Bush Team Should Explain Aggressive China Policy," *Seattle Times*, July 3, 2001.

Pincus, Walter, "U.S. Nuclear Proposals Envision Sharp Cuts in Missiles, Bombers," *Washington Post*, May 26, 2001, p. A1.

_____, "U.S. Aims for 3,800 Nuclear Warheads: Cold War Strategy Is Being Replaced," *Washington Post*, January 10, 2002, p. A9. Online at http://www.washingtonpost.com (as of March 22, 2002).

Platt, Kevin, "U.S. a Threat? Just Ask China," *Christian Science Monitor*, October 28, 1999, p. 6.

Pomfret, John, "China Ponders New Rules of 'Unrestricted War,'" *Washington Post*, August 8, 1999, p. A1.

Pry, Peter Vincent, *War Scare: Russia and America on the Nuclear Brink*, Westport, Conn.: Praeger Publishers, 1999.

Quinlivan, James T., and Glenn C. Buchan, *Theory and Practice: Nuclear Deterrents and Nuclear Actors*, Santa Monica, Calif.: RAND, P-7902, 1995.

Rhodes, Richard, *The Making of the Atomic Bomb*, New York: Simon & Schuster, 1986.

_____, *Dark Sun: The Making of the Hydrogen Bomb*, New York: Simon & Schuster, 1995.

Richter, Paul, "U.S. Works up Plan for Using Nuclear Arms," *Los Angeles Times*, March 9, 2002, pp. A1, A16.

Roman, Peter J., *Eisenhower and the Missile Gap*, Ithaca, N.Y.: Cornell University Press, 1995.

Rosenberg, David Alan, "The Origins of Overkill: Nuclear Weapons and American Strategy, 1945–1960," *International Security*, Vol. 7, No. 4, Spring 1983, pp. 3–71.

Rosenthal, Andrew, "U.S. to Give up Short-Range Nuclear Arms: Bush Seeks Soviet Cuts and Further Talks," *New York Times*, September 28, 1991, pp. A1, A4–5.

Rowley, Storer H., interview with Robert McNamara, Former Secretary of Defense, *Chicago Tribune*, February 18, 2001.

Rumsfeld, Donald H., "Foreword," in Department of Defense, *Nuclear Posture Review Report*, Washington, D.C., January 2002.

Sagan, Scott D., and Kenneth N. Waltz, *The Spread of Nuclear Weapons: A Debate*, New York: W. W. Norton & Company, 1995.

Sakharov, Andrei, *Memoirs*, New York: Knopf, 1990.

Savage, David G., "Nuclear Plan Meant to Deter," *Los Angeles Times*, March 11, 2002, pp. A1, A15.

Schelling, Thomas C., *Arms and Influence*, New Haven, Conn.: Yale University Press, 1966.

Shanker, Thom, and David E. Sanger, "White House Wants to Bury Pact Banning Tests of Nuclear Arms," *New York Times*, July 7, 2001, p. A1.

Smith, R. Jeffrey, *The Strategy of Conflict*, New York: Oxford University Press, 1963.

_____, "The Dissenter," *Washington Post Magazine*, December 7, 1997, pp. 18–21, 38–45.

Trachtenberg, Marc, *History and Strategy*, Princeton, N.J.: Princeton University Press, 1991.

_____, *A Constructed Peace: The Making of the European Settlement 1945–1963*, Princeton, N.J.: Princeton University Press, 1999.

Turnbull, Peter C. B., et al., *Guidelines for the Surveillance and Control of Anthrax in Humans and Animals*, 3rd ed., Geneva: World Health Organization, WHO/EMC/ZDI/98.6, 1998.

Turner, Stansfield, *Caging the Nuclear Genie: An American Challenge for Global Security*, Boulder, Colo.: Westview Press, 1997.

U.S. Air Force, *New World Vistas*, Munitions Volume, 1995.

U.S. Department of Defense, *Nuclear Posture Review*, Washington, D.C., 1994.

U.S. Senate Foreign Relations Committee, Hearings on The Terrorist Nuclear Threat; Focusing on Dirty Bombs and Backyard Nukes, hearings, March 6, 2002.

Warner, Edward L., III, "Statement of the Honorable Edward L. Warner III, Assistant Secretary of Defense for Strategy and Threat Reduction, Before the Senate Armed Services Subcommittee on Strategic Forces Hearing on Nuclear Deterrence April 14, 1999," Washington, D.C.: Federal Document Clearing House, Inc., 1999.

Werrell, Kenneth P., *Blankets of Fire: U.S. Bombers over Japan During World War II*, Washington, D.C.: Smithsonian Institution Press, 1996.

Wohlstetter, Albert J., "The Delicate Balance of Terror," *Foreign Affairs*, January 1959, pp. 211–234.

Chapter Eight

COUNTERING WEAPONS OF MASS DESTRUCTION AND BALLISTIC MISSILES

Richard F. Mesic

The attractiveness of both ballistic missiles and weapons of mass destruction (WMD) to various actors throughout the world represents a serious threat to U.S. security. These weapons and delivery systems are popular in part because they are highly visible symbols of military prowess. As such, they provide significant influence on regional balances of power, even among friends and allies. Furthermore, as the ranges of available delivery systems increase, proliferants will eventually be able to reach, and therefore threaten, the U.S. homeland. Such weapons also give at least the appearance of being unstoppable, even by active defenses. All these features together make such weapons highly attractive as asymmetric counters to potential U.S. regional power projection. Increasing proliferation over the next few years is likely to change the very nature and scope of conflict.[1]

Ironically, the various U.S. policies and initiatives that attempt to stem proliferation and to delegitimize these weapons simply enhance potential proliferants' beliefs in and quests for these weapons. And there are questions about the helpfulness of various defenses against ballistic missiles. What, then, can the United States do to respond to these threats?

Ballistic missile systems themselves are relatively well understood, but there seem to be significantly different views of the nature and

---

[1]Thanks to RAND colleagues Ted Warner, Joel Kvitky, Greg Jones, David Vaughan, Gary McLeod, and Myron Hura for their thoughtful reviews of the material in this chapter. Any residual errors are, of course, the author's.

significance of WMD. A ballistic missile armed solely with conventional high explosives is inefficient and expensive compared to one armed with WMD. This is so even though WMD-bearing missiles may be "unusable" except for coercion or in extreme circumstances. What makes these warheads so valuable is the enhanced terror potential they provide.

The next section elucidates this potential by considering the three distinct classes of WMD—chemical, biological, and nuclear—separately. This is particularly important because the nature, scope, and significance of these classes are dissimilar. For example, as this chapter will explain, chemical agents are the most "usable" but the easiest to counter; biological agents may be the least suitable for ballistic missile delivery but are the most difficult to counter; and nuclear weapons are the most expensive and challenging but are the least likely (one hopes) to be used.[2]

After that discussion, the chapter postulates a wide range of candidate U.S. responses and identifies those that may be most compatible with the roles, missions and capabilities of the Air Force of the 21st century. The following two sections then focus on possible Air Force counterforce attacks and active defenses—theater missile defense (TMD) and national missile defense (NMD). Finally, the chapter concludes with a brief summary.

WMD CHARACTERISTICS AND SCENARIOS

The proliferation of WMD threats is worrisome for a number of reasons we will attempt to summarize in this section. The term *WMD* is in many ways an unfortunate aggregation of very dissimilar threats, unfortunate because the differences in nature, scope, and significance could yield profoundly different effects, not just for Air Force operations, but on the entire force structure, power-projection strategy, and warfighting tactics.

---

[2]Of course, the word *use* has different meanings in practice. This qualitative assessment of relative likelihood of WMD use pertains to *actual use* for WMD effects rather than, for example, veiled or explicit "use" as a coercive instrument. The absolute likelihood of nuclear and biological use is, obviously, impossible to estimate. The relative likelihood, however, reflects an analysis of differences in the range of conceptual opportunities and risks this chapter outlines.

Biological weapons have been a part of warfare throughout recorded history.[3] They are sometimes referred to as the "poor man's nuclear weapons" both because of their potential, pound-for-pound, to inflict human casualties on a scale comparable to nuclear weapons and because their costs could be less than the costs of even modest nuclear arsenals by a factor of ten or more. They are of increasing concern to national security policymakers and warfighters.

Chemical weapons were widely employed in World War I. Since then, defenses against them have played an important role whenever deployed U.S. forces have faced adversaries possessing chemical weapons (e.g., Iraq and North Korea).

The nuclear weapons dropped on Hiroshima and Nagasaki helped end World War II but spawned an arms race between the United States and the Soviet Union that ended with the demise of the latter. Nuclear weapons continue to be a serious proliferation concern.

WMD, particularly biological and nuclear weapons, could have a *revolutionary* impact on warfighting. But the United States has yet to address these complex counterproliferation problems adequately—and this brief summary certainly cannot do so. The threat WMD pose may be a forcing function for revolutionary changes. If so, a much broader consideration of WMD and their implications in the overall strategic context is certainly warranted.

The following pages describe biological, chemical, and nuclear weapons and outline the pros and cons of employing them as

---

[3]Christopher et al. (1997) provides an excellent historical overview of the use of microorganisms and toxins as weapons. This well-referenced article from the Journal of the American Medical Association discusses several historical cases, including the Tatar's 14th-century use of plague in laying siege to Kaffa (now Feodossia, Ukraine). Reportedly, the Tatars turned a plague epidemic among themselves into an offensive opportunity by catapulting their own plague cadavers into the city. Plague subsequently broke out in the city, and Kaffa's defenders retreated, but whether this outbreak was a result of the Tatar "biological attack" is uncertain. This uncertainty, which is typical of many actual or alleged uses of biological weapons over the ages, is due to the difficulty of differentiating naturally occurring diseases from biological weapon-caused epidemics, the challenges of gathering convincing forensic evidence, and allegations of biological attack for political or propaganda purposes. Examples of the last point of confusion include allegations the Soviets, Chinese and North Koreans made against the United States during the Korean War and the "yellow rain" allegations the United States made against the Soviets and their proxies in Laos (1975–1981), Kampuchea (1979–1981), and Afghanistan (1979–1981).

warfighting instruments. This sets the stage for assessment of the utility of various Air Force offensive and defensive responses and will help in rationalizing investments. Table 8.1 summarizes the characteristics of these weapons.

Background

Biological Weapons. While the potential effectiveness of biological agents is well understood (and very frightening) in purely clinical terms, their strategic and tactical military potential is much less clear.

The significance of disease-producing agents in warfare is unquestionable: Over the centuries, armies have suffered more casualties from natural disease than from man-made weapons. But with very few significant and/or well-documented exceptions, biological agents have not been *effectively* used *on purpose* to disable or kill enemy soldiers or civilians.[4]

During World War II, Japan's Unit 731 aggressively explored the potential use of biological weapons in modern warfare, to the extent of testing experimental agents and delivery systems against prisoners of war and against civilians in Chinese cities. Japanese attempts to use biological weapons on the battlefield in China may, however, have backfired: Japanese soldiers were inadequately protected and reportedly suffered 10,000 casualties after a cholera attack on Changteh in 1941.

The United States, Great Britain, and the Soviet Union also pursued biological weapon capabilities during and after World War II. Indeed, after the war, the U.S. programs were able to tap into the knowledge and experience of Japanese scientists to some extent. But with the emergence of nuclear weapons as the ultimate in mass destruction, and possibly because of some of the problems associated with biological weapons that I will outline below, the United States abandoned research and development (R&D) on offensive biological

---

[4]Christopher et al. (1997) notes one of the most notorious cases: the use of smallpox-infected blankets and handkerchiefs to "reduce" Native American tribes hostile to the British during the French and Indian War (1754–1767).

Table 8.1
Weapon Characteristics

Biological Weapons	Chemical Weapons	Nuclear Weapons
Antipersonnel, not antimateriel.	Antipersonnel, not antimateriel.	Can produce antipersonnel and/or anti-materiel effects.
Biological agents produce a delayed effect (kicking in minutes to hours to days or even weeks after exposure).	The initial danger of chemical exposure in military use scenarios is by skin contact or inhalation as the aerosolized agent cloud drifts by in the wind and the liquid agent "rains out."	Nuclear weapons produce devastating, prompt, and long-term effects.
The primary danger of infection in military scenarios using biological weapons is by inhalation as the aerosolized agent cloud drifts by in the wind—once the cloud is past, the danger is largely over.	Some chemical agents remain lethal for substantial periods of time (hours to days), that is, they produce a persistent effect.	
The extent of the exposed region depends on the agent type, release conditions, weather (winds, temperature, humidity) and time of day.	The longer-term, persistent effects depend on agent type, release conditions, and weather (winds, temperature, humidity) and are caused by contact with the liquid agent that has settled out on the ground or on equipment or by inhalation of vapors from evaporation of these deposits.	Nuclear weapons produce a wide range of effects that are relatively predictable given specific target characteristics.
The release of agent may or may not be associated with observable phenomena (such as hostile aircraft flyby, bombs, artillery or missile attack).	The release of chemicals will probably be associated with observable phenomena (such as flyby of hostile aircraft or attack with bombs, artillery, or missiles).	Nuclear weapons will most likely be delivered by ballistic missiles.

Table 8.1—Continued

Biological Weapons	Chemical Weapons	Nuclear Weapons
Warning that a biological attack is under way or has recently occurred is critical—warning makes very effective defensive responses possible.	Warning that an attack is under way is critical—if warned, very effective defensive responses are possible.	Warning (minutes) will likely result in only modest survivability benefits.
Biological weapons will be viewed as strategic assets with operational limitations due to political taboos.	Chemical weapons may be viewed as tactical assets.	Nuclear weapons will be viewed as strategic assets with operational limitations because of the small number of weapons likely to be available.
Details about the size and character of an adversary's biological weapons are critical (e.g., specific agents and strains) but will likely be unknown.	Specific details about the character of an adversary's chemical arsenal (e.g., specific agents and delivery mechanisms) will likely be uncertain (but less critical than the biological weapon uncertainties).	Specific details about the size and character of an adversary's nuclear arsenal (e.g., numbers and design particulars, such as yield) will be important but likely will be unknown.

weapons and destroyed its limited arsenal of weaponized agents in the late 1960s and early 1970s.

The Biological Weapons Convention (BWC), signed in 1972, repudiated biological weapons by internationally prohibiting their manufacture, stockpiling, and use. Despite the known violations of one signatory, the Soviet Union, the BWC has succeeded in stigmatizing offensive biological weapon systems. Biological weapon use is inconsistent with generally accepted norms of "civilized" warfare. No country today openly admits it has an offensive biological weapon program but many, particularly relatively poor third-world countries, are suspected of covert programs. The "poor man's nuclear weapons" may therefore play a role in a future conflict involving the United States with these potential biological weapon proliferants.

The covert nature of potential biological threats exacerbates the problems the United States faces in anticipating and countering them but may also limit the ways biological weapons can be used and their risks. Intelligence collection may point to specific threat agents and delivery means, but the enemy's intent, as reflected in his strategy and doctrine, is less likely to be well understood. Further, the need for covertness limits the proliferant's abilities to explicitly incorporate biological weapons in his overall campaign plan and to train and exercise his troops to use them. This could limit his tactical flexibility and/or confidence in the effectiveness of the weapons at the tactical level, unless biological operations could be conducted independent of other military operations. At the strategic level, the proliferant risks becoming an international pariah if he uses biological weapons overtly. Although biological weapons put awesome destructive potential within the reach of many states, actually using them would entail awesome operational and political risks. The scenario issue we will explore below is whether the military benefits outweigh the risks.

Chemical Weapons. Chemical agents made their modern military debut in World War I. These weapons were effective in trench warfare because they could penetrate fortifications, causing causalities and demoralizing the stalemated troops. Nonlethal chemical agents have been used since then for a variety of purposes, from crowd control to defoliation. The potential effectiveness of modern lethal and nonlethal chemical agents, in purely clinical terms, is well understood, but their strategic and tactical military potential in the very dynamic and complex battlefields of the future is more uncertain, although probably not as uncertain as the effectiveness of biological weapons.

Chemical weapons also have a negative image relative to more "humane" and discriminating instruments of warfare. The Chemical Weapons Convention, which more than 140 states signed in 1993 (the United States has not yet ratified it), bans the development, production, and possession of chemical weapons and reinforces the international norm against their use.

Nuclear Weapons. The relatively crude approximately 15 kt implosion and gun-type atomic weapons dropped on Hiroshima and Nagasaki ended World War II and introduced the world to the awesome power and effects of nuclear fission devices. In a relatively short time, the development of thermonuclear weapons ("H-bombs") rewrote the book on destructive potential—with modern weapons producing the power equivalent of about 1,000 tons of high explosives per pound of warhead (a 2,000,000-to-1 increase in efficiency over high explosives). While multimegaton (millions of tons of high-explosive-equivalent power) weapons were created, the primary effect of thermonuclear weapons technology was to allow weapons with 10 to 100 times the power of the 1945 weapons to be made small enough to pack several on long-range ballistic missiles.

In the proliferant context, the assumption is that current "entry-level" weapons would be more sophisticated than the 1945 U.S. devices but would still be limited to single-stage (fission) yields on the order of a few to a few tens of kilotons. Such a first-generation nuclear weapon could be made small and light enough for a Scud missile to deliver (that is, 1,000 to 2,000 pounds).[5]

Characteristics of WMD Affecting Their Use

Characteristics of Biological Weapons. Biological agents are unlike conventional weapons in that they affect only living organisms. Anthrax can kill a human but will not harm a tank. But unlike an antitank gun, the bacteria, viruses, or, to a lesser extent, the toxins,

---

[5]There is also the problem of "loose nukes." Presumably, a proliferant might be able to buy or steal all-up complete weapons from a nuclear state, in which case the single stage fission limitation here would not pertain.

take a relatively long time before their effects are observed. These qualities lead to the first two observations about biological weapons: Biological agents are antipersonnel, not antimateriel, and they produce a delayed effect (kicking in minutes to hours to days or even weeks after exposure).

Most, though not all, of the principal agents are highly infectious but not contagious; smallpox is one that is contagious, while anthrax is not. Noncontagion is obviously a desirable property to control the spread of disease beyond the targeted group. The principal means of becoming infected with a biological agent is by inhaling a vapor containing the organisms or toxin. This mechanism is effective only if the aerosol particles are in the 1 to 10 μm range. Particles this size are invisible to the naked eye and are suspended in the air in Brownian motion. They diffuse and drift with the wind. To infect people, of course, these winds have to be near the surface of the earth, in the mixing layer. This means that the aerosol must be released in this region (typically below 1,000 m, depending on meteorological conditions and time of day). Once released, the agent begins to lose its potency from drying out or exposure to ultraviolet light. Some agents, such as anthrax, can live for years as spores under the right conditions, but most biological agents are relatively short lived (hours to days). Also, the agent cloud disperses as it drifts away from the release point, reducing the density of the agent, eventually below concentrations needed to achieve the desired effect (whether incapacitation or death). Particles that settle out of the cloud and come to rest on surfaces, such as the ground, no longer constitute a serious threat since reaerosolization in sufficient quantity seems unlikely (although this is a somewhat controversial area in which additional tests are planned).

This illustrates two more observations: The primary danger of infection in military scenarios using biological weapons is by inhalation as the aerosolized agent drifts in the wind—once the cloud is past, the danger is largely over. The extent of the exposed region depends on the agent type, release conditions, weather (winds, temperature, humidity), and time of day.

Delivery means could range from slightly modified commercial handheld agricultural sprayers to spray tanks on high-performance aircraft or ships to sophisticated munitions delivered in gravity bombs or in warheads on cruise and ballistic missiles. Water and food contamination is also possible but is much less significant in military actions than in terrorist actions against urban populations. Weaponizing biological agents is probably the most challenging aspect of developing an effective biological weapon capability. The process of aerosolization is hard on the agent, and achieving a particle size appropriate for effective distribution is also difficult. Delivery means are generally relatively inefficient, either killing much of the agent or allowing it to fall out prematurely because the particles are too large. Therefore, despite the theoretical potency of very small quantities of an agent, military effectiveness requires releasing amounts larger than theory would suggest.

Another characteristic of these agents is that their release may or may not be associated with observable phenomena (such as flyby of a hostile aircraft or a bomb, artillery or missile attack). Once dispersed, the agent cloud is generally invisible. People in its path will not see it coming and will not notice when the cloud engulfs them. Warning systems are being developed to detect and characterize biological aerosols both at a distance (e.g., light detection and ranging [LIDAR] on helicopters and detectors on unmanned aerial vehicles [UAVs]) and locally, as the cloud passes by (e.g., the Biological Integrated Detection System [BIDS] equipment mounted in Army trucks).

These warning and agent identification systems are important because, with adequate warning, very effective defensive measures can be taken. These measures include moving out of the cloud's path, if possible, or, if not, donning simple, inexpensive face masks for partial protection. Note that, unlike the case of chemical agent threats, which require full-body protection (e.g., using mission oriented protective posture [MOPP] suits), which can significantly impair operations for an extended time, the biological agents can be mitigated simply with a mask. The mask that is included as part of the standard MOPP ensemble will work, but so will less-elaborate masks that simply cover the mouth; nose; and, just to be on the safe side, the eyes.

Even if there is no warning, so that troops are exposed to the biological cloud before they can don masks, there may still be some hope. Prior inoculation against the specific agent is one possibility, assuming the individual does not inhale enough agent to overwhelm whatever immunity the inoculation provided. The last is, however, a big "if," for a number of reasons. Further, identifying the agent before observable symptoms develop and promptly beginning an appropriate treatment (e.g., with antibiotics or vaccines) may prevent adverse effects (other than psychological effects). Even when antibiotics are not effective, as with smallpox, vaccination immediately after exposure may be beneficial. However, once the agent's symptoms appear, it is generally too late to prevent the disease from running its course. This means that warning that a biological attack is under way or has recently occurred is critical—warning makes very effective defensive responses possible.

In summary, biological agents can be delivered by a variety of means (some exceedingly stealthy or covert); a small amount can cover a large (but "controllable") area (e.g., a few kilograms of anthrax could cover a city, airfield, or port); the effects are delayed; and warning and attack assessment are technologically and operationally challenging. However, with adequate warning of an attack, very effective treatment is possible.

Characteristics of Chemical Weapons. Chemical agents are unlike conventional weapons in that they primarily affect living organisms. The nerve agent VX can kill a human, but it will not harm an airplane (although some agents are corrosive). But the direct causalities in a chemical attack may be considered a bonus to the primary military effects, which are generally indirect: Chemical agents are antipersonnel, not antimateriel.

Agents are characterized as lethal or nonlethal and by their physiological effects (e.g., blister, blood, nerve agents). They are also characterized by their persistence, that is, the length of time that an exposed area remains dangerous after agent release. Persistence is a function of the chemical properties of the agent and environmental factors (e.g., temperature, moisture, winds). Most modern agents are delivered as liquid aerosols—not as a gas. Their effects can be produced by skin contact (percutaneously), but causalities will also result from inhalation or eye contact.

The initial danger of chemical exposure in military use scenarios is by skin contact or inhalation as the aerosolized agent cloud drifts by in the wind and liquid agent "rains out." Some chemical agents remain lethal for substantial periods of time (hours to days), that is, they produce a persistent effect. The longer-term, persistent effects depend on agent type, release conditions, and weather (winds, temperature, humidity) and are caused by contact with the liquid agent that has settled out on the ground or on equipment or by inhalation of vapors from evaporation of these deposits.

Delivery means could range from slightly modified commercial handheld agricultural sprayers to spray tanks on high-performance aircraft or ships to sophisticated munitions delivered in gravity bombs or in warheads on cruise and ballistic missiles. Water and food contamination is also possible but is much less significant in military actions than in terrorist actions against urban populations. Weaponization of chemical agents is relatively straightforward.[6] The release of chemicals will probably be associated with observable phenomena (such as flybys of hostile aircraft or attack with bombs, artillery, or missiles).

Warning systems are important because adequate warning allows very effective defensive measures to be taken. Protection against chemical threats requires total-body MOPP gear, which can significantly impair operations for an extended time. If troops are exposed to the agent before they can be warned and don protective (MOPP) gear, there may still be some hope. If chemical symptoms appear, the soldiers can inject themselves with chemical antidotes, such as atrophine. Thus, warning that an attack is under way is critical.

In summary, chemical agents can be delivered by a variety of means; a small amount can cover a large (but "controllable") area; and, with adequate warning, very effective defenses are possible.

Nuclear Characteristics. The principal characteristic of nuclear weapons in the proliferant context is that very few are likely to be available because the critical nuclear materials (plutonium and/or highly enriched uranium) required to fabricate these weapons are difficult and expensive to make. Of course, if this process is circumvented by buying or stealing the materials or even complete weapons—the so-called "loose nukes" problem—this may not be a limiting factor.

---

[6]An important exception is fusing. The dispersal of chemical (and biological) weapons is most efficient if the agent can be dispersed before it hits the ground. Achieving a reliable, optimal airburst is a technical challenge that few potential proliferants have yet managed to achieve.

This brings us to the first three observations regarding nuclear weapons: Nuclear weapons will be viewed as strategic assets with operational limitations because of the small number of weapons likely to be available. Specific details about the size and character of an adversary's nuclear arsenal (e.g., numbers and design particulars, such as yield) will likely be unknown. Finally, the details about the adversary's nuclear capabilities may be important (e.g., do they have one weapon, a few, or a few tens?).

While one tends to think of the blast and radiation effects of a ground or low airburst nuclear weapon and the resulting mushroom cloud and downwind radioactive debris fallout, these weapons can also be used in other ways to produce tailored effects. For example, they can be detonated at high altitudes or in space (a) to disturb radio frequency propagation; (b) destroy unhardened missiles and satellites promptly with long-range X-rays; (c) degrade the effective lives of satellites through trapped radiation effects; or (d) expose aircraft and terrestrial systems to high-energy magnetic pulse (HEMP), which can upset or damage electronic components.

The specific effectiveness of nuclear weapons used in these various ways depends on device design specifics, but proliferants would, typically, already understand the effects of nuclear weapons and how to best achieve any desired damage mechanisms. In particular, even limited-yield, first-generation weapons can produce significant HEMP effects.[7] So, nuclear weapons can produce antipersonnel and/or antimateriel effects, and the effects are devastating, prompt, and long term.

WMD Scenarios

Biological Weapon Scenarios. As we noted above, there is very little historical precedent or published doctrine that the analyst can rely on when postulating how a proliferant might use biological weapons. Further, intelligence breakthroughs in this area are more a matter of

---

[7]While the physical effects of the nuclear detonation are relatively well understood, the response of specific systems exposed to these environments is much more contentious. Significant testing (some unintentional, as occurred in Hawaii during the high-altitude U.S. nuclear tests conducted in the South Pacific in the mid-1950s) was carried out during the Cold War years, but most of today's military and commercial systems have not been hardened or tested against HEMP.

luck than effort (e.g., the right defector at the right time). So, it is necessary to fall back on common sense, letting the characteristics above suggest how biological agents might most reasonably be used against the United States or its regional allies. This amounts to "informed speculation." This structuring of the problem of anticipating the potential use and effectiveness of biological weapons is not predictive, and measures of "confidence" cannot reasonably be provided; however, the logical framework should be helpful in stimulating further community consideration of these threats to help identify and prioritize Air Force response options.

An adversary might consider two generic types of biological weapon "use":

1. coercive threats of biological weapon use

2. military use (covert and overt).

Because of the significance of biological weapons as an instrument of terror (e.g., potentially massive casualties among unprotected urban populations), the existence of even a *suspect* capability will cast a shadow over future regional crises. The use of or even overt threats of the use of biological weapons would violate generally accepted international norms of proper behavior, thanks to such instruments as the BWC.

Therefore, it seems that the most likely form of coercive use would be rather unspecified, veiled threats of "dire consequences" if the aggressor's demands were not met. The demands could involve specific actions by the states under attack or could be aimed at the United States. The principal concerns are that these threats will do one of the following:

1. cause the victim state (a friend or ally of the United States) to acquiesce to the aggressor's demands because the risks of resisting seem too high

2. deter the United States from intervening successfully to protect both its short-term and long-term national interests.

The bottom line in each of these issues is the potential cost of resisting the aggressor versus the cost if the United States or our allies acquiesce. "Costs" can be measured in terms of blood or treasure or can be more abstract, involving a either a loss of stature and influence in the region (or even globally) or a weakening of the moral underpinnings and rule of law that govern interstate relations.

These coercive threats will be of concern in three scenario contexts:

1. in a crisis, before the start of military hostilities

2. during the course of a military campaign

3. after the war has come to an end.

The crisis scenarios would involve the attempts of a would-be regional hegemon to compel his neighbors to do his bidding "or else." The threat of using biological weapons—whether explicit, or, more likely from our perspective, veiled—may introduce an asymmetry in the regional balance of power. If the threatened states cannot protect themselves adequately, are relatively weak militarily, and have no credible punitive threats with which to deter the use of biological weapons, the coercive threats may succeed. Of course, the United States could address these deficiencies to some extent through security commitments, including an "extended" deterrent policy and posture. The credibility of U.S. commitments and capabilities will depend, of course, on its ability to deal militarily with the biological threats, which will be discussed in more detail below.

During the course of a regional conflict in which the United States is engaged, there is a chance that biological threats will be used to attempt to limit U.S. war aims through escalation potential. A possible threat, for example, might be: "If the Great Satan advances to within 250 km of Baghdad, the wrath of God will be visited upon the Infidels." Such threats could be punctuated by limited "demonstrations" of the potential of biological weapons, although the decision to cross the "use line" is not likely to be taken lightly. Perhaps the more serious concern would be the enemy's threats to, in effect, take his opponents with him in defeat. If the United States were prevailing militarily, as would be expected, the enemy might at some point believe that all is lost. At this point, threats to lash out in revenge with biological weapons would become much more worrisome.

After the end of the war, a defeated aggressor may still pose a biological threat to the world by sponsoring terrorist organizations or reconstituting an offensive biological weapon capability to regain influence in the region.

Actual use of biological weapons for military purposes (i.e., other than purely for revenge, as discussed above), would seem to pose two distinct possibilities:

1. that limited, covert (or possibly even overt) use could tip the military balance significantly in the opponent's favor
2. that unlimited, overt use could win the war.

If possible, successful covert use would be preferable to overt use from the enemy's perspective for two principal reasons:

1. It would reduce the risk of punitive responses (military and political).
2. It could increase the effectiveness of the biological attack.

Here, *covert use* refers to the release of biological agents through unconventional means that deny timely attack warning and leave no "smoking guns" to implicate the user. The military goal would be to confuse and demoralize the forces attacked and to cause them to take defensive precautions that reduce their military effectiveness so that the aggressor's conventional forces would have a better chance of prevailing. The intent would not be to inflict extensive casualties. In fact, the goal may be to limit casualties to a very small number to minimize the likelihood of a "disproportionate" punitive response and to help mask the true nature of the attack.[8] Many biological agents occur in nature. So, for example, it is possible for troops to contract anthrax after stirring up the soil where sheep or camels have been grazing. If the agent were delivered in very low concentrations and even during daylight hours (as opposed to night, which would be the preferred time to maximize agent coverage area), the release might be difficult to detect, and the casualties might be appropriately small.

The problems with covert use from the enemy's perspective are the same problems conservative military planners have with all types of

---

[8]An example "biological sucker punch" scenario might go as follows: Attempts to coerce the United States have failed, and the United States is in the theater. Special operations forces (SOF) teams have covertly dispersed nonlethal biological agents (e.g., Brucellosis, Q-fever, VEE, SEB) over a wide area. As symptoms appear and chaos sets in, the aggressor launches massive combined arms strikes. The United States asks for a cease-fire to withdraw and treat casualties.

"special operations." First, the unpredictability of the results puts this kind of use into a "bonus" category—if it works, so much the better, but the user cannot count on it working to achieve the military objectives. This "bonus effect" might be desperately needed in some situations, but the conditions would seem to argue for "more aggressive" measures, such as those described below. Arguing against covert operations is the essentially incalculable risk of exposure as the initiator of a biological attack, which could, at best, embarrass the perpetrator or, at worst, compromise his conventional military plans and capabilities and/or cost him dearly in response. If the adversary state is discovered to have used biological weapons against allied forces, the allied war aims and strategy will likely change dramatically. This may hasten the user's defeat by making subsequent use of biological weapons much less effective (the opponent has been warned) and rallying international opinion against him. Such use would also likely increase the consequences of defeat (e.g., in war crimes trials).

The point of *overt use* (other than as the last desperate act of revenge of a doomed regime) would likely be to attempt to achieve both warning denial and widespread serious effects. Surprise would be important to minimize defensive actions and to give the biological agents time to take effect before executing other elements of the strategy. Widespread attacks would likely be planned to maximize the shock value of the use of biological weapons and to get the greatest possible effect from the initial use, since subsequent tactical uses may be less effective because of the defensive actions that would be invoked upon warning. But how could the dual objectives of warning denial and widespread attacks be met?

This sort of widespread, intense biological attack, whether as a preemptive move in a crisis or as an escalatory move during the war, would be difficult to plan and execute. If conventional delivery means were used, surprise would surely be impossible to achieve. For example, ballistic missiles armed with biological weapons could be salvoed to hit targets across the theater nearly simultaneously. However, the launch signatures would almost surely provide enough warning for troops in the target areas to don their MOPP gear in anticipation of a chemical or biological attack (of course, the forces might let their guard down if there were a number of false alarms). If this reasonable planning assumption were accurate, a ballistic missile attack with biological weapons would have little chance of even modest success (and a good chance of prompting a massive, punitive response). Similarly, aircraft-delivered biological attacks would be difficult to coordinate for undetected, theaterwide, nearly simultaneous release. If one or more aircraft were intercepted and found to be carrying biological weapons, the rest of the attack would likely fail.

Perhaps the best chance of both denying warning and striking theaterwide targets near-simultaneously with biological weapons would be to attack with SOF. These stealthy troops would infiltrate and, using handheld or vehicle-mounted spray devices, would release the agent upwind of their targets at a prearranged time or signal. Of course, local winds are variable and unpredictable over the interval of weeks to days to hours (or even minutes in certain situations) between the drafting of the attack plan and its execution, so some real-time adjustments would need to be made. Conservative military planners will have some difficulty endorsing important but risky unconventional operations, such as this, but they are conceivable, especially if they can be rationalized as a bonus if they work.

Whether or not massive near-simultaneous surprise attacks across a theater succeed, it is conceivable that localized biological attacks for specific tactical purposes also will be attempted. All such attacks must also achieve some level of surprise if they are to succeed. While biological aerosol clouds may be difficult to detect, it may be possible to detect the aircraft or missiles that release the agent, thereby providing warning to downwind troops. There is much less chance that more covert, unconventional delivery means (e.g., handheld agricultural sprayers) will be detected.

Chemical Weapon Scenarios. Chemical weapon use is more likely to be overt than the discussion above suggests biological weapon use would be. This is in part because the amount of agent required for a given effect is a factor of ten (or more) greater with chemical agents than with biological agents and in part because the principal effect is a reduction in the warfighting effectiveness of the exposed troops, whether measured in terms of aircraft sortie generation or the ability of soldiers to fight in MOPP gear.

Surprise may be a goal in chemical attacks to inflict causalities among unwarned troops, but once the chemical use threshold has been crossed, operational tempo and efficiency are the primary causalities. Chemical attacks can be used to attempt to shut down operations on airfields and ports, to deny territory or slow advances, and to create battlefield asymmetries (e.g., allied forces in MOPP gear but enemy forces not exposed or protected).

Nuclear Scenarios. Given the likely shortage of weapons and their strategic value, it seems unlikely that battlefield nuclear use of the sort that was envisioned in NATO versus Warsaw Pact scenarios is the greatest concern. Using nuclear weapons against large concentrations of forces and support personnel at such locations as airfields and ports would yield impressive local devastation (i.e., within a radius of about a mile or so from a 10-kt airburst). But there are likely many more such potential targets than there are available weapons. So, while a few "high-value" facilities might be attacked with nuclear weapons delivered by aircraft (for which defense penetration is an issue) or ballistic missiles (for which reliability is an issue), this application may not be considered sufficiently "strategic."

With a very small nuclear arsenal, if an adversary wanted to use part of this force for direct military effect, it could be argued that the most effective use (beyond simply brandishing them) would be to detonate one or two at high altitude over U.S. forces to produce the wide-area effects of HEMP. The risks are minimal: The lack of direct enemy causalities reduces the likelihood of a disproportionate nuclear response, and even if the detonations did not have much effect on the equipment exposed to HEMP, this might still be considered a highly dramatic warning shot and show of resolve. On the positive side, the HEMP effects could cause widespread disruptions in critical electronic systems (e.g., on Joint Surveillance and Target Attack Radar System [JSTARS] and Airborne Warning and Control System surveillance and control aircraft), which could impair U.S. warfighting abilities and allow the enemy's conventional operations to be more effective by exploiting our reduced capabilities.

Finally, of course, the adversary would likely maintain a "strategic reserve" of a few nuclear weapons to hedge against the prospect of a catastrophic defeat. This reserve would most likely be used to threaten cities in the region to convince the United States to halt the war before all was lost. At the extreme, nuclear weapons could be used as the last act of revenge by a dying regime.

Implications of These WMD Scenarios

The above WMD scenarios are somewhat one-sided. Other than alluding to U.S. "punitive" responses and the role passive defenses could play given direct or indirect warning, the scenarios do not really play both sides of the interaction. How can the United States mitigate the threats of WMD use outlined above? Several kinds of response should be considered in four areas: policy, intelligence, operations, and R&D.

Policy. This area presents several knotty issues. Active pursuit of nonproliferation policies and further reinforcement of the stigma against WMD development and use are good starting points, but more is needed to add credibility to our deterrent posture. Specifically, the United States needs to articulate its policy for deterring WMD use against the United States or our allies. Such statements as those made during the drive to renew the Non-Proliferation Treaty (NPT), to the effect that the United States would not use nuclear weapons against a nonnuclear state, may have been sensible in the NPT context but are unfortunate, or at least too explicit and/or premature, in the larger context of preventing WMD use.

An extended deterrent posture, such as the one NATO adopted during the Cold War, would seem to be an unlikely policy to adopt in the context of possible theater wars against WMD-equipped adversaries. Since the U.S. government has disavowed chemical and biological weapons, threatening preemptive first use of nuclear weapons to destroy hostile WMD systems would be contrary to its nonproliferation goals and to NPT-related commitments. However, a controlled nuclear *response* following enemy use of WMD, particularly if that use were against civilian areas and caused massive casualties, probably should not be ruled out.

More practically, what is needed is a policy statement that might go something like the following:

> The United States is dedicated to the worldwide elimination of all weapons of mass destruction (WMD). The U.S. has destroyed its chemical and biological weapons and is negotiating nuclear reductions with the other nuclear powers. The community of nations cannot be held hostage by WMD-armed aggressors. Assuring that regional predators cannot use WMD to coerce or defeat their neighbors is a vital U.S. national interest. The reasoning is simple: Non-

proliferation and WMD disarmament are critical because WMD have no moral justification other than to deter the use of WMD by others. If WMD were used to coerce or destroy an enemy, the user of WMD for offensive purposes might win a short-term victory, but the long-term implications would be dire. Allowing a proliferant to prevail would set a precedent that would encourage further WMD proliferation (either in defense or to serve aggressive ambitions), thereby increasing the risk of further WMD use. This is a downward spiral that civilization may not be able to survive. Consequently, the United States will not let a WMD-armed aggressor prevail: No matter what the U.S. national interests are in a regional conflict, the introduction of WMD will transcend the other interests. The United States will use whatever means necessary to deny the aggressor his prize and to punish him appropriately.

Simply stated, the argument is that it should be U.S. policy (as well as that of other nations and international organizations) not to tolerate offensive (first) WMD use. The issue, of course, is making this policy credible—where are the teeth? And can this policy be enforced without creating the same sort of long-term proliferation problems the policy attempts to prevent. For example, what responses to WMD use are "appropriate" or "proportionate," and might they, of necessity, involve the United States or its allies using nuclear weapons?

This general concern suggests that U.S. warfighting capabilities should emphasize (a) intelligence collection to help counter WMD threats politically and militarily; (b) responsive operational concepts and systems, such as counterforce precision strike and defenses (passive and active) against WMD attack to limit damage and, hence, the scale of "proportionate" responses; and (c) aggressive R&D to support the development of effective conventional response capabilities.

Intelligence. Several priorities are clear for this area. Intelligence should tell policymakers and warfighters if countries show an interest in WMD, have R&D programs under way, or have deployed such weapons: Which WMD are of concern? Where are they being developed? stored? What are the characteristics of the delivery systems? What testing has been conducted? Is there any indication of enemy intent or its doctrine for WMD use? Are they training for "combined arms" battles that include WMD use? Are SOF forces, agents, or paramilitary forces training for WMD delivery?

Specific threat details that will likely be very difficult to collect will be important. The more specific the details on these weapons (e.g., production methods, strain, and delivery means for biological agents), the better, because the details help determine which responses would be most effective (e.g., the correct vaccines to be produced and used, the best postexposure treatment of casualties, and the appropriate focus for warning systems that detect and identify agents). In fact, the forensic proof that an enemy has produced and/or used specific WMD agents may be very important both during and after the conflict; the intelligence community should play a critical role in associating the enemy with a WMD attack. Details on the WMD infrastructure will be important in planning counterforce strikes, particularly with respect to limiting collateral damage from any agent that might be released by attacks on WMD production and storage facilities. Information on adversary doctrine, exercises, and any other sources would have obvious value to the regional commanders in chief who seek insights on how an adversary might employ WMD.

Operations. U.S. forces can take several actions operationally, including preventive strikes on WMD facilities; preemptive strikes on WMD stocks and delivery means; active defenses, such as airborne laser (ABL), Theater High-Altitude Area Defense (THAAD), and Patriot; and passive defense measures, such as nuclear hardening, inoculations, antibiotics, warning systems, masks, and collective shelters. Of these, the most important are passive defense measures.

The success of counterforce attacks to destroy the WMD before they can be used will be unpredictable because of inherent uncertainties in locating and destroying these weapons. Active defenses will not be leak-proof, and there are enemy countermeasures that are compatible with the delivery of biological and chemical weapons that may defeat or avoid defenses, such as the proliferation of submunitions can defeat Patriot and THAAD TMD systems, and handheld agricultural sprayers can avoid air defenses. Therefore, a prudent commander in chief will not count on counterforce and active defenses to completely neutralize the WMD threat.

Antibiotics work only against bacteria. Vaccinations work against many bacterial agents (e.g., anthrax), some viruses (e.g., smallpox), and some toxins (e.g., botulinum). Even proven vaccines, however, are strain dependent and may be overwhelmed by large agent concentrations. Inoculation is a logistical burden.[9] Moreover, adverse side effects are always a risk, particularly with multiagent vaccination "cocktails." Because of these problems, it may be preferable to treat troops after they have been exposed to certain agents, if the exposure can be detected and if the agent can be identified quickly (that is, before symptoms appear). For example, victims who receive vaccinations in parallel with antibiotics immediately after being exposed to anthrax may never experience any symptoms. Even when antibiotics are not effective (for example, with smallpox), administering vaccinations immediately after exposure may be beneficial.

R&D. Masks and shelters can be highly effective with minimal logistical and operational burdens, but effective use will require warning. Warning is a greater technological and operational challenge for biological weapons than for chemical weapons. The biological warning systems currently in R&D have limitations that a knowledgeable opponent could exploit.[10]

For example, BIDS currently can detect and identify only four specific agents, and the methods are very sensitive to operator skill and motivation. Furthermore, because of the sensitivity setting adopted to minimize false alarms, agent concentrations that can produce casualties can pass by a BIDS system without triggering an alarm. There will only be one BIDS platoon (seven vehicles) per Army corps. The long-range (50 to 100 km) biological standoff detection system currently under development using an eye-safe LIDAR on UH60 helicopters cannot identify agents but has pattern-recognition software that allows it to recognize aerosol clouds with characteristics (size, shape, particle size, and density) consistent with biological threats. These aircraft will be able to monitor suspect regions, but the operational concepts have not yet been fully developed. Other devices for timely identification of biological and chemical agents are in R&D, including the spectroscopic excitation and classification of trace effluents (SPECTRE) system.

---

[9]Usually, a series of shots is required over time, with periodic boosters. Thus, at any one time, some troops will be unprotected because of the introduction of new recruits into the service and departure of older troops.

[10]The need for transparency in activities related to biological weapons means that most of these systems are unclassified, which makes exploitation easier.

Table 8.2 summarizes what I believe to be the projected capability of the United States to respond to WMD threats. These assessments reflect the author's experience and judgments but are admittedly qualitative. Others may emphasize different aspects of the threat and response, which could lead to different assessments. While the particulars might be arguable, however, the overall impression this matrix gives is reasonably representative of the current state of affairs.

Deterrence is an important element of a U.S. response to the more severe biological and nuclear threats, but as discussed above, deterrence based on punitive response may not be appropriate, and hence not effective, in some biological and nuclear use scenarios. The chemical threats are serious but can be managed through prudent investments in such areas as collective protection. Much the same can be said for proliferant nuclear threats, although the potential for massive casualties is very sobering. The biological agents that were the focus of much of the discussion above are, in many ways, the most troubling.

RESPONDING TO THE WMD THREAT: POTENTIAL AIR FORCE INITIATIVES

The remainder of the chapter investigates possible Air Force responses to these threats by exploring two related but distinct contexts: power projection and homeland defense. The following discussion will touch briefly on many of these potential initiatives but will focus on counterforce and active defenses.

Table 8.2

Summary of WMD Threat Implications

Priorities	Chemical	Biological	Nuclear
Deter WMD Use	Modest	Modest to Substantial	Modest to Substantial
Deny WMD Use	Poor	Poor	Modest
Defend Civilians			
Passive	Modest	Modest to Poor	Poor
Active	Modest	Modest to Poor	Modest
Defend Military			
Passive	Substantial	Modest to Poor	Modest
Active	Substantial	Modest to Poor	Modest

Generic WMD and ballistic missile responses relevant to U.S. power projection scenarios will involve a mix of approaches in which the Air Force may be a significant participant:

- *nonproliferation* measures include limiting technology transfer, delegitimizing systems, and offering security guarantees to non-proliferants

- *counterproliferation* measures include providing deterrence, passive defenses, counterforce attacks, and theater missile defense (TMD)

- *other response* measures involve doctrine and force structure, such as relying more on long-range operations with bombers).

Homeland defense has related but distinctly different elements:

- *nonproliferation* measures include limiting technology transfer and delegitimizing systems

- *counterproliferation* measures include providing deterrence, counterforce attacks, and national missile defense (NMD).

Some specific candidate Air Force activities would support the generic counterproliferation initiatives above,[11] such as the following:

- *power-projection measures,* including force protection (warning, hardening), threat avoidance (counterforce, camouflage, concealment and deception; mobility; dispersal; sanctuary basing; and operations)

- *homeland defense measures,* including strategic warning, preemptive counterforce, attack warning and characterization, NMD cueing, air and cruise missile defense (close other threat doors), red teaming.

As the discussion above notes, counterforce attacks and active defense are responses to WMD delivered by theater ballistic missiles (TBMs). The next section will get more specific about Air Force TMD initiatives, both ongoing and conceptual, describing candidate con-

---

[11]These activities range from the conceptual state to being programmed and/or deployed. The programmatic details change over time and are not critical here and so will not be addressed explicitly.

cepts and discussing their pros and cons. The following section will then focus on NMD responses as a natural extension of the theater defense initiatives.

POTENTIAL AIR FORCE TMD INITIATIVES

The purpose of U.S. TMD efforts is straightforward: to address the emerging threats that TBMs pose for U.S. military power-projection operations. The Gulf War introduced crude Scud threats with military and political implications disproportionate to the Scuds' actual effectiveness. In contrast, the emergence of future TBM threats—most notably missiles with precision guidance, advanced conventional submunitions, or WMD—will have the potential to profoundly alter the course and conduct of major theater wars (MTWs). For example, TBMs increase the vulnerability of forward-positioned forces and equipment. TBMs can slow and complicate (politically and operationally) access to ports and air bases making rapid deployment of decisive forces to a theater more difficult. If casualties and collateral damage are an issue, an opponent could create response dilemmas by using TBMs against U.S. and/or allied forces and populations. If the United States depends on swiftly meeting its defined warfighting objectives, TBM use may change these objectives and lengthen and/or expand the conflict.

Although TMD *may* have some deterrent value, the primary purpose of acquiring and deploying TMD systems (and, by extension, as longer-range threats emerge, NMD as well), is to ensure that the United States will not be deterred from taking the appropriate actions to defend its primary security interests in an MTW fought under threatened or actual WMD strikes on the United States or our allies.

To achieve this purpose in "WMD shadowed MTWs," acceptable counterproliferation strategies will require both TMD (and, eventually, NMD) capabilities to support U.S. *offensive* campaign strategies that seek to deny or minimize the damage from possible TBM strikes against the United States, our allies, and our forward-deployed forces.

TMD offers some protection both against enemy missile delivery of advanced conventional munitions and against WMD warheads. The two types of warheads pose qualitatively different cases. Advanced conventional warheads, even with future precision guidance, are much less challenging than WMD warheads, both in terms of the threat they pose to defended areas and the difficulty of designing defenses to blunt this limited threat potential. However, the WMD case is very challenging because the price of U.S. failure to respond in some manner to WMD could be high in political, if not military, terms. Since these risks are very real but incalculable, it is impossible to calculate analytically what should be expended on the development and deployment of TMD systems, as well as on directly related intelligence, surveillance, and reconnaissance, in an attempt to mitigate the risks, so value judgments must be made. Fortunately, there is only a limited set of candidate "solutions" from which the United States can pick and choose. As these decisions are made, it is likely that TMD efforts will be focused on countering WMD threats.

The United States will attempt to field capabilities that can *deny* damage from WMD threats, but it is not possible to create a perfect defense.[12] Force developers will therefore want to create TMD systems that are good enough to make the entry price for potential opponents high enough (at best) to discourage them from developing a missile delivery capability in the first place or (at least) to increase their uncertainty of success and risk of failure, should they contemplate using WMD. The TMD system should provide "enough" protection from a conservative defense viewpoint that the United States is not deterred by the prospect of missileborne WMD attacks. Unfortunately, these complex judgments must be reviewed periodically and adjustments must be made to our TMD efforts as the political situation, emerging threats, operational experience, and technological opportunities unfold. Once the United States commits itself to pursue substantial TMD capabilities, the dynamic (and potentially very costly) action-reaction process with a determined adversary could be prolonged.

---

[12]But what will decisionmakers and military planners *believe* about the effectiveness of their defenses? Unfortunately, these "beliefs" are unlikely to be well founded in careful, "real-world" tests and exercises but almost surely will not be the result of real wartime experience. This inherent uncertainty in the effectiveness of complex, scenario-dependent systems, such as TMD and NMD, has a dangerous aspect: Might our overconfidence in defenses make us more prone to taking risks we should not? What if we came to depend on defenses and they failed us? What is our fallback strategy to win this war? What about the next war—will we have to completely reshape and rebuild our military forces?

TMD Concepts of Operation

All TMD architectures have four components:

1. **active defenses,** of several distinctly different types:

    — terminal defense, in which intercepts occur primarily within the atmosphere (e.g., PAC-3, located near the targets to be defended)

    — midcourse defense, in which intercepts occur after booster burnout, but before warheads reenter the atmosphere (e.g., THAAD or Navy Theater-Wide [NTW], which covers a large area)

    — boost-phase or ascent-phase intercept defense, which would, for example, require a system of forward-operating aircraft with high-energy lasers or capable of launching hypersonic interceptors at a missile with its warhead still attached to the burning rocket motor during its boost phase or intercept of the separated warhead post boost in its ascent phase before it reaches apogee.

2. **counterforce operations** (e.g., air-to-ground or ground-to-ground systems to attack TBM infrastructure and transporter-erector-launchers [TELs]) *before, during, or after the launch of missiles*)

3. **passive defenses** (e.g., hardened aircraft shelters and revetments, suits, masks, and collective protective shelters to protect against the use of chemical and biological agents, mobility, counterforce, camouflage, concealment, and deception)

4. **battle management and command, control, communications, computers, intelligence, surveillance, and reconnaissance (BM/C⁴ISR).**

As pictured in Figure 8.1, the overall TMD concept involves an integrated joint-service "system of systems" comprising Army, Navy, and Air Force elements. The following paragraphs will focus on the Air Force components of this system.

The Air Force is currently pursuing counterforce operations and active defenses, as outlined in Figure 8.2. Pre- and postlaunch counterforce concepts of operation (CONOPs) build on dedicated non-

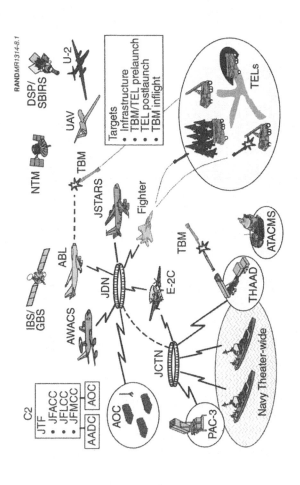

Figure 8.1—Joint-Service TMD Architecture Programs for Dealing with the TMD Problem

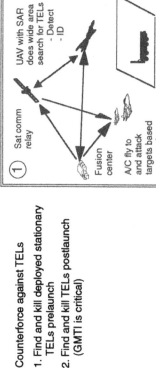

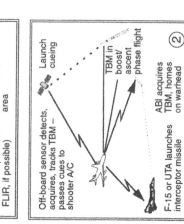

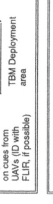

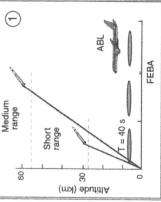

RAND*MR1314-8.2*

Counterforce against TELs

1. Find and kill deployed stationary TELs prelaunch

2. Find and kill TELs postlaunch (GMTI is critical)

Active defense against TBMs

1. Kill missiles during powered flight with a high-power laser

2. Kill missiles/warhead(s) in boost/ascent with a kinetic kill vehicle

Figure 8.2—A Broad Range of Potential Air Force TMD CONOPs

TMD Air Force systems in the field or under development with the support of improved C⁴ISR systems and operations. The ABL system is an Air Force active defense system currently under development, while the airborne interceptor (ABI) is a conceptual extension of air-to-air systems and capabilities, again with appropriate supporting C⁴ISR advances, but ABI is not currently under development within the Air Force.

Counterforce Operations. Prelaunch counterforce CONOPs involve sensors on various platforms—satellites, stand-off aircraft, and UAVs—to find, identify, track, and target mobile TELs used to carry and launch TBMs. It is assumed that the enemy will keep his TELs hidden (in caves, warehouses, under bridges, etc.) until just prior to launch. The TELs will leave their hides and/or resupply bases and move to a safe distance to set up for launch. After launch, the TELs will move off as quickly as possible to a hide. Total exposure times for this cycle during prelaunch activities might range from 30 minutes to a couple of hours, with similar exposures postlaunch.[13] There will likely be no radio frequency emissions at any point when the TEL is in the open.

Within the brief prelaunch exposure period, the surveillance systems must be able to search the entire deployment area (which could be thousands of square kilometers, depending on intelligence preparation of the battlespace (IPB) for moving or stationary TELs, detect and identify them, and pass the information to the attack systems, which must then respond within minutes—a very challenging set of tasks. It implies multisource data fusion; close coordination and cueing between ground moving target indication (GMTI) and all-weather, day-and-night imaging systems, such as synthetic aperture radars (SARs); very high search rates; near-continuous deep dwell; and advances in moving target identification and high-resolution (<1 m) imagery with automatic target cueing (ATC) and automatic target recognition (ATR) processing to achieve a high probability of

---

[13]There are, of course, several feasible variations on this generic enemy TBM concept of execution in which the timelines could be more or less stressing. For example, in North Korea, the TELs could be hidden in specially fortified caves (of which there are several hundred) and could be launched from just outside the portal blast doors. This would eliminate TEL travel exposure entirely, but the Koreans would have to be confident that the TELs and reload TBMs inside the caves were safe from U.S. attack after the initial launches compromised the locations.

TEL detection and recognition with manageable false alarm rates.[14] The current state of the art is not yet up to this challenge, other than under admittedly very optimistic assumptions (good IPB localization, open desert terrain, etc.). This discussion will be more explicit and quantitative about the admittedly optimistic assumptions it will use to provide an interesting point of departure in the sensitivity analyses to follow.

Postlaunch counterforce operations can take advantage of the cue from the missile launch detected by the Defense Support Program infrared satellites or by its follow-on, the Space-Based Infrared System–High (SBIRS-High). This will allow operators to immediately focus intelligence, surveillance, and reconnaissance and attack assets on a very limited area. Of course, this area might be deep in enemy territory, so the sensors, attack platforms, and command links will have to reach deep in a timely manner and/or have appropriate survivability enhancements (e.g., stealth). GMTI and SAR capabilities will need to have improved resolution and ATC/ATR capabilities for this mission, as well as for the prelaunch mission.

Active Defenses. The ABL concept employs a multimegawatt chemical laser with a lethal range of several hundred kilometers. An advantage of the this concept is that destruction of the booster during the boost phase will cause the warhead to fall short of its target—possibly impacting in the adversary's territory. In addition, engagement during the boost phase can defeat the use of decoy warheads and many other countermeasures, multiple warheads, and submunitions (a particular concern with chemical or biological threats and a real challenge to midcourse and terminal intercept defense systems, such as NTW and PAC-3).

The ABL concept consists of placing a laser-armed 747-400 in an airborne orbit at a safe standoff distance from enemy territory and above the cloud layers (which are typically below 10 to 15 km in altitude). During engagements, the ABL tracks the booster's plume signature, slews the laser in the direction of the rising TBM, and maintains the laser spot on the booster for several seconds, which heats

---

[14]Hyperspectral imaging (HSI) is promising technology in this regard. Of course, the HSI sensors must be on suitable platforms (e.g., UAVs), and integrated into an effective overall C⁴ISR architecture (including near–real time sensor tasking, data fusion, analysis, and dissemination).

the pressurized missile body, resulting in a catastrophic stress failure that explodes the booster.

The ABL fuel magazine is planned to be sufficient to negate about 10 to 20 missiles, depending on the dwell time required for booster destruction. Currently, a prototype 747-400 ABL system is planned to demonstrate track, dwell, and kill for a boosting missile body in 2003.

Responsive threats to an ABL defense include tactics and technology that exhaust the laser magazine or increase the attack time needed to destroy each booster. By far the most probable response is a saturation missile attack, or near-simultaneous multiple attacks. Figure 8.3 shows a notional ABL effectiveness against multiple launch threats for a laser magazine that could generate intercept attacks lasting a total of 100 seconds. The number of boosters killed is shown as a function of booster-to-booster launch delay and total laser attack times of 5 and 10 seconds. For typical Scud engagement parameters, the attack time for the first missile would be about 12 seconds; the

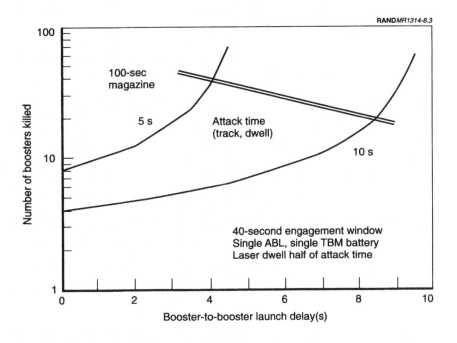

Figure 8.3—Notional ABL Effectiveness Against Saturation Attack

time would be 8 seconds for successive missiles, since course detection and turret slew times are not repeated. A 10-second attack time with a 100-second magazine can result in about 10 TBM kills.

The inherent antisatellite capability of the ABL requires a careful accounting of the ephemeris data of friendly and third-party satellites that are potentially susceptible to unintended exposure. This would be necessary to avoid accidental collateral damage and to avoid accusations that the United States conducted an antisatellite attack. This would also require keeping records of lasing directions and times to provide a format for assessing the impact of ABL operations.

Another defense concept is to employ an ABI, a missile that can be launched from a fighter or an unmanned tactical aircraft (UTA).[15] The objective would be to destroy the TBM during boost or early ascent (after burnout but before payload separation) with a kinetic kill interceptor, which has a standoff range of hundreds of kilometers. A potential advantage of this approach is that the ABI might be directed to impact and destroy the warhead; the ABL concept, in contrast, targets the pressurized booster. The intercept is, of course, slower than the speed-of-light laser, but the ABI's extended engagement window combined with the laser's requirement to lase the booster for several seconds may offset the difference.

The effectiveness of the ABI depends on achieving a high average intercept velocity of about 3,000 to 4,000 m/s to reach the target quickly for a kinetic energy impact. Offboard sensors are needed for initial detection, for acquisition, and for tracking the TBM before passing the tracking cues to the ABI platform. The performance of the ABI will depend on the type of acquisition system used, such as radar or infrared search and track sensors. A potential advantage of this concept is that the ABI platforms may be stealthier and smaller than the ABL and could thus be fielded in much larger numbers. ABI platforms will almost certainly be better able to operate within threat areas, where the large ABL aircraft could not fly safely, and therefore may be able to exploit threat localization to allow effective shorter-range intercepts. While the technology currently exists for develop-

---

[15]See Vaughan, Isaacson, and Kvitky (1996). Presently, such vehicles are called *unmanned combat aerial vehicles* (UCAVs).

ing an ABI, no effort to develop this defense concept is currently being funded, to the best of our knowledge.[16]

TMD Effectiveness Analyses

Counterforce limitations were clearly demonstrated during the Gulf War. From today's perspective, it seems clear that high levels of robust counterforce effectiveness will, with one significant exception, require advances in sensors and sensor processing that may not be achievable. Effective *prelaunch* TEL kills will be a matter of effective all-source (particularly human) intelligence data collection, data fusion, and luck. But the good news is that even a little prelaunch counterforce capability can be significant over the course of a war.

For example, assume that

1. IPB has localized the TBM threat to an area, A, of 2,500 km².
2. The search rate, R, is 1,000 km²/hr (against a stationary target).[17]
3. The TEL is stationary in the open for $T = 0.5$ hr.
4. P_d x P_{ID} is 0.8; R_{fa} is the false alarm rate (number per km²), e.g., 0.05.
5. If cued, the shooter can effectively engage the target ($P_k = 1.0$).
6. R_w is the rate targets can be attacked, e.g., 12 per hour.

Then, *prelaunch attrition* will be

$$P_A = \min\left[\frac{1,(R \times T)}{A}\right] \times \left(P_d \times P_{ID}\right) \times P_k\left[\frac{1, R_w}{\left(R \times R_{fa}\right)}\right]$$

$$= \left[\frac{1,000 \times 0.5}{2,500}\right] \times \left(0.8\right) \times 1.0 \times \left[\frac{12}{500 \times 0.04}\right]$$

$$= 0.1$$

---

[16]The sticky technologies tend to be homing and terminal guidance. While the state of the art in individual technologies is not inconsistent with a successful ABI (or any hit-to-kill system, for that matter), systems integration challenges are daunting.

[17]If search were for a *moving* TEL the negative exponential would apply: $P = 1 - \exp(-(R \times T) / A)$. (Note that for small values of X, $1 - \exp(-X) \approx X$.)

This example uses very optimistic assumptions (particularly for IPB and search rates), but each time a TEL comes out of its hide to attempt to launch a missile, there is still only one chance in ten that it will be found and killed before it launches. As will be discussed later, even this limited prelaunch kill capability may be significant over the course of an entire campaign.

The exception to our generally pessimistic projections regarding counterforce effectiveness is *postlaunch* TEL kills. This is an exception simply because the game starts with a very observable event—the launch of a missile—that is impossible to mask. Within several tens of seconds of a launch, U.S. forces in the theater will likely be able to determine fairly precisely where the launch occurred, and tactical controllers, if they act promptly, may be able to direct intelligence, surveillance, and reconnaissance assets and/or shooters to find and track and/or destroy the TEL.

If we expand counterforce attack operations to include postlaunch kill attempts, as well as prelaunch attacks, Figure 8.4 presents the resulting counterforce effectiveness for the example case outlined in the box on the left.[18] Most of the kills in this example occur as the TEL runs to its hide after missile launch.

Because of this, and because the only way U.S. forces currently have to identify moving vehicles is to get fairly close to them (e.g., from a few kilometers up to about 10 km) and use a forward-looking infrared system or to have the pilot visually identify the vehicle, these results are sensitive to the total number of moving vehicles (TELs and non-TELs) in the threat area. Figure 8.5 shows this sensitivity for two representative road densities as a function of the traffic on the roads. The concepts and technologies for moving target exploitation may thus be as critical as ATC/ATR.

Since counterforce operations are unlikely to be 100-percent effective (an understatement, given our Desert Storm experience), a prudent planner must expect that a substantial number of missiles will

---

[18]Note that "MTID" refers to the ability to identify moving targets. Currently MTID is limited to electro-optical and infrared systems. The JSTARS MTI radar can, however, do a rough moving target *classification* between tracked and wheeled vehicles by virtue of the double Doppler returns from the tracks.

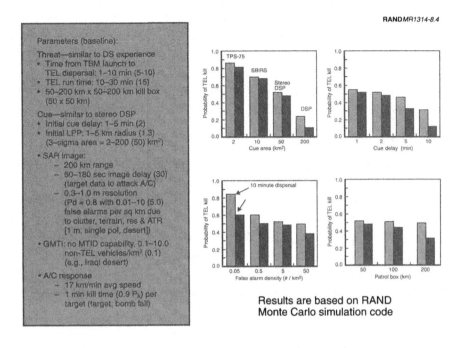

Figure 8.4—Sensitivity of Combined Prelaunch and
Postlaunch Attack to C^{4}ISR Parameters

be launched. These TBMs become a problem for active defenses. The Air Force's ABL should get the first shot during the boost phase. If enemy salvos saturate the ABL system, the "leakers" that get through must be intercepted by midcourse and terminal defenses, such as the Aegis-based NTW and Navy Area Defense (NAD) systems, THAAD, and Patriot PAC-3.

How effective might the integrated TMD elements be? This is difficult to predict, but the potential value of even rather modest (though optimistic) capabilities can be dramatic. Figure 8.6 provides a quantitative example of the potential combined effectiveness of Air Force counterforce (pre- and postlaunch) and ABL operations. The baseline case presumes a 0.1 prelaunch TEL/TBM kill probability (per launch attempt). The postlaunch TEL kill probability is assumed to be 0.3 (per launch). In addition, we have assumed that the ABL can

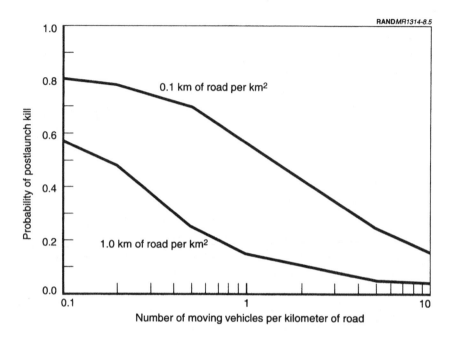

**Figure 8.5—Sensitivity of Postlaunch Attack
Operations to Moving Vehicle Clutter**

engage up to 10 missiles launched in a salvo, achieving a 0.9 kill probability on each.

The net effect of these capabilities in a hypothetical MTW that includes TBM attacks by the adversary is shown in Figure 8.6. This discussion assumes that the enemy has 50 TELs, each with several reload TBMs,[19] and is able to launch one missile a day from each TEL. It is also assumed that these launches would occur in a salvos in an attempt to saturate the defenses and increase the number that get through to strike their targets.

The solid curve in Figure 8.6 shows the number of TBMs launched each day, assuming a maximum effort by the enemy. The first day, only 45 were launched because 10 percent of the initial 50 TELs were

---

[19]The intelligence community thought that Saddam had less than two dozen TELs and over 800 Scuds.

killed by prelaunch counterforce operations as they prepared to launch. Of the 45 launches, the ABL could engage only ten but killed nine. Therefore, 45 − 9 = 36 TBMs made it through the layered Air Force TMD systems (and would, therefore, have to be engaged by the Army and Navy active defense systems not modeled explicitly here). So far, this does not seem like much of an Air Force missile defense accomplishment, but the situation improves as the war progresses. As irreplaceable TELs are killed, salvo rates go down, and fewer TBMs get through. In fact, by day five, the ABL can no longer be saturated. Within a week or two, expected leakage has decreased to very low levels (e.g., <0.01 per day). In effect, all the remaining reload TBMs have been grounded, ending the TBM threat to coalition forces.

Note that this primarily resulted from successful counterforce attacks, both prelaunch and postlaunch. How sensitive are these results to the assumed capabilities? Figure 8.7 provides quantitative answers. In these graphs all the baseline parameters and scenario assumptions are the same as in Figure 8.6, except for the parameter

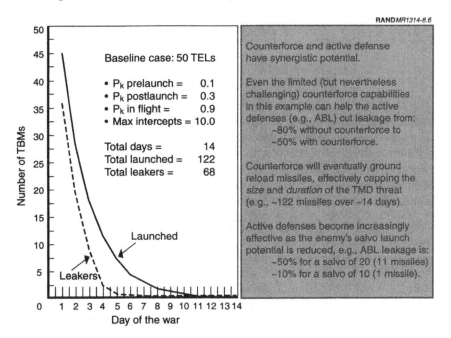

Figure 8.6—The Significant Synergistic Potential of Air Force TMD CONOPs

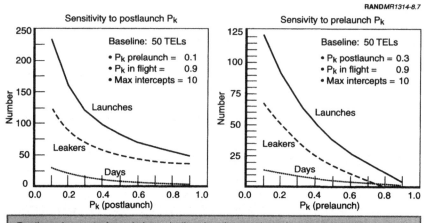

Postlaunch counterforce is less challenging than prelaunch counterforce due to (robust) potential to exploit TBM launch signatures.

If the threat includes multiple TBM reloads per TEL, feasible postlaunch counterforce capabilities may be much more significant than feasible prelaunch capabilities.

C⁴ISR approaches to enhance postlaunch counterforce are better understood than for prelaunch—progress can be achieved quickly at the margin rather than only after unpredictable technological breakthroughs (e.g., ATC/ATR, multispectral data fusion).

**Figure 8.7—The Effects of Small Improvements in
Counterforce Capabilities**

on the abscissa—either the prelaunch or postlaunch probability of kill (P_k). The measures of effectiveness (MOEs) here are (a) the total launches over the entire campaign, (b) the total leakers, and (c) the length of the campaign (time to achieve <0.01 expected leakers per day). Note that, for the baseline case, these values were (a) 122 successful launches, (b) 68 leakers, and (c) 14 days.

The good news from Figure 8.7 is that improvements *either* to postlaunch P_k *or* to prelaunch P_k can improve campaign MOEs dramatically. If prelaunch effectiveness proves too difficult (or too costly), the Air Force might be well advised to focus on postlaunch operations, which most likely can be more effective and robust at affordable investment levels. The costs, of course, have to be determined, but the *relative* difficulty between pre- and postlaunch missions suggests that a given level of postlaunch capability should be less costly than the same level of prelaunch capability.[20]

As noted above, postlaunch counterforce effectiveness is sensitive to the ability to locate the moving TEL in the presence of other moving ground clutter. Figure 8.8 shows the sensitivity of integrated defenses to moving vehicle clutter. Clearly, as with needed ATC/ATR advances in imagery exploitation, advances in moving target exploitation are uncertain, but their potential value suggests that aggressive research programs are warranted (as ideas permit).

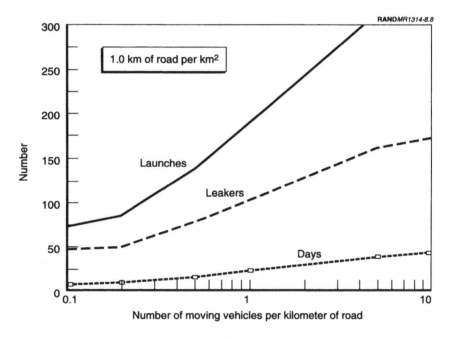

Figure 8.8—The Strong Effect of Moving Vehicle Clutter on Overall TMD Effectiveness

[20]Of course, the value of postlaunch kills depends to a great degree on the ratio of reload missiles to TELs (high for Scud systems). If TELS become one-shot "throw-aways," the main value of postlaunch kills is their negative effect on the morale and willingness to fight of missile crews. These postlaunch operations may also help us get a better idea of TEL operations, which may improve our prelaunch kill effectiveness.

Figure 8.9 looks at sensitivities to ABL performance parameters. In this case, the MOE is total TBM leakers over the campaign. The two curves are for the baseline salvo handling potential (10) and an enhanced capability (20). Note that, for either of these curves, the sensitivity to P_k (in flight) is not very dramatic within reasonable ranges.[21] For example, the top curve illustrates that the difference in total leakage between a P_k of 0.9 and 0.7 is 68 to 78, or just a 15-percent increase in leakage for a 22-percent reduction in P_k.

Finally, note that the notional layered Air Force contribution to the composite U.S. TMD system has not been able to enforce the extremely low leakage levels that might be a goal in a WMD-shadowed war. But it has greatly reduced the threats that the mid-course and terminal defenses must counter, significantly increasing the potential for near-zero leakage from the multilayered, system-of-systems TMD architecture.

As with the ABL, the effectiveness of other active defenses is enhanced as the adversary's salvo launch capability is reduced, such as by counterforce actions. The ABL system can provide early launch warning and can estimate launch and target impact points to support other TMD elements. This would include handing over track files for the missiles that get through and damaged booster and warhead debris maps to terminal defenses, such as THAAD, PAC-3, and NTW/NAD.

The ABI concept could complement the ABL defense approach by helping to fill gaps in the ABL coverage and could therefore increase the overall boost- and ascent-phase intercept potential, increasing salvo handling potential. Multiphenomenology active defenses would also improve robustness to countermeasures. The C[4]ISR considerations for the ABI are similar to those for the ABL.

This layered defense theme is displayed in the trajectory chart in Figure 8.10. The box for each period briefly describes (a) the CONOPs, (b) the platforms and weapons, and (c) the C[4]ISR support.

---

[21]"Reasonable" should not be construed in this context as "achievable." These effectiveness numbers are "reasonable" design goals. Actual performance will depend on technological, engineering, and operational ingenuity, as well as on the creativity and effectiveness of responsive enemy countermeasures.

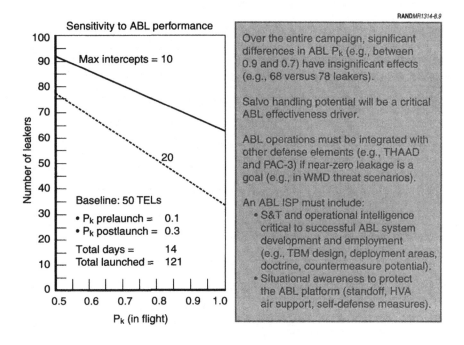

Figure 8.9—The Sensitivity of ABL Performance Salvo Handling and Lethality

Current TMD capabilities are very limited. In the near term, the principal thrust will be the addition of Army and Navy surface-based active defenses, with appropriate C⁴ISR support to integrate the systems into a multilayer defense. In the far term, these defenses may be augmented with Air Force boost-phase intercept defenses and improved counterforce capabilities (as technology permits), greatly increasing the effectiveness and robustness of the overall TMD architecture.

Table 8.3 summarizes some first principles for TMD. Prelaunch counterforce is very challenging. It will require new enabling C⁴ISR capabilities, such as deep dwell GMTI/SAR systems with enhanced ATC/ATR processors on survivable platforms with search rates and resolution much greater than currently achievable (for "reasonable" assumptions about IBP effectiveness), in combination with CONOPs

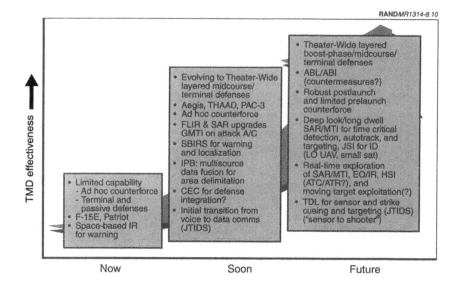

**Figure 8.10—Air Force Capabilities for Shaping the
Potential TMD Trajectory**

that allow controllers to bring attacks to bear while the TEL is in the
open. However, commitments to specific prelaunch counterforce
CONOPs and system elements may not be appropriate until required
advances in the state of the art (e.g., ATR) have been demonstrated
under realistic operational conditions.

The same caution, of course, applies to postlaunch counterforce
CONOPs, but the near-term potential is greater there. The Air Force
does not yet have well-developed postlaunch counterforce operations, but has fielded or is fielding some critical C4ISR investments
(e.g., Attack and Launch Early Reporting to Theater, SBIRS-High). In
addition, similar, but more modest, GMTI/SAR improvements than
for prelaunch counterforce above will be required for effective and
robust postlaunch attack operations. A significant unknown at this
point is the moving target clutter problem, which may necessitate
advances in moving target exploitation that are difficult to forecast.

The ABL development program is well under way. It is essential to
develop C4ISR architectures to support the ABL, recognizing that it

Table 8.3

New C⁴ISR Capabilities Needed for TMD

	Capabilities	What the analysis has shown:
Counterforce against TELs	Find and kill deployed TELs prelaunch	This is very challenging.
		"Hiders and finders" favors hiders.
		Key C⁴ISR issues are deep dwell/increased search rate at <1-m resolution and orders of magnitude advances in ATC/ATR Pd/Pfa and near-real-time shooter redirection (Joint Tactical Information Distribution System [JTIDS]. SAR/GMTI/HSI sensors on low-observability UAVs or satellites? Unattended ground sensors [UGSs]?).
	Find and kill deployed TELS postlaunch	This is less challenging.
		Enemy cannot hide launch signature.
		Key C⁴ISR issues are time-urgent exploitation of SBIRS cues using deep dwell SAR/GMTI sensors to image, identify, and track TELs, redirect sensors and shooters (JTIDS+, SAR/GMTI/HSI sensors on observability UAVs or satellites? UGS?).
Active defense against TBMs	Kill missiles during powered flight with a high-power laser	The ABL will require significant C⁴ISR support.
		Intelligence (science and technology on TBM design and signatures, countermeasures?).

Table 8.3—Continued

Capabilities	What the analysis has shown:
Kill missiles/warheads(s) in boost/ascent with a kinetic kill vehicle	Operations support (e.g., IPB and situational awareness interface (Cooperative Engagement Capability–like) with other defense elements (THAAD, PAC-3 and Aegis).
	ABIs may complement ABL.
	C^{4}ISR issues probably less stressful than for ABL (CONOPs and systems to be determined).

will be part of a layered TMD architecture. The ABL concept can engage a limited number of TBM boosters *autonomously* but cannot operate effectively in *isolation*. To be effective, the ABL needs to be integrated with the other TMD elements—particularly if the overall multitier TMD system seeks to achieve near-zero leakage with high confidence. The ABL system requires significant C⁴ISR support in science and technology areas and in IPB. For science and technology, it is important to know the technical characteristics of the threat missiles, such as burn time, booster material and thickness, plume characterization (relative to the booster body), and possible countermeasures. IPB includes intelligence on potential locations of TBM launchers and air defense threats to the ABL to allow establishment of a safe but effective standoff distance and orbit parameters. As with other weapon systems under development, these issues must be addressed early for the ABL in an evolving intelligence support plan.

ABI concepts are less well developed than those for the ABL. If ABI systems are pursued, it would seem that the required C⁴ISR support will be similar to that for ABL, but differences may emerge due to (1) different lethality mechanisms (kinetic kill of the missile payload versus laser kill of the booster), (2) different defense platform characteristics and operations,[22] and (3) different IPB and situational awareness needs (associated with intercept range and overflight differences).

These TMD programs have exciting potential, but this potential can only be achieved through focused developments in enabling C⁴ISR systems.

POTENTIAL AIR FORCE NMD INITIATIVES

Deploying NMD, also referred to as antiballistic missile (ABM) defense, to protect the United States from long-range ballistic missile attack[23] seems, on the surface, to be a good idea. In fact, many

---

[22]ABI may be based on a relatively small and survivable fighter aircraft or UCAV rather than on the large ABL aircraft. but ABI will likely depend on offboard battle management, cueing, and intercept support. ABI may overfly hostile territory (depending on the size of the country and the location of the TBM launchers), but ABL will stand off.

[23]These missiles may be launched intentionally, accidentally, or without leadership authorization by a rogue commander. Of these possibilities, this chapter focuses on the first.

Americans evidently believe that this capability already exists. When told the actual situation, most favor developing and fielding an NMD system. Unfortunately, however, as this section tries to show, the NMD decision is much more complex than "the conventional wisdom," as expressed by polls, would seem to suggest.

It is, of course, a basic purpose of governments to protect their people from foreign threats. Countries, like individuals, have a "moral imperative" for self-defense. Many see the proliferation of WMD and long-range delivery systems, particularly ballistic missiles, as a threat to our national security, if not now, then surely in the future. So, in turn, many see NMD as a moral imperative.

But an NMD system is only one of several approaches to countering WMD threats to the United States. Not all approaches are compatible or certain to succeed. All compete for resources. This section outlines some of the NMD and related TMD issues, culminating with some impressions on the nature of possible solutions to these problems—solutions that will require the United States to recognize that NMD and TMD issues cannot successfully be resolved "on their own" but rather must be considered in a broader national security context, which will require debate on matters that are much more troublesome and morally uncertain than the self-defense imperative.

Background: The Cold War

One aspect of the moral dilemma was clear in the Cold War years. As delivery systems advanced from bombers to cruise missiles and then to ballistic missiles with increasing ranges, accuracies, and explosive yields, the offensive nuclear threat reached the point that the United States and the Soviet Union could "destroy" each other several times over.

The destructive potential of thousands of megatons of nuclear yield is difficult to grasp—there is obviously no historical precedent. But our limited vision was clear enough: Unconstrained nuclear war between the United States and the Soviet Union would surely be a catastrophe of unprecedented scale. While much intellectual capital was expended debating concepts of nuclear deterrence and the nuclear balance, survival, warfighting, and even "victory," it became clear that the most fundamental common goal had to be avoiding war. Given the ideological differences and inherent tension between the protagonists, war, in particular nuclear war, was avoided through deterrence, which, when stripped bare of the "details," came down to a very understandable concept: Mutual deterrence came from both sides having the assured capability to inflict massive, "unacceptable" damage on the other, even after absorbing a would-be disarming first strike. Most people considered this standoff, with both sides vulnerable to "assured destruction," to be morally justified because its goal was to avoid war—if it failed, there would be hell to pay.

The primary strategic tension in the Cold War was between defensive and offensive concepts.[24] The offense won because the defenses conceivable at the time could not fundamentally change the vulnerability of both sides to devastating retaliation under any circumstance of war initiation. This was so despite the fervent desire President Ronald Reagan articulated in 1983 to make nuclear-armed ballistic missiles "impotent and obsolete." To see this, however, defense planners had to take the long view. That is, the development and deployment of NMD systems was not something that could, in effect, eliminate the nuclear balance of terror between two determined and capable adversaries. It takes time to develop and field NMD systems. The adversary will react to any change in a way that avoids putting himself in a position of inferiority.

Thus, the quest to develop and deploy a highly effective NMD should be viewed as part of a strategic chess game, which suggests that strategists and planners should look ahead for as many moves as they can before acting. When leaders on both sides did this during the Cold War, they saw dangerous instabilities that might emerge and the potential for an offensive *and* defensive arms race, both of which seemed to increase the risk that deterrence might fail. Moreover, the offense-defense balance seemed to continue clearly to favor the offense, so heavy investments in defenses would not only consume vast resources but also might prove counterproductive in the

---

[24]Defenses included civil defense measures (e.g., warning, evacuation plans, and fallout shelters), air defenses, and eventually limited BMD. Offensive capabilities were also referred to as "damage limiting"—a kind of "defense" that reflects the old adage that the best defense is a good offense (unfortunately, to work best, the offense has to preempt, which is not comforting from the perspective of stability in a crisis). Note that these various elements of a defensive strategy (passive, active, offensive, and BM/C^{4}ISR) are the so-called "four pillars" that the Strategic Defense Initiative Office (SDIO), Ballistic Missile Defense Organization, has discussed in their architectural efforts, which have, so far, emphasized the active defense pillar.

long run. Since deterrence was the U.S. strategic nuclear corner-stone, anything that could conceivably weaken it (no matter how well intentioned or "moral," e.g., "defensive") was rejected. So, while technologists continued to study strategic defenses in the hopes that "advanced technology" could change the conclusions about the inadequacy of NMD relative to offensive threats, the policy focus shifted toward arms control solutions with the conclusion of the SALT I agreement and the ABM Treaty of 1972. Since then, the SALT II, START I, and START II agreements have clearly reduced the risk of superpower nuclear war. Today, no one questions the cliché that "the Cold War is over." But that does not mean that the lessons the competition between offensive and defensive capabilities taught have been overturned.

Post–Cold War Issues

Setting aside, for now, such residual Cold War issues as "minimum" deterrence, the ABM Treaty, and the risks of accidental or unautho-rized launch, as well as the lessons learned, let us turn to the con-temporary rationale for the pursuit of a limited NMD capability: proliferation of WMD and intercontinental ballistic missiles (ICBMs) to deliver them.

Again, there seems to be a moral imperative connecting NMD to TMD: How can the United States justify the expense of protecting its troops, friends, and allies with TMD while not providing protection to its own homeland with NMD that can cope with a relatively lim-ited long-range missile attack? The NMD arguments in this context, pro and con, tend to be driven by assessments of the *urgency of* (rather than the *rationale for*—which is seldom debated) deploying a limited NMD, which, in turn, is directly related to assessments of "the threat."[25] The threat, of course, is the seemingly inevitable acquisition of the appropriate hardware systems (e.g., WMD mated to ICBMs) by potentially hostile so-called "rogue states" or "states of

---

[25]The "threat" discussions tend to focus on hardware rather than how and why that hardware might be used. This is natural because hardware is tangible; intent is not. Nevertheless, while some elements of intent can be inferred from the hardware (e.g., missile range determines potential targets), the "threat" descriptions and uncertain-ties here should be treated much less simplistically—the missiles, per se, are not "the threat."

concern" (e.g., Iran and North Korea).[26] The pro-NMD supporters recommend getting started now, while the people who argue against NMD deployments say that, since the threat is still a ways off, the United States should pursue R&D efforts to assure that it will be able to deploy the most cost-effective limited NMD possible when the time comes.

One side argues that NMD is most certainly not "an ABM system" in the old Cold War sense, but the opposition wonders whether effective NMD deployments against proliferation threats can be consistent with the requirements of nuclear deterrence, given the deep reductions both sides are debating in informal "START III" discussions.

What about the ABM Treaty? Can it be modified to accommodate layered NMD systems that protect all 50 states? Can treaty-compliant NMD concepts be responsive to emerging threats? Must the United States withdraw from the treaty, or can the treaty be modified to permit the deployment of a limited NMD system with nationwide coverage that can address these emerging security needs? Many NMD proponents see the old Cold War problems and new proliferation problems as separable and as having solutions that are not necessarily incompatible.[27] The opponents of near-term NMD deployments may agree about this but, in any event, do not see the urgency. Certainly, some who argue for a "go slow" approach to NMD see the potential for conflict between historical lessons learned and new concerns about proliferation. However, as time has passed since the breakup of the Soviet Union, the Cold War issues have seemed less relevant and the new emerging threats have become more pressing for people on all sides of the debate. The United States must do *something* about proliferation—why not NMD?

It may be that the essence of the current NMD rationale is, in actuality, indistinguishable from the TMD rationale. The two will, at some

---

[26]Note that this discussion does not include the "established" long-range missile threats from Russia and China.

[27]The United State's French allies have strong feelings about these matters. They see even limited NMD as potentially undermining their "minimum deterrent" nuclear forces. This should be a reminder that the NMD and TMD issues are distinctly multipolar and are hence even more complex than the ABM issues were in the bipolar Cold War context.

point in the development of these "regional threats," become logically inseparable as the reach of the rogue states becomes intercontinental. This simple point needs no further development (there is still the timing issue noted above, however).

It is paradoxical but conceivable that the old Cold War instability argument that was used *against* substantial U.S. and USSR missile defenses will be the principal argument *for* defenses in the future. The Cold War concern was that ABM defenses were destabilizing (bad) because, while they could not offer substantial protection against a well-conceived and executed massive first strike, they might be very effective against a much smaller, more ragged retaliatory strike. Therefore, in a crisis, the argument went, ABM defenses tended to favor the superpower that struck first—not a desirable situation from a crisis-stability perspective.

It could be argued that current U.S. moves toward the deployment of a limited NMD to counter proliferation of long-range missiles armed with WMD is turning this logic on its head. At least for the foreseeable future, the United States is not threatened with annihilation by a WMD-armed rogue proliferant. In a future regional crisis in which vital U.S. interests are at stake, the United States need not fear a preemptive strike by a proliferant armed with a handful of ICBMs.[28] It is much more likely that the rogue state will try to use its missileborne WMD to *deter us* from intervention in "its region" by threatening *retaliatory* strikes against us. Of course, for the same reason that it seems incredible that the rogue state would strike first against us, it also seems incredible that it would attempt to retaliate against us— its fate would surely be sealed by such a move, while the United States would survive (albeit somewhat bloodied). But unlikely as it seems in absolute terms, this sort of retaliation is *relatively more likely*, either as a last, vengeful act before being destroyed in a war against a U.S.-led coalition, or as an attempt, by marching up the

---

[28]The underlying assumption is that the United States will have the physical means and the national will to respond "appropriately" and that the potential attacker appreciates this. There is no doubt, as history demonstrates, that, if provoked, the people of the United States can be ruthless in response. Nevertheless, this area deserves much more careful thought and preparation so that the country can live with its actions (the moral issues) and so that the actions do not create a legacy of hatred, revenge, and alienation. The United States might be able to destroy a regional aggressor but may not be able to control the long-term repercussions.

escalation ladder, to stop that coalition short of total victory (even though its first step might be its last). If the rogue state's leadership were desperate enough or mad enough, it could strike out. Unfortunately, there is a good chance that the United States and its allies would be forced to put the rogue state into this position.

The logic behind this assertion runs as follows. If the United States backs down in a crisis with a rogue proliferant, the proliferant wins. If the rogue wins, proliferation will likely increase. If proliferation increases, risks of WMD use will increase. Clearly, the United States and other status quo powers would like to avoid that. So, if the vital interests of the United States are even *perceived* to be at stake, it cannot back down—it must prevail.

The United States could prevail in several ways. First, it could avoid the problem if its nonproliferation policies succeed, but that seems highly unlikely. Second, the United States could deny potential enemies a WMD capability through "preventive" attacks,[29] but this seems even more unlikely. Finally, the most likely scenario is that a would-be regional aggressor would provoke the United States, which would in turn apply decisive force to defeat him on the battlefield. But if the United States pushes such countries too far into a corner, they may strike back. Limited NMD is designed to be our hedge against desperation attacks.

NMD Systems Implications

If this perspective rings true, the kinds of limited NMD systems the United States might need to defend against attacks on the homeland by regional aggressors may resemble what the United States needs for TMD—in fact, some of the same systems *may* be able to do double duty.[30] The NMD system(s) need not be deployed and on high

---

[29]"Preventive" attacks include such actions as the air strikes the Israelis carried out against the Iraqi Osirak nuclear reactor to prevent them from using it to produce nuclear weapons. The United States is most unlikely to take serious *overt* preventive military actions, although reputable senior people have discussed such actions (most recently with respect to the North Korean nuclear program). *Covert* actions may be another matter, however.

[30]The defense architecture to protect both theater and continental U.S. targets from attack could consist of conventional ground or ship-based radars and interceptors

alert every day, because a surprise attack is not credible. The NMD system also need not be sized to defeat "large" missile threats.[31] The issue of balance of defenses is less troublesome in these scenarios than it was during the Cold War and than it would be if the United States were concerned about surprise or "terrorist" attacks.[32] Mobile missile defense systems (which are not permitted under the ABM Treaty) would be desirable (assuming different threats had distinct coverage limits). Finally, the edge the United States has in technology and our relative wealth make an offensive-defensive arms race very unlikely—the United States ought to be able to keep up easily with any conceivable regional adversary. Of course, this may change at some point, which should be reason to think more seriously about "preventive attacks" than U.S. planners evidently have to date.

Returning to nuclear deterrence issues, missile defenses, both TMD and NMD, can conceivably be different enough in character, operation, size, and performance from robust NMD systems to allow superpowers to continue reducing offensive arms, whether negotiated or accomplished through parallel unilateral cuts. This is the problem that the United States should be pursuing in any ABM Treaty discussions with the Russians, instead of drawing technical distinctions between TMD and NMD, which are, at best, artificial and, at worst, seriously constraining. The most worrisome issue is China, since the Chinese are likely to attempt to increase their regional influence and may even have superpower pretensions. Their

---

deployed in the theater and in or near the continental United States to provide layered engagement opportunities (e.g., midcourse, high-end, and terminal). Alternatively, TMD and/or NMD capabilities could, to use an example from colleague Glenn Kent, be provided by advanced boost-phase intercept systems deployed as a barrier over the threat region.

[31]This could be one way out of the strategic deterrence conflict, but the multipolar issues and concerns about verification and inherent potential for abuse (e.g., breakout) make success far from certain.

[32]The deficiencies in current U.S. defenses against cruise missiles, manned bombers, and even "suitcase bombs" relative to proposed NMD capabilities should be less an issue for the most worrisome retaliation scenarios discussed above. If ballistic missile threats exist and if the United States can react to them, it probably should, even though a vengeful enemy might also be able to plant nuclear "suitcase bombs" (or other, more likely, terror weapons) in U.S. cities. The underlying issue, of course, is affordability and balance in defense expenditures. There is no avoiding the need for difficult decisions that, in the end, must be based on informed judgments rather than precise calculations.

current strategic nuclear arsenal is quite small (two orders of magnitude below the Russian and U.S. arsenals, even at START II levels). That will make the distinction between theater and strategic ballistic missile defense more difficult to sell to the Chinese, possibly fueling an arms race like that of the Cold War. The Bush administration seems to be trying to link NMD with deep cuts in U.S. offensive nuclear forces. For China, the appeal of further force reductions as a NMD sweetener would seem to be minimal. Let us look at a simple example "calculus." To illustrate this point, consider the following.

In the current situation, the United States has several thousand nuclear weapons that threaten China, but no NMD. China has a couple tens of weapons that could reach (some part of) the United States. Here, the bottom line is that neither side seems terribly troubled by this "correlation of strategic nuclear forces." While China would probably not mind if the numbers were a bit more balanced, it does not seem to be working very hard to achieve this sort of "essential equivalence," so its leaders must not be too concerned about the status quo. In this scenario, it is assumed that China considers 20 or so nuclear weapons enough to hold the United States at risk, for the sake of deterrence.

But what if in, say, 20 years, the United States has made significant reductions (e.g., much less than half of today's nuclear force) and has deployed a "thin" (e.g., 100 interceptor) NMD system. While China would certainly not mind the reductions in U.S. offensive forces, the reductions would probably not affect China's strategic planning greatly one way or another. After all, U.S. offensive forces would still be much larger than those of China, without a dramatic Chinese buildup. But in this scenario, maintaining the same level of deterrence in the face of U.S. NMD as today would require the Chinese to have about five times as many long-range missiles. That would be a reasonably significant increase if China really believes that 20 would otherwise be sufficient. This means that, with a limited U.S. NMD and if China is self-consistent, it will instead need to be able to throw about 120 weapons at the United States. The first 100 would either hit the targets the Chinese care about or would exhaust the defenses. The next 20 would finish the job.

So, China ought to be expected to complain about NMD. Even unilateral reductions of U.S. offensive arms (as long as it is not down to hundreds) would not matter to the Chinese. They will have to increase their offensive capabilities significantly—but not, it seems, enough for the United States to really be concerned.

But what if China has bigger plans, such as a vision of reaching some sort of strategic nuclear parity with the United States within, say, 20 to 30 years.[33] In this hypothetical situation, China ought to welcome U.S. NMD and force reduction initiatives. In an "arms race," defensive investments are a loser. A modest 100-interceptor system could surely be overtaken by threat technological advances and numbers in 20 years. So, in the Chinese view, such an NMD system would, at worst, sponge up a few percent of their much larger parity force. At best, it would allow them to catch up with the United States offensively years earlier than they might have otherwise.

NMD would not seem, therefore, to be much of a problem as far as China is concerned. If it builds up its forces in response, to restore the previous status quo, the net result is a wash. NMD, per se, is not likely to spawn an "arms race" between the United States and China (other things might, but it would seem the United States would have to be fairly inept to make that happen). Does the United States really care whether China can throw 20 or 120 weapons at it (particularly if, in the latter case, the United States has a chance of shooting some of them down)? But if an arms race is inevitable between the United States and China, NMD probably will not matter much one way or the other. And the United States should not worry too much about the "growth potential" of near-term deployments because NMD (for all the reasons found during the Cold War) will rapidly become unattractive (at least with respect to China) once the arms racers leave the starting line.

Protection against accidental or unauthorized launches of strategic missiles from Russia or China is impossible to address with any precision. The kinds of NMD the United States may want to support its regional power projection capabilities could, however, fill this niche at the price of adopting a day-to-day alert posture for the limited NMD system. This would, however, require additional costly capabilities, particularly sensors, as well as BM/C⁴ISR that would provide the capability to defend around the clock, day in and day out. If the

---

[33]Although, from a nuclear perspective, just why they might want to do this is hard to imagine, considering the small benefits and high costs and risks relative to just maintaining something like the current status quo.

additional costs could be kept low enough, the U.S. national leadership may decide that NMD is a prudent investment.

SUMMARY

The potential for the use of ballistic missile–delivered WMD warheads is likely to shadow any future major regional conflict in which the United States is involved. WMD threat scenarios can be conceptualized (with varying degrees of credibility), as this chapter did earlier. But these WMD threat systems are largely untested in battle, and exactly how and to what extent they will change the future of warfare is highly uncertain. As shown earlier, some of the WMD lessons learned during the Cold War no doubt still apply, but the WMD calculus of the proliferant rogue states could also be quite different from the bipolar Cold War calculus in significant ways—particularly with respect to the potential value of defenses. This chapter has argued that

- The primary purpose of U.S. TMD and NMD systems is not only to deter *enemies* from striking the United States or its allies with WMD but also to *prevent the United States from being deterred* in a future regional crisis fought under a WMD shadow.

- To achieve this purpose, acceptable counterproliferation strategies could require both TMD and, eventually, limited NMD capabilities to support U.S. *offensive* campaign strategies, to minimize the damage from possible retaliatory strikes against the United States that cannot be suppressed completely by offensive actions.[34]

- The distinctions between TMD and NMD systems may be artificial. Some of the same hardware systems might even satisfy both missions; examples include space-based launch detection, tracking, and kill, as well as use of an ABL or an ABI for boost-phase interception.

---

[34]The key word here is *eventually*. We would argue that this eventuality should be delayed for as long as we can comfortably delay it because of the technological, engineering, and operational challenges involved in making an NMD system work and because of the continuing long-term commitment (and expense) implied by a deployment decision. Zealots should reflect on lessons learned from Safeguard, which was operational for a *very short time* (accounts vary from days to weeks) before it was decommissioned.

- TMD and/or NMD systems can be deployed on strategic warning of a theater war with a state possessing WMD-equipped TBMs or ICBMs. These defense systems need not be maintained on high alert every day, since surprise attacks or even preemptive attacks by proliferants on the United States are highly unlikely.

- Effective TMD and NMD systems (in limited numbers) need not conflict with the *intent* of the ABM Treaty or with continuing progress in superpower strategic arms reductions, but there is work to be done here. TMD and NMD are now multipolar issues, and the effects on China, in particular, may be significant.

If these views are accepted (and many are admittedly quite contentious), they suggest that the effects on the Air Force and its role in addressing these threats could range from relatively minor to quite significant. Critical issues for the Air Force to consider include

- developing CONOPs and systems initiatives to hold time-critical mobile targets, such as TBM TELs, at risk[35]

- further developing and integrating the ABL (and, possibly, an ABI) into the TMD system-of-systems architecture

- determining the Air Force's role in NMD to support or augment the roles played by the Army and Navy.[36]

Over the longer term, if missileborne WMD threats evolve beyond the relatively manageable levels the discussions in this chapter imply, Air Force responses may have to go beyond defensive overlays to current power projection concepts and systems. Sanctuary basing, long-range force application, migration of surveillance systems from aircraft to space, and even space-based weapons (e.g., space-based laser) may be required to counter more-aggressive WMD threats. The implications for doctrine and force structure could be substantial.

---

[35]These will involve the sorts of BM/C^{4}ISR capabilities and weapon developments described in the TMD effectiveness analyses above.

[36]At a minimum, the Air Force satelliteborne launch detection and tracking systems and terrestrial radars will provide warning and cueing. At the other extreme, the Air Force may be able to offer a backup or growth option (e.g., Minuteman—possibly employing a nuclear intercept) for the Ballistic Missile Development Office's kinetic kill systems, if technical or operational problems emerge (e.g., threat object discrimination).

BIBLIOGRAPHY

Christopher, George W., Theodore J. Cieslak, Julie A. Pavlin, and Edward M. Eitzen, Jr., "Biological Warfare: A Historical Perspective," *Journal of the American Medical Association,* Vol. 278, No. 5, August 6, 1997, pp. 412–417.

Hura, Myron, et al., *Investment Guidelines for Information Operations: Focus on Intelligence, Surveillance, and Reconnaissance,* Santa Monica, Calif.: RAND, 1998. Government publication; not releasable to the general public.

Vaughan, David R., Jeffrey A. Isaacson, and Joel S. Kvitky, *Airborne Intercept: Boost- and Ascent-Phase Options and Issues,* Santa Monica, Calif.: RAND, MR-772-AF, 1996.

Part III

SUPPORTING FUTURE FORCES

Chapter Nine

PROVIDING ADEQUATE ACCESS FOR EXPEDITIONARY AEROSPACE FORCES

David Shlapak

OVERTURE[1]

> Sadat's decision to ask for a cease-fire (in the Yom Kippur War) was greatly influenced by the effectiveness of the American resupply operation.
>
> —*Chaim Herzog*[2]

> For generations to come, all will be told of the miracle of the immense planes from the United States bringing in the material that meant life for our people.
>
> —*Golda Meier*[3]

Saturday, October 6, 1973. It is Yom Kippur, the most solemn holiday of the Jewish calendar, and Israel is at prayer. Her leaders, however, have other pressing matters on their hands. That morning, fresh intelligence confirmed that Syria and Egypt were about to attack, embroiling Israel in the fourth full-scale war of its brief 25-year exis-

---

[1]What follows is in large part derived from and based upon work reported on in Shlapak et al. (forthcoming). I am indebted to my colleagues in that endeavor, John Stillion, Olga Oliker and Tanya Charlick-Paley, for much that the reader finds here. A number of other recent RAND studies have addressed the access and basing question from a regional perspective. See, for example, O'Malley (2001) and Khalilzad et al. (2001). For a more conceptual overview, see Killingsworth et al. (2000).

[2]From Herzog (1982), p. 377.

[3]Quoted in Boyne (1998), p. 59.

tence. Denied authorization to launch a preemptive air attack on Syrian forces—which pose the greater threat, since their jump-off positions are barely 70 miles from Tel Aviv—Israeli Defence Forces (IDF) General David Elazar orders a partial call-up of reserves and waits for the blow to fall. He does not have to wait long.

At 1400 hours local time, 2,000 Egyptian artillery pieces opened fire on IDF positions along the so-called "Bar-Lev" line of fortifications in the Sinai Desert. More than 10,000 shells fell on the defenders in the first minute as 240 Egyptian aircraft screamed overhead to assault airfields, surface-to-air missile sites, and other targets in the Israeli rear area. Meanwhile, on the Golan Heights, Syrian armored forces, which in some key sectors outnumbered Israeli tanks ten to one, moved forward under the cover of a 50-minute artillery barrage (Herzog, 1982, p. 241). The Yom Kippur War had begun, and Israel was already fighting for her life.

The U.S. airlift to Israel in October 1973—dubbed Operation Nickel Grass—was a historic event. Not only did it provide the IDF with badly needed weapons and supplies at a critical juncture, but it had a coercive impact as well, convincing Egyptian President Anwar Sadat, the mastermind of the Egyptian-Syrian offensive, that the war could not be won. Without firing a shot, the United States had established its preeminent role in the politics and strategy of the Middle East, inflicted a bitter if symbolic defeat on its Soviet rival, and demonstrated its ability—even while mired in the early stages of its post-Vietnam "funk"—to bring power rapidly to bear wherever and whenever it deemed necessary.

But Nickel Grass almost did not happen. As the U.S. Air Force's Military Airlift Command (MAC)[4] prepared to execute President Richard Nixon's order to "send everything that can fly" to support Israel, it became clear that this would be no routine operation. One after another, key U.S. allies refused diplomatic clearance for MAC's C-5s and C-141s to use their bases or overfly their territory. Fearful of Arab reprisals, Great Britain, Greece, Italy, Spain, and Turkey all denied access to the United States. Only Portugal agreed to cooperate; after

---

[4]The predecessor to today's Air Mobility Command.

much persuasion, Lisbon allowed U.S. airlifters to use Lajes Field in the Azores as a stopover on their way to and from Israel.[5]

Without this "discreet" support from Lisbon, Nickel Grass would have been crippled before it got off the ground. A nonstop flight from the U.S. east coast to Israel would have been impossible for MAC's primary airlifter, the Lockheed C-141A, which was not equipped for aerial refueling. Structural problems with its wings meant that the huge Lockheed C-5A also could not be refueled in the air, although it was theoretically capable. And, while the *Galaxy* could have flown from the United States to Tel Aviv without stopping for fuel, its payload would have been reduced to 33 tons, less than half what it carried on average during Nickel Grass.[6] Without access to bases on foreign soil, the emergency airlift so vital to Israel would have been, for all practical purposes, an impossible undertaking.

ACCESS YESTERDAY, TODAY, AND TOMORROW

A Troublesome Track Record

The airlift to Israel is one striking example of the importance of overseas access—basing and overflight—for U.S. power projection, but it is by no means unique.[7] In 1986, lack of support from NATO allies again complicated a U.S. military operation. In April of that year, President Ronald Reagan ordered air strikes on a number of targets in Libya in retaliation for alleged terrorist activities. Operation El Dorado Canyon was complex enough to begin with, involving as it did F-111 and EF-111 aircraft flying from Great Britain and U.S. Navy jets operating from two carriers in the Mediterranean Sea. Difficulties multiplied when both Spain and France refused to allow the F-111s to fly over their territories during the mission. This substantially lengthened the flying times for the F-111s, which had to start the trip to their targets in the southeast by flying southwest over

---

[5]A single C-5 sortie was allowed to move spare artillery tubes from Germany to Lajes (Comptroller General, 1975, pp. 10, 47).

[6]John Lund described the refueling restrictions in 1990 in an unpublished RAND manuscript, "The Airlift to Israel Revisited." Range and payload data are from Comptroller General (1975), pp. 10, 30.

[7]Despite the technical inaccuracy of the practice, this chapter will use "overseas" as a synonym for "outside the territory of the 50 United States."

international waters opposite the French and Spanish coastlines, then slipping through Gibraltar and across the Mediterranean. Having followed this tortuous course on their inbound journey, the crews were expected to avoid strong Libyan air defenses, deliver their weapons (subject to extremely stringent rules of engagement), then turn around and make their way home back the way they came.

The prolonged trip necessarily took a toll on men and machines. By the time the F-111s made it to Libya, numerous aircraft had difficulties with their sensitive targeting systems that either prevented them from dropping the bombs they had carried such a distance or resulted in the weapons being delivered well off target. Tired aircrews also made errors that resulted in improperly aimed ordnance. While the attack can at least be arguably assessed as a success on a strategic level, the strikes achieved significantly less tactically than planners had hoped. At least some of the blame for the disappointing performance must be assigned to the excruciating mission profile, which stressed aircrew and, especially, aircraft, well past the bounds of their normal operations.

More recently, U.S. attempts to punish Saddam Hussein's regime in Iraq have frequently been hampered by a lack of cooperation from friends and allies:

- September 1996: Baghdad perpetrated a gross violation of the terms of the Gulf War cease-fire, launching a large-scale ground attack against Kurds in northern Iraq. Neither Turkey nor Saudi Arabia permitted the United States to fly combat missions against the Iraqi troops. Jordan, too, denied the United States the use of its airspace.[8] Deprived of the use of its land-based airpower, the U.S. response was limited cruise missile strikes against air defense command and control facilities in southern Iraq.

- November 1997: Iraq expelled six U.S. members of the United Nations Special Commission (UNSCOM) weapon inspection team. Saudi Arabia denied permission to launch attacks from its bases and did not allow any additional forces into the country.

---

[8]This despite the fact that a USAF aerospace expeditionary force (AEF) had recently been deployed there.

Turkey was not asked for permission to conduct strikes from its territory but made it clear that, if asked, it would refuse.

- January 1998: The unresolved crisis flared again when Saddam blocked weapon inspectors from inspecting presidential palaces and "sensitive sites." Because of extraordinary U.S. arm-twisting, Kuwait and Bahrain gave assurances of cooperation in military operations.[9] Even under such pressure, however, Saudi Arabia declined to support strikes on Iraq; not only did Riyadh deny the use of U.S. aircraft based in Saudi Arabia, but it would not allow the aircraft to be moved to a neighboring country to conduct attacks from there. Faced with such unequivocal Saudi opposition, first Bahrain, then Kuwait backed away from their initial support of the United States. Qatar and the United Arab Emirates also refused to allow the use of their territory, and Jordan, Turkey, and Egypt all expressed opposition to any U.S. air strikes.

- November 1998: Another incident occurred when Iraq announced an end to cooperation with the UNSCOM inspectors. Although many Arab governments were markedly more critical of Iraqi actions than previously, they remained unsupportive of U.S. military action against Baghdad. Prominently, Saudi Arabia again refused the U.S. access to its facilities for offensive operations.

- December 1998: Finally, UNSCOM reported that Iraq had not complied with UN demands that Baghdad dismantle its programs for developing and producing weapons of mass destruction. U.S. and British air forces attacked Iraqi military units, installations, and facilities suspected of being related to weapons of mass destruction. Both Saudi Arabia and Turkey—where the United States had its largest concentrations of deployed assets—again denied the use of their bases.

So, while some may argue that access problems "have never stopped an operation to which the United States was seriously committed," it is clear that such difficulties have adversely affected important USAF actions in many instances.[10]

---

[9]First, Secretary of State Madeline Albright visited the region, followed closely by the Secretary of Defense, William Cohen.

[10]"The Access Issue" (1998).

The Current Context of Military Access

The United States has many friends and allies around the world and relies on them to facilitate its overseas military operations. Broadly speaking, the United States enjoys three kinds of worldwide access:

- *Permanent bases hosted by allies whom the United States is committed by treaty to defend.* Though such bases often serve as focal points for U.S. military operations overseas, host nation approval for use of these facilities in missions not directly related to their intended purpose—defense of the host's territory—is by no means automatic.[11]

- *Substantial presence in support of ongoing military missions.* When the mission ends, the U.S. troops expect to depart. The current deployments in Saudi Arabia supporting Operation Southern Watch are an example of this *mission presence*. As with permanently based forces, having the troops in place does not betoken any right to use them how and when the United States wants.[12]

- *Visits to a number of countries each year to assist in training, for exercises, or to take part in contingency operations.* On each occasion, of course, this *limited presence* is subject to the invitation and/or approval of the host.

Specifics of the situation will often drive a country's response to U.S. access requests, and the possible permutations are manifold. Responses can vary with the type of mission being undertaken—an ally may routinely grant overflight and transit basing to airlifters, for example, while prohibiting combat aircraft on or over their territory. Or the nature of the contemplated operation may be decisive, with the government permitting air-defense missions but interdicting offensive operations. Restrictions can range from vexing preconditions (Riyadh's demand that mine clearing equipment being airlifted from Israel during Desert Shield be flown first to Cyprus before

---

[11]So, for example, NATO ally Turkey did not allow the use of U.S. forces stationed at Incirlik to counter Iraqi intervention in the Kurdish civil war in 1996. And there are concerns that U.S. forces in Japan might not be permitted to participate if the United States should decide to actively support Taiwan in a struggle with mainland China.

[12]As noted earlier, the Saudis have repeatedly vetoed air strikes on Iraq when they have not shared the U.S. assessment of their necessity.

entering Saudi territory, for example) to outright refusal and every step in between. Uncertainty dominates.

So the first, obvious principle of any access calculations: *There is no such thing as "assured" access to the territory of any other sovereign power.* Regardless of what treaties, agreements, or understandings may be in place, nations retain ultimate control of their territory and airspace. No matter how friendly or closely aligned, a foreign government will consider its own interests first; if the proposed U.S. action is not in accord with them, it is unlikely to receive unequivocal support. This may represent the closest thing to a durable general principle yet discovered in this complex political arena: *The host's perceptions of the interests at stake matter at least as much as does the U.S. perspective.* Shaping these perceptions, not just in crisis but from day to day, is key to increasing the likelihood of the parties' developing a common frame of reference and thereby reducing both the chance and magnitude of any disagreement when access issues arise.

The United States faces different access challenges (and opportunities) across the many regions where it wishes to maintain its ability to project power. Thus, a successful USAF global access strategy will emphasize different approaches from place to place, capitalizing on strengths and compensating for weaknesses as necessitated by the dynamics of each environment.

Europe. In many ways, Europe is the United States' gateway to much of the rest of the world. The United States has relied on its substantial forward presence in Europe not only for local missions but also to support operations in the Middle East and Africa. Europe's tremendous infrastructure, modern economies, and strong historical ties to the United States have made it an obvious choice to support and facilitate a wide range of combat, peacekeeping, and humanitarian operations, a situation that is likely to continue.

As noted earlier, however, even these closest of long-standing alliance ties—such as those between the United States and Great Britain—have from time to time failed to guarantee the kinds and levels of cooperation the United States desires. Nonetheless, the situation in Europe continues to be broadly favorable, if only because the options are so plentiful and diverse that occasional setbacks are fairly easily overcome. This does not mean that new options should not be pursued, however, as the experience of being but one country away from mission failure, as happened with Nickel Grass, is always a possibility. Focusing attention on and building ties with Partnership for Peace states and, to a lesser extent, Russia may provide just the opportunity the United States will need at some point in the future, as well as potentially increasing its overall reach further east.

The Persian Gulf and Southwest Asia. Access in the Gulf region, on the other hand, has always been limited and case-by-case, with few formal agreements in place between host countries and the United States.[13] Because the Gulf War ended with Saddam Hussein still in power, there has been a willingness in the region to permit some U.S. forces to stay, particularly since that presence has had the sanction of the United Nations. Insofar as actual combat operations against Iraq have been concerned, however, the Saudis and some of their neighbors have not felt and do not feel comfortable serving as bases for what many in the region perceive as continued harassment of Iraq. While military action to defend them—as had been taken in 1987[14] and 1991—was acceptable in Gulf capitals, these more recent strikes have not been seen as advantageous to the host states. They have instead been perceived as irritating and angering Baghdad in a region where grudges are long lasting. "You Americans will eventually go home," the Gulf countries in essence say, "leaving Saddam's regime (or an equally revanchist successor) intact and us, his neighbors, vulnerable to retribution." It should not be completely surprising then that the Saudis and their neighbors have concluded that they have little or nothing to gain from supporting these ongoing and inconclusive American attacks.[15]

The overall Gulf and Middle East environment for access is therefore problematic. The lack of strong alliance ties creates a great deal of uncertainty that continues to plague operations and planning.

---

[13]In fact, only Oman has a formal defense arrangement with the United States that predates Desert Storm. Kuwait, Bahrain, and Qatar have since acceded to various defense ties with the United States.

[14]Operation Earnest Will, the reflagging and escort of Kuwaiti tankers during the Iran-Iraq war.

[15]The most notable exception to this attitude, unsurprisingly, has been Kuwait, which continues to support the majority of U.S. actions. Kuwait, of course, continues to feel the greatest threat from Iraq and thus has the greatest security dependence on the United States.

Asia. The United States maintains a strong and sizable presence in northeast Asia, and there is a fair amount of regional consensus that the United States should remain involved in this part of the world. Although it has been quite some time since anything other than exercises and the occasional humanitarian operation has actively involved U.S. forces here, the U.S. presence is widely regarded as stabilizing.

At the same time, however, most countries in the region wish to avoid inflaming tensions in what is seen as a reasonably stable and quite prosperous period of Asian history. So, for example, few are willing to openly avow support to the United States if it comes to the aid of Taiwan in a possible war with the Peoples' Republic of China. These attitudes make risky any predictions of how countries would respond to U.S. calls for support; the response is likely to be highly variable and sensitive to the details of the specific scenario.

In the Asian arena, the USAF's biggest problem may be lack of adequate basing in the South China Sea and in Southeast Asia. Addressing these shortfalls promises to be a difficult and long-term problem. While Guam is a valuable chunk of sovereign U.S. territory in East Asia, the island is distant from most likely conflict locations. Similarly, U.S. forces in Korea and Japan, while well-situated for their primary mission of deterring North Korean adventurism, are based quite far away from the Taiwan Strait and the South China Sea.[16]

Given this sparse set of existing basing arrangements and the uncertain political dynamics of the region, the presence and security ties that the United States enjoys in East Asia leave a dangerous level of uncertainty. The only viable solution appears to be to diversify and to hedge, maintaining and building as wide a network of ties as possible to increase the odds of access and thus facilitate whatever operations may be necessary in the future.

This is even truer in southeast and south Asia, where existing U.S. relationships are far less developed than in the East, even as the region grows increasingly volatile. The United States has limited access in Thailand and Singapore and long-standing close relations with Australia. Further to the west, U.S. ties with Pakistan were severely curtailed first in the 1970s and again in 1990 because of U.S.

---

[16]A useful survey of Asian basing issues can be found in Khalilzad et al. (2001).

concerns over Islamabad's nuclear ambitions. Pakistan's 1998 testing of a nuclear weapon confirmed these worries and further strained relations with the United States. The military coup in October 1999 further complicated matters. India, the region's other nuclear power, has never been an ally of the United States. Contacts were beginning to develop in the mid 1990s, with some exchanges of high level visits having taken place, when they were derailed by India's atomic testing.[17]

An alternative may lie somewhat to the north. Through the Partnership for Peace and bilateral cooperation programs, connections are being built with several of the post-Soviet Central Asian states, notably Kazakhstan and Uzbekistan. These actors, although carrying considerable baggage of their own, could provide considerable infrastructure and may be worth exploring as potential operating locations should need and opportunity intersect in this area.[18]

Latin America. The handover of the Panama Canal Zone and associated military infrastructure has dramatically altered the access situation in Latin America. Counternarcotics operations have already been impeded by the loss of Panamanian access, and while some arrangements have been made to compensate, these appear far from permanent (see Abel, 1999; Farah, 1999, p. A19; and Grossman, 1999).

Also, while there has been a great deal of successful interaction between the United States and several Latin countries in counter-drug activities, the refusal of such states as Venezuela to cooperate on some fronts is indicative of a general ambivalence that many in the region feel toward the United States. The danger that U.S. actions will be interpreted as imperialistic requires taking particular care in engagement in this region. Transparency of goals and strategies is important here, but it is equally important to strengthen ties in

---

[17]Relations with both Pakistan and India are undergoing potentially dramatic changes as a result of Operation Enduring Freedom. Relations between the two subcontinental powers, meanwhile, continue to be fraught with tension.

[18]The war on terrorism has prompted the United States to dramatically increase its level of engagement with Uzbekistan, Tajikistan, and Kyrgistan. Russia, too, may prove a useful partner in Asia. Military contacts between the Russian Far Eastern forces and U.S. Pacific Command, for example, have developed substantially over the past few years.

peacetime and to build economic relationships that can help foster trust. Military ties alone will likely not be sufficient to alleviate local concerns and may backfire in the long run.[19]

Africa. Like South Asia, Africa is something of a void for U.S. engagement. This lack of U.S. involvement has also created something of a vacuum in understanding the complex political realities that drive relations between states there.

What the United States, or anyone else, can do in Africa is significantly constrained by the abysmal infrastructure and the dearth of sophisticated local forces to contribute to operations, humanitarian or otherwise, on the continent. While South Africa maintains a highly modern and effective military, it is located at the very southern tip of the continent and has shown little inclination to cooperate with the United States in such ventures as the building of a U.S.-sponsored African peacekeeping force (Heyman, 1999).

The current outlook for Africa suggests that U.S. operations there will focus on either peace enforcement, response to humanitarian disasters, or both. However, large-scale operations of these kinds could be very difficult to execute, given the region's woeful infrastructure and the long distances between where U.S. forces would come from and where they would need to go.

Future Demands Require Expanded Access. In considering the U.S. experience with access around the globe, it is worth noting that the kinds of contingencies that crop up in the next decade or two will likely very often occur in areas where the United States faces sizable access uncertainties. Europe—where the United States enjoys a history of close security relationships, an enduring alliance superstructure, and a plethora of potential basing options—may continue to witness limited conflicts on its southern and eastern fringes. However, the probable foci for large-scale warfare lie in regions of problematic access: Southwest Asia, the Taiwan Strait and South China Sea, South Asia, and Africa all loom large as possible hot spots.

Future Threats Add Complexity. The spread of advanced weaponry and military technology will complicate access and basing in the

---

[19]Particularly given the somewhat murky history of U.S. cooperation with and support to the repressive military-controlled regimes that in many cases were overthrown or at least compelled to step aside by the current governments of many Latin states.

future. It seems increasingly likely that many future adversaries will deploy ballistic and cruise missiles—some equipped with nuclear, biological, or chemical (NBC) warheads—that could credibly threaten many regional targets, including close-in USAF air bases. Stillion and Orletsky (1999, especially Chapter Two) painted a dramatic picture of the threat that could be posed to aircraft parked in the open, critical maintenance facilities, and the "tent cities" housing aircrew and other personnel by a fleet of fairly crude GPS-guided cruise missiles equipped with a simple submunition-dispensing warhead. Ballistic missiles with similar warheads, while more technically challenging to develop, would represent a still more dangerous threat because of their long range, rapid flight times, and the difficulties of defending against them.

Figure 9.1 shows how such weapons could affect the basing options available to U.S. forces. Each solid circle represents the "threat ring" presented by a *No-Dong* class, 700 nmi–range missile based at one of three locations in Iran; each dotted circle does the same for a 500 nmi–range cruise missile similar to those Stillion and Orletsky proposed. Much of the Arabian peninsula and Turkey—including most of the main operating bases (MOBs) the USAF would use—are within range of one or both weapons.

Confronted with such a situation, a future commander would face a difficult choice: bed down his air forces at bases closer in, thus maximizing their combat power while putting them in the bullseye for the enemy's missile forces, or base them further out, trading sortie generation for added protection.[20]

Passive and active defenses can mitigate the risks enemy missiles, as well as such other threats as special forces or terrorist attacks, present. Unfortunately, it may be difficult to provide adequate protection to forces deployed in a truly expeditionary mode. Hardened shelters for aircraft, equipment, and people may be unavailable at many if not most future operating locations. Other RAND analysis suggests that it would be possible to develop deployable shelters that

---

[20]Our work suggests that roughly doubling the one-way distance from base to target from 500 to 1,100 nmi results in about a 40-percent drop in combat power for a deployed force of about 40 A-10s, F-16s, and F-15s. See Shlapak et al. (forthcoming), especially Chapter Three.

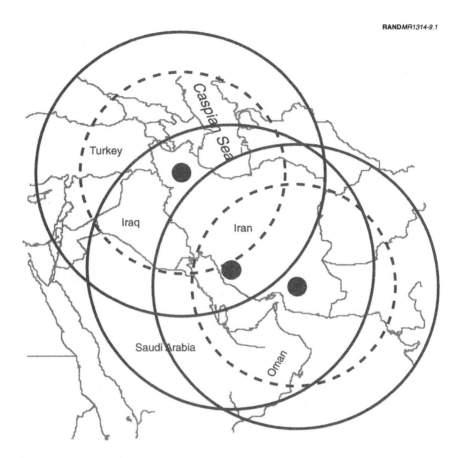

RANDMR1314-9.1

Figure 9.1—Notional Missile Threat Rings

could provide protection against cluster munitions, mortar frag-
ments, and small-arms fire. However, even such minimal protection
carries an enormous penalty in terms of the weight and volume of
material to be moved in support of a deployment and would probably be practical only in scenarios in which substantial prepositioning
is possible (Stillion and Orletsky, 1999).

Active defenses against cruise and ballistic missiles suffer from similar problems. Land-based systems, such as Patriot and, in the future,
Theater High-Altitude Area Defense, are large and heavy and consume large amounts of airlift to deploy.

Sea-based theater missile defense (TMD) systems, such as the proposed Navy Theater-Wide, are self-deploying. However, limitations on both the number of vessels to be equipped with the system and the number of TMD rounds available on each make it likely that a sufficiently well-endowed and determined attacker could overwhelm the defense with sheer numbers. Boost-phase defense systems, such as the USAF's proposed airborne laser (ABL), could deploy rapidly to a distant theater. However, the ABL will require a considerable degree of friendly air control as a precondition for its most effective usage, creating something of a bootstrapping problem: How will the theater commander protect the air superiority assets needed to create the environment in which the ABL can happily operate? And, of course, a TMD architecture that is sufficiently effective to adequately protect the full range of critical theater targets—not just air bases—is many years in the future.

And it is not just the dangers to U.S. forces that can affect access calculations. An adversary fielding ballistic or cruise missiles armed with NBC warheads could choose to use them to intimidate potential host countries by threatening to attack them if they cooperated with the United States. Less well-equipped opponents could employ terrorism against U.S. security partners in a similar effort to convince them to withhold assistance. An ability to help protect regional friends against these threats—whether with rapidly deployable TMD capabilities or intelligence and counterterrorist units—may become a precondition for U.S. access in many future scenarios.[21]

ACCESS OPTIONS: FROM "PURE" STRATEGIES TO A PORTFOLIO

The past, then, contains numerous examples of USAF operations that have been adversely affected by difficulties with access. The geopolitical picture presents complications and uncertainties galore, and likely future threats may impose difficult choices between operational effectiveness and force protection. How should the Air Force confront this complicated set of demands as it confronts its future as an expeditionary force? Clearly, it must plan, organize, equip, and train itself according to a new set of principles suited to a world that

---

[21]Thanks to RAND colleague John Peters for this insight.

demands frequent, short-notice deployment and employment across an entire spectrum of conflict virtually anywhere in the world. And it must do so in the face of grave uncertainties—driven by ineluctable political and military realities—regarding where, how, and when it will be able to operate. Projecting power effectively and reliably under such circumstances requires a robust global strategy for managing access, one that can cope with the unavoidable and sometimes dramatic ebb and flow of operational demands and political dynamics.

Five "Pure" Strategies

Our work identified five broad alternative approaches for improving access and basing in the future[22]:

- Expand the number of overseas MOBs to increase the likelihood that forces will be present where and when they are needed.

- Identify one or more "reliable" allies in each region of the world, and count on them to cooperate when asked.

- Proliferate security agreements and alliances to broaden the set of potential partners in any given contingency.

- Negotiate and secure long-term extraterritorial access to bases, as was done with Diego Garcia.

- Rely on extended-range operations from U.S. territory.

Each of these strategies is, in and of itself, insufficient to ensure adequate access. The following subsections briefly discuss each.

Expand Overseas MOBs. Proliferation of permanent presence overseas has a historical pedigree; USAF forces were at one time stationed at dozens of locations around the world. With the end of the Cold War, that base structure was substantially reduced. Why not rebuild a larger and more robust array of overseas MOBs to support the USAF's power-projection mission?

At least three serious objections can be raised to this approach:

---

[22]A sixth strategy, that of imperial conquest as the British Empire of old, was after brief consideration eliminated a priori as a viable option.

- There appear to be no popular constituencies, either domestic or foreign, for such an expansion.

- Unless host countries pick up all or part of the tab, foreign MOBs are expensive propositions. Freeing up money to build, or reopen, these facilities would be extremely difficult.

- Having forces stationed on another country's territory does not in itself guarantee that they can be used how and when desired. Spain, Saudi Arabia, Turkey, and others have demonstrated this time and again over the past 30 years.

Rely on the Reliable. Great Britain has proven to be a particularly stalwart friend to the United States by, for example, enabling the 1986 raid on Libya. It and Turkey are the only countries that continue to share the burden of policing the no-fly zones in Iraq. Can the United States not identify one or more "Britains" in other parts of the world—partners whose reliability will be such that they will rarely, if ever, be uncooperative? Unfortunately, our analysis suggests not.

First, candidates are few and far between. Britain and the United States have enjoyed a mutually beneficial "special relationship" since the 1940s. It began with Lend-Lease, was solidified through the war against Hitler and British participation in the Manhattan Project, and was set on firm ground with continued cooperation on postwar nuclear matters. Too, the United States has a strong cultural attachment to and affinity for Britain that is deeply rooted in their common history. Looking around the world, it is difficult—one is tempted to say "impossible"—to find another country that shares a similar range and depth of connections with and similarity of perspective to the United States. This is especially true in Asia and the Greater Middle East, the regions where access promises to be especially problematic in the near term.[23]

---

[23]Israel might represent a plausible candidate for a "special relationship." Its somewhat shadowy status amongst its neighbors could, however, impose very great limitations on its utility as a point of access to the region. Should these circumstances change for the better, this assessment could also change. Australia may appear to be a possible "England" in the Western Pacific. However, Canberra's regional and global perspectives are not identical to those of the United States, and a significant portion of its people are likely to oppose greatly expanded defense ties with the United States. Furthermore, Australia's location makes it less than ideally suited to support USAF operations outside of its immediate Southeast Asian vicinity.

Also, it again bears noting that even "reliable" Britain has at times asserted itself by refusing to cooperate with the United States. London's failure to support the Nickel Grass airlift to Israel in 1973 is probably the most notable example.

This is not to say that the United States should not try to nurture close and robust relationships with other countries, only that it would be imprudent to rest an overall access strategy on this single leg.

Expand Security Agreements and Alliances. Another option would be to greatly expand the existing network of alliances and other security arrangements that bind other countries to the United States and vice versa. Indeed, NATO's recent expansion and the success of the Partnership for Peace program has in fact opened new doors for USAF access.[24] Two points, however:

First, as with the idea of expanding the number of USAF overseas MOBs, it is difficult to identify the political constituencies that would support a wide-ranging extension of U.S. alliance guarantees. Domestically, support for NATO expansion may have been a one-off affair, based more on public familiarity with the Atlantic Alliance's long-time role in U.S. security than any desire to see the American security umbrella more broadly spread. And, while there is little doubt that America will remain an engaged and active power on the international scene, the persistence of the isolationist strain within the national political debate may indicate that these are not the most propitious times to advocate such an expansion.[25]

Second, our review of the historical record suggests that much of the payoff in terms of cooperation from enhanced security arrangements may come during the early stages of a relationship when the prospective partner is anxious to prove its value to Washington. A desire for improved relations with the United States may motivate a new friend to be more cooperative than it will be when, secure in its status, those improvements are cast in stone.

---

[24]As witness Hungary's cooperation with NATO during Allied Force.

[25]Isolationist voices have been somewhat muted in the aftermath of the terrorist attacks on New York and the Pentagon, but history suggests that the silence is only temporary.

"Rent a Rock." The value of Diego Garcia in supporting U.S. operations in the Persian Gulf and Afghanistan leads to the question: Are there opportunities to make similar arrangements elsewhere in the world? To help improve U.S. access in the area around Taiwan, for example, might it be possible to lease from the Philippine government one of the many desolate, uninhabited islands in the archipelago and build a MOB there? It is an intriguing and potentially powerful idea.

Of course, extraordinary circumstances are typically needed to induce a country to cede sovereignty over part of its territory; Britain granted the lease on Diego Garcia under the strains of the Cold War and in the context of an intimate preexisting security relationship with the United States. It is certainly imaginable, though, that some set of incentives might prompt Manila, say, to agree to a similar arrangement with the United States. Filipino perceptions of rising hostility from Beijing, for example, could drive them to pay a high price for U.S. protection. The idea should therefore not be dismissed out of hand. There are, however, at least two reasons that it is not a complete solution for future USAF access needs.

First, these arrangements are and will certainly remain rare indeed. The United States enjoys such extraterritorial status at Diego Garcia and Guantanamo Bay in Cuba; the first was acquired from a close friend facing a common foe, and the other was a remnant of a colonial past.[26] Assuming that Washington will be able to acquire such privileges anywhere else, let alone at multiple locations, would be foolhardy.

Second, it is likely that only utterly uninhabited locales could even come under discussion as candidates for such an arrangement. And, such places are typically without people for good reason: A pestilential climate, lack of livable real estate, or the absence of fresh water are three common explanations. Each of these would represent a significant barrier to establishing a major military installation as well. None is necessarily strictly prohibitive; swamps can be drained; mountains can be flattened; and salt water made fresh through the sufficient application of ingenuity and cash. However, the up-front costs of such an undertaking are likely to be very high, and the real-

---

[26]A third such concession, the Panama Canal Zone, has recently passed into history.

location of resources within the Department of Defense to provide for them would be extremely painful.[27]

Project Power from U.S. Territory. A final option is to reduce reliance on overseas access by resorting increasingly to employing airpower from sovereign U.S. territory. The success of long-range bomber raids from bases in the continental United States (CONUS)—B-52s carrying cruise missiles from Louisiana to Iraq and B-2s attacking targets in Serbia and Afghanistan from Missouri—gives this idea great credibility. And clearly, the rapidly improving conventional capabilities of the USAF's heavy bomber fleet earmark them for a more prominent role in future conflicts. Two factors, however, will limit the extent to which these sorts of operations can, at least in the near to mid-term, dramatically reduce the need for overseas access across all contingencies.

Sheer weight of numbers is the first factor. The USAF currently fields over 2,000 fighter and attack aircraft but only about 150 bombers and has no plans for further procurement of long-range strike platforms for at least 20 years.[28] Thus, about 90 percent of the Air Force's combat aircraft cannot and will not be able to operate effectively from U.S. territory in any but the most exceptional scenarios.

This quantitative difference looms even larger when taking into account the productivity difference between a bomber based on U.S. territory and a fighter that is in theater. Heavy bombers flying 30- to 40-hour CONUS-to-CONUS missions must obviously generate less than one sortie per aircraft per day. In fact, for analytic purposes, it is typically assumed that a realistic sortie rate may be one every two or three days, and this appears broadly consistent with what has been achieved thus far in practice. An F-15E, on the other hand, can achieve an average of between 1.5 and two sorties per day when based within 1,200 nmi or so of its targets.[29] And although the bomber's heavy payload makes up somewhat for the disparity in sortie rates, the limited number of bombers available—in compari-

---

[27]Costs are also a major factor militating against a higher-tech variant of this approach: the construction of large floating air bases. Another strike against such platforms is that, unlike England, islets, and atolls, they lack inherent unsinkability, making them potentially very lucrative targets for a capable adversary.

[28]USAF force numbers as of July 1999 (Mehuron, 2001).

[29]See Figure 3.10 in Shlapak et al. (forthcoming).

son to the number of fighters and attack aircraft—further reduces the heavy force's relative impact, as the illustrative numbers in Table 9.1 show.[30]

Again, this does not denigrate the value of the heavy bomber force; indeed, RAND strongly advocates that the USAF consider the near-term development of a new long-range strike asset. However, enthusiasm for the role bombers can play in power projection must be tempered by the real limitations of their capabilities.

The second problem with operating mainly from U.S. territory is that, for some missions, it is simply not practical. Consider a complex operation other than war in Central Africa.[31] The problem here is not putting ordnance on target but supporting complicated and intensive operations on the ground in the heart of Africa. It is difficult to conceive how that could be accomplished absent access to numerous countries in the region.

U.S. territory should become an increasingly important launching pad for overseas operations; however, it does not appear to be a complete solution to the access problem.

Table 9.1

Illustrative Comparison of Weapon-Delivery Potential

Aircraft	Payload	Daily Sortie Rate	Weapons Delivered per Day	Weapons Delivered in 10 Days
1 x F-15E	3 x GBU-24	1.75	5.25	52.5
24 x F-15E	3 x GBU-24	1.75	126.00	1,260.0
1 x B-2	16 x JDAM	0.33	5.00	50.0
16 x B-2	16 x JDAM	0.33	85.00	850.0
1 x B-2	16 x JDAM	0.50	8.00	80.0
16 x B-2	16 x JDAM	0.50	128.00	1,280.0

---

[30]This very rough comparison ignores a host of operationally important factors, not least of which is the value of the B-2's low-observable configuration. Nonetheless, it does, we believe, present a reasonably valid comparison of capabilities along one important dimension.

[31]A more in depth discussion of such a scenario can be found in Chapter Four of Shlapak et al. (forthcoming).

Embracing Uncertainty with a Mixed Strategy

If pure strategies are not adequate to cope with the challenges to come, a hybrid approach must be called for. The USAF might consider a metaphor from the financial world, and treat the construction of an appropriate access and basing strategy as a problem in *portfolio management*. The analogy appears sound along several dimensions:

*   As on Wall Street, the environment facing USAF planners is one dominated by *uncertainty*. It is not possible to predict where the next contingency will erupt, what form it will take, or how the geopolitical stars will align to facilitate or restrict the level of international cooperation the United States will receive. In such a "market," a well-hedged portfolio is the best path to success.

*   Managing risk and exploiting opportunity require *diversification*. No single investment can ensure maximum financial success, and no single-point solution can provide a sufficiently robust guarantee of adequate access. Success will depend on having a range of contingency options, plans, and capabilities.

*   *Information flows* are critical to good decisionmaking. Just as a competent broker must match the needs of buyers and sellers, the United States must remain informed and aware of its partners' sometimes-divergent goals, strategies, and interests. *Engagement* and *transparency* play pivotal roles.

This approach recognizes that the ultimate question of access—will country X let the United States do what it wants to do when it wants to do it?—is in many ways outside the USAF's control and prepares the Air Force to operate effectively in a variety of access contexts.

In constructing its portfolio, the USAF may want to think in terms of three broad classes of future access and focus its efforts on options that enhance its ability to operate across more than one of them.

In the first, the USAF would operate its expeditionary forces from close-in, well-protected forward operating locations. This mode, which is similar to the USAF's current preferred operational style, maximizes the productivity of the Air Force's fleet of short-legged fighters while accepting some additional risk of enemy attack. This tactic would be most viable in well-developed theaters where the United States was fighting from long-standing MOBs (as in a Korea scenario) or in a region where prepositioning and cooperation with the host permitted the construction of hardened facilities and the quick establishment of robust TMD and other defenses (such as, perhaps, Saudi Arabia).

The second mode would require the Air Force to operate from more distant bases outside the range of the adversary's offensive weapons. This model would significantly reduce the force-protection challenge at the cost of a loss of operational efficiency. As theater missiles and NBC weapons proliferate, this may become by necessity the default for most expeditionary operations against a reasonably sophisticated adversary. Even in such theaters as the Persian Gulf, where the United States has many friends and large stocks of propositioned materials, political uncertainties may impose this manner of operation on the USAF.

Finally, the USAF should plan for situations where it is compelled, for whatever reason, to project power exclusively from facilities on *de facto* U.S. territory.[32] It is difficult to conjure up a large number of contingencies that would impose such severe strictures on U.S. basing options; a confrontation between China and Taiwan may be the most likely.[33] However, given the uncertainties that color every aspect of the access issue, it would be prudent for the Air Force to hedge against this unlikely but very challenging circumstance.

BUILDING THE PORTFOLIO: EIGHT RECOMMENDATIONS

A global access strategy for the Air Force will consist of many components that, like a good financial portfolio, balance risk and opportunity and hedge against the many uncertainties of the marketplace. The following eight subsections propose a series of elements for such a portfolio.

---

[32]De facto U.S. territory includes the 50 states; U.S. possessions, such as Puerto Rico and Guam; facilities that are leased to the United States on a long-term basis, such as Diego Garcia; and U.S. warships operating in international waters.

[33]It is certainly possible that Japan and other U.S. friends in East Asia would opt to sit out a war between China and Taiwan, particularly if it could be plausibly argued that the island republic "provoked" the conflict (perhaps by making what Beijing interpreted as moves toward de jure independence). Under such circumstances, of course, Taiwan itself could offer to host U.S. forces; for a variety of political and military reasons, the United States might be wise to decline that opportunity.

Retain Existing MOBs

The first and most obvious element is that the United States maintain its current array of overseas MOBs in Europe and Asia. These installations are fairly secure and reliable footholds that can serve as points of entry to virtually every region of possible interest. These bases have in the past been critical for rapidly responding to contingencies around the world and should continue to play that role into the indefinite future.

Build Forward Support Locations

The USAF should establish a small number of forward support locations (FSLs) worldwide. Much discussed under a variety of names, an FSL is essentially a "mega-MOB" intended to support power projection.[34] Spares, equipment, and munitions could be prepositioned at these locations, which should be built where access is either guaranteed or highly likely. FSLs could also host repair facilities for key components, such as engines and critical avionics units, and would serve as both strategic and intratheater airlift hubs when the situation demanded it.[35] Extensive RAND analysis strongly suggests that properly located and outfitted FSLs offer significant leverage in enabling both rapid and sustainable expeditionary operations.[36]

Figure 9.2 shows five possible locations for FSLs. It demonstrates that even a small number of such installations could provide broad coverage of likely contingency locales. Note that of the five FSLs shown, three are in U.S. territory (Alaska, Guam, and Puerto Rico); a fourth is on de facto U.S. territory (Diego Garcia, at least until 2039), and the fifth is on the territory of America's most reliable ally, Great Britain. Taken together, these locations put most of the world within C-130 range of a permanent center of U.S. power projection capability.[37] Extreme southern South America and southwestern Africa are left

---

[34]RAND has worked extensively on the FSL concept. See, for example, Killingsworth et al. (2000). Also see Galway et al. (1999).

[35]For a discussion of the kinds of maintenance facilities that might be put at FSLs, please see Peltz et al. (2000).

[36]See Chapter Ten.

[37]These locations also have the virtue of being outside the range of the bulk of any likely adversaries' probable offensive capabilities.

uncovered, but much of the rest of the world's landmass can be served from two different FSLs.

Plan for Uncertainty

The USAF should develop tools and procedures that support the rapid creation and modification of flexible deployment and employment plans. A world of fluid political arrangements, unpredictable contingencies, and evolving threats will not be at all kind to rigid and doctrinaire planning processes. The Air Force should exploit the power of widely available information technologies and the skills of its highly trained people to ensure that its forces are prepared to respond quickly and adaptively to rapidly changing circumstances. For example, the ability to adjust deployment plans on the fly as circumstances evolve could prove invaluable. Current processes that revolve around intricately synchronized and hard-to-

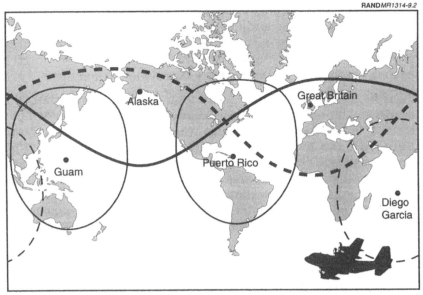

NOTE: The circles on the map have radii of 3,000 nmi; the AF Pamphlet 10-1403 planning factor for a C-130 is 3,200 n mi with a 12-ton payload. Solid circles denote an FSL on U.S. territory, while dotted ones are drawn for foreign ones.

Figure 9.2—Coverage Available from Five FSLs

change time-phased force deployment lists may not be adaptive enough to meet such demands.

It is especially important not to overlook flexible logistics planning. Rapid resupply and effective maintenance are critical to maximizing the capabilities of future AEFs, and the USAF logistics community must prepare to operate amidst the same uncertainties that bedevil their operational brethren. Current initiatives to increase real-time visibility into demands, inventories, and flows will likely be very valuable in this regard. Robust, well-stocked FSLs will certainly be a major asset but must be combined with careful planning that covers a wide range of possible contingencies and circumstances, and these plans must be exercised to increase the chances of smooth execution if and when necessary.

Build AEFs with Flexible Configurations

In addition to planning, force packaging must also be very responsive to possible access constraints. Otherwise, basing and access limitations could impose significant penalties on expeditionary operations. Any one or a combination of threat, politics, and infrastructure limitations could compel the USAF to operate from bases located at a considerable distance from the forces' main area of operations; in fact, this is likely to be a major mode of future operations. Analysis indicates that the capabilities of the fighter and attack aircraft in the USAF inventory—again, about 90 percent of the warfighting forces—are subject to fairly rapid and dramatic reduction as these distances grow.

RAND work has identified several steps that can improve this situation. Figure 9.3 shows the effect of adding additional tankers (about a half-dozen) to and doubling the effective aircrew ratio of a deployed fighter task force of about 40 jets. These two measures buy back many of the sorties lost when the force is based further away from its operational area.[38] Note how fuel consumption per sortie increases by almost 50 percent in the "standoff" versus the "close-in" case; this additional logistics burden must also be taken into consideration when planning operations from extended ranges.

---

[38]The detailed analysis behind this figure may be found in Shlapak et al. (forthcoming).

Develop Improved Active and Passive Defenses

The USAF should work to maintain its ability to operate from close-in bases even in the face of modest threats from adversary missiles, special operations forces, and terrorists. Therefore, the USAF needs to develop and deploy improved active and passive defenses for deployed units. These could include gun systems for defense against low-flying cruise missiles, improved automatic sensors and small unmanned aerial vehicles for protection against threats on the ground, and protective "blankets" to shield high-value aircraft from shrapnel and small submunitions.[39] To facilitate rapid mobility, these assets—many of which are low-tech and should be relatively

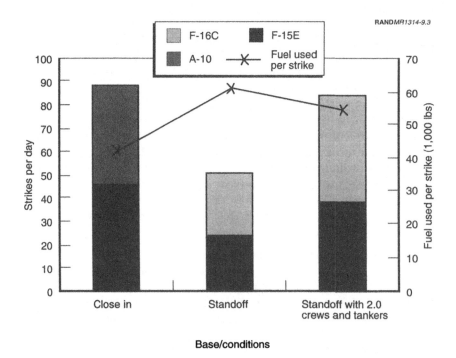

Figure 9.3—Impact of Adding Aircrew and Tankers on Sorties and Fuel Use

---

[39]These recommendations echo some that can be found in Stillion and Orletsky (1999) and Shlapak and Vick (1995).

inexpensive[40]—should be widely prepositioned at FSLs (from which they could be airlifted via C-130 to forward operating locations and thus not compete for strategic airlift resources) and afloat.

Expand Contacts with Potential Partners

While history shows that engagement does not necessarily equate to "assured" access, close security relations with potential hosts do appear to facilitate cooperation. Contact between the United States and its partners helps develop common perspectives on key issues and encourages the pursuit of joint goals and objectives. This shaping—while not guaranteeing anything—does help lay the foundations for cooperation in the event of a crisis or conflict. Therefore, the United States should seek to maintain and expand its contacts with key security partners worldwide. While there appears to be no need to pursue additional formal defense ties as a means of shoring up prospects for access, consistent engagement is of great value. Training exchanges, joint exercises, and temporary deployments help establish relationships—both formal and, perhaps equally important, informal—that can prove of great value in a crisis. And because U.S. deployments for training and exercises often include engineering undertakings—repairing runways and parking aprons, improving fuel storage and delivery facilities, and so forth—they offer opportunities to enhance infrastructure as well as relationships. Finally, these interactions serve to foster the strategic transparency that is invaluable for helping shape partners' perceptions in ways that facilitate future cooperation.

These military-to-military contacts could prove especially valuable in regions where access appears problematic: Asia outside of Korea and Japan, Africa, and Latin America. Laying the groundwork now could produce substantial payoffs in some future crisis.

Adjust the Force Mix

The USAF should aim at fielding a modified mix of short- and longer-range platforms that has improved capabilities to operate effectively

---

[40]For example, Stillion and Orletsky (1999) suggest that a standard 0.50-caliber M-2 heavy machine gun fitted with a night sight and mounted on a simple tower could be an effective weapon against simple, low-and-slow flying cruise missiles.

when based further from target areas. This does not necessarily mean buying additional B-2s or developing another heavy bomber. Options could include procuring a new generation of highly accurate long-range munitions to extend the effective ranges of existing platforms or taking advantage of the soon-to-be-fielded family of "small, smart" weapons to build the 21st-century equivalent of an F-111.

By way of example, an aircraft with a 2,000-nmi unrefueled range could, with minimal tanker support, be able to cover much of the world while operating strictly from the five FSLs shown in Figure 9.2. Using supercruise capability to maintain a fighterlike sortie rate, the aircraft would combine speed and high operating altitudes to increase survivability. If it carried a payload of 2,500 pounds (roughly ten small diameter bombs), preliminary calculations suggest that the platform would have an empty weight somewhere between 33,000 and 55,000 pounds, putting it between an F-15 and an F-111 in size. Even if the per-pound cost of the aircraft were two to three times that of the F-16, it should be sufficiently affordable that the price tag for a force of one or two wings (100 to 200 airframes) need not be prohibitive.

In any event, given the very real uncertainties governing future access arrangements, the USAF should consider carefully whether it is best served by focusing its next round of combat aircraft development solely on platforms whose effective combat radii are no greater than those of its currently fielded fighter force.

Explore New Options

Finally, Air Force long-range planners should conduct ongoing explorations of "outside-the-box" options. Initial ideas could include attempting to identify candidate islets for long-term lease ("rent-a-rock") in the Western Pacific. If one or more are found, some thinking should be done on what kind of facilities might be called for, how they might be built, and what the costs might look like. Then, the United States will be prepared to act if a situation should arise in which the theoretical possibility of such a deal is, for whatever reason, transformed into a real opportunity.

The USAF should also begin evaluating the potential contributions of new access partners. In case of a crisis involving China or even Iran, for example, Kazakhstan and its neighbors could have great utility as hosts for USAF forces. Similarly, Mongolia, Malaysia, and even Vietnam could help support U.S. actions in Asia, while Israel and the former Soviet republics in the Caucasus could be useful in a Southwest Asia (SWA) contingency.

Summary of Recommendations

To show how these recommendations might help to facilitate access in the future, Table 9.2 racks them up against key regions of the world according to our estimate of the value of each in that region. It is worth noting that many of the steps described here serve Asia and SWA—two areas in which U.S. interests are deeply engaged, the potential for conflict is relatively high, and existing access arrangements are uncertain—well.

CONCLUDING REMARKS

Like a good financial portfolio, a global access strategy for the USAF will strike a careful balance between short- and long-term payoffs, and between risk and opportunity. Core investments, such as maintaining the existing overseas base structure, establishing and stocking FSLs, and increasing planning flexibility, represent the heart of the portfolio. Hedges against risk—flexibly configuring AEFs to allow increased tanker support and higher crew ratios, improving active and passive defenses for deployed AEFs, and increasing the long-

Table 9.2

Effects of Recommended Steps Across Regions

	Europe	SWA	Asia	Latin America	Africa
Retain MOBs	√		√		
Build FSLs		√	√		
Flexible plans		√	√	√	√
Flexible AEFs		√	√	√	√
Improve defenses	√	√	√		
Expand contacts		√		√	√
Adjust force mix		√	√		
New options		√	√		

range capabilities of the USAF force mix—provide a cushion against the proverbial rainy day. And, marginal investments prospecting for new opportunities, such as renting a rock in a key area or reaching an access agreement with a new partner, offer the possibility of substantial future returns.

There is no panacea solution to the challenge of overseas access, no "silver bullet" waiting to be discovered. Old problems, like the vagaries of international politics, will persist, and new ones—dozens or even hundreds of accurate, long-range missiles aimed at U.S. bases—will emerge. Furthermore, nothing comes free: Real costs, monetary and opportunity, are associated with any course of action the USAF might take to deal with potential problems in this area. This is the bad news.

On the other hand, that observation should not suggest that this is nothing but a tale of woe. The problems that exist are manageable, and a well-thought-out global access strategy can minimize even those that cannot be foreseen—always the most worrisome. The strategy this chapter suggests calls for increased flexibility and pays off in enhanced robustness against the inescapable uncertainty that characterizes this problem. To sum up in a sentence: Access is not a problem to be solved—it is a portfolio to be managed.

REFERENCES

"The Access Issue," *Air Force Magazine*, October 1998, p. 42.

Abel, D., "Holes Open in US Drug-Fighting Net," *Christian Science Monitor*, July 28, 1999, p. 2.

Boyne, W. J., "Nickel Grass," *Air Force Magazine*, December 1998, p. 54.

Comptroller General of the United States, *Airlift Operations of the Military Airlift Command During the 1973 Middle East War: Report to the Congress*, Washington, D.C.: U.S. General Accounting Office, LCD-75-204, April 16, 1975.

Farah, D., "Handover of Panama Base Hinders Anti-Drug Efforts," *Washington Post*, May 30, 1999, p. A19.

Galway, L., et al., "Expeditionary Airpower: A Global Infrastructure to Support EAF," *Air Force Journal of Logistics*, Vol. XXIII, No. 2, Summer 1999, pp. 2–7, 38–39.

Grossman, E. M., "Commander Says Pact for Base Access in Ecuador Is Close at Hand," *Inside the Pentagon*, September 23, 1999.

Herzog, C., *The Arab-Israeli Wars: War and Peace in the Middle East*, New York: Random House, 1982.

Heyman, C., ed., *Jane's World Armies*, Jane's Information Group, 1999.

Khalilzad, Z., et al., *The United States and Asia: Toward a New U.S. Strategy and Force Posture*, Santa Monica, Calif.: RAND, MR-1315-AF, 2001.

Killingsworth, P. S., et al., *Flexbasing: Achieving Global Presence for Expeditionary Aerospace Forces*, Santa Monica, Calif.: RAND, MR-1113-AF, 2000.

Mehuron, T., ed., "Equipment," *Air Force Magazine*, May 2001, p. 55.

O'Malley, W., *Evaluating Possible Air Field Deployment Options: Middle East Contingencies*, Santa Monica, Calif.: RAND, MR-1353-AF, 2001.

Peltz, E., et al., *Supporting Expeditionary Aerospace Forces: An Analysis of F-15 Avionics Options*, Santa Monica, Calif.: RAND, MR-1174-AF, 2000.

Shlapak, D. A., et al., *A Global Access Strategy for the U.S. Air Force*, Santa Monica, Calif.: RAND, forthcoming.

Shlapak, D., and A. Vick, *"Check Six Begins on the Ground": Responding to the Evolving Ground Threat to U.S. Air Force Bases*, Santa Monica, Calif.: RAND, MR-606-AF, 1995.

Stillion, J., and D. T. Orletsky, *Airbase Vulnerability to Conventional Cruise-Missile and Ballistic Missile Attacks: Technology, Scenarios, and U.S. Air Force Responses*, Santa Monica, Calif.: RAND, MR-1028-AF, 1999.

Chapter Ten

A VISION FOR AN EVOLVING AGILE COMBAT SUPPORT SYSTEM

Robert Tripp and C. Robert Roll, Jr.

The development of Expeditionary Aerospace Force (EAF) operations requires rethinking many Air Force functions, among them the combat support system. To a large extent, success of the EAF depends on having an increasingly agile support system. The Air Force has thus designated Agile Combat Support (ACS) as one of the six essential core competencies for Global Engagement and has begun the transformations necessary to achieve an ACS system.[1]

Continuing development of this system will require making hard decisions on allocating limited resources to make it capable of meeting a wide range of uncertain scenarios. ACS requirements will vary with the scenario, and each scenario will require unique trade-offs, such as that between speed and cost or, more generally, between different characteristics the Air Force values. These trade-offs will change as support technologies, policies, and practices change.[2] As a result, ACS planning must be continuous and must evolve toward a more agile logistics infrastructure making the best use of resources and information.[3]

This chapter offers a vision of what the future ACS system might look like and how it could help the Air Force meet EAF operational goals.

---

[1] A Logistics Transformation Team comprising Air Force and KPMG personnel is leading much of this transformation work. The team has also previously been known as the Agile Logistics Team and, earlier, as the Lean Logistics Team (Menendez, 1999).

[2] For a detailed discussion of how changing technology affects one part of the support system, see Peltz et al. (1999).

[3] For a more general discussion of this point, see Tripp et al. (1999a).

This vision draws from ongoing RAND research evaluating how ACS design options affect EAF effectiveness and efficiency.[4] The system will have to support EAF operations ranging from major theater wars (MTWs) to small-scale contingencies to peacekeeping missions. This chapter describes basic elements for a system we believe would meet EAF operational requirements.

The ACS system will need to be a global network comprising the following:

- forward operating locations (FOLs), with resource allocations that support differing employment timelines

- forward support locations (FSLs), with differing support processes and resources

- continental United States (CONUS) support locations (CSLs).

These infrastructure elements will need to be connected by a logistics command and control system and by a very responsive distribution system to ensure that support resources arrive as combat commanders need them.

ACS DECISIONS AND THEIR "TRADE SPACE"

The Air Force has recognized the need to transform its support system to meet the needs of the EAF. Some of the system's elements and processes were Cold War relics designed to support the needs of a large overseas force in simultaneous major conflicts occurring in Central Europe and Northeast Asia. Under the old scenario, FOLs received specific resources for waging combat in "known" places. Planners assumed that the resources needed for MTWs would suffice for all lesser conflicts. There was less uncertainty to consider in such a planning environment.

Today's support resources must meet the needs of a smaller force facing a wide variety of scenarios in uncertain locations. The new planning environment also has limited resources for supporting

---

[4]Some related RAND reports include Tripp et al. (1999b); Galway et al. (2000); Peltz et al. (2000); Tripp et al. (2000); Amouzegar, Galway, and Geller (2002); and Feinberg et al. (2002). See also Kumashiro (2001).

multiple areas of responsibility (AORs). This means that the support system needs to be flexible enough to move resources across AORs.

The old assumption was that a deploying aviation unit would need to take along enough supplies to be self-sufficient for 30 days. Units meet the new EAF demand for more-rapid deployment by carrying fewer supplies but then require resupply almost immediately. This has shifted the supply emphasis from keeping large stockpiles of resources at FOLs to relying on fast resupply to replenish small stockpiles.

More generally, support resources must be considered strategically rather than tactically. In the past, support requirements have been determined by calculating the specific resource requirements to meet specific "planned" wartime scenarios. Now these requirements must be determined for a wide range of scenarios and for a wide range of variation within scenarios. The resulting resource mix may not be the best for any one particular scenario, but it may be the one that is most robust against the entire range of scenarios or that holds up best in the face of uncertainty. Thus, the future ACS system must be agile, with logistics processes in place to determine how to move limited resources from one place to another to meet rapid deployment, employment, sustainment, and reconstitution needs.

Some key variables affecting ACS system design include

- operational options for force composition, employment timeline, and operational tempo to achieve the desired effects

- FOL capabilities, including infrastructure and resources, as well as the political and military risks associated with prepositioning resources at specific locations

- technology options affecting performance, weight, and size of test equipment, munitions, support equipment, and other support resources and processes

- resupply time, particularly as it affects initial operating requirements (IORs) and follow-on operating requirements (FORs)

- alternative support policies, such as conducting repair operations at deployed or consolidated support locations

- strategic and tactical airlift capacity.

These and other variables form a rich array of possible decisions from which Air Force leaders will choose in designing the ACS system. Generally, there are no right or wrong answers. System trade-offs will be required.

ACS design decisions depend on how Air Force leaders value different criteria. Some system needs, such as rapid employment time-lines, high operational tempos, and airlift constraints, favor forward positioning of resources. Others, such as the cost and risk of positioning resources at FOLs, favor consolidating resources.

Figure 10.1 depicts the general trade-offs. Investment costs (black wedge) are higher for an extensive support structure positioned at numerous forward locations. These costs decline as the number of

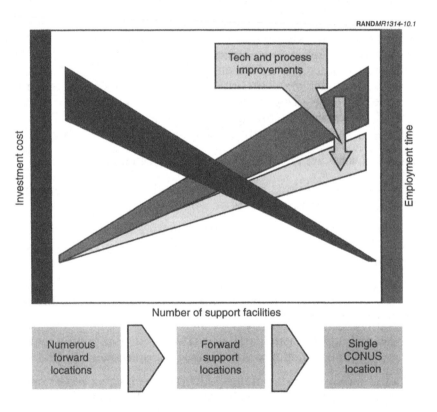

Figure 10.1—General Decision Trade Space by Location

support locations declines. Employment time (dark gray) is lower for an extensive support structure with numerous forward locations. The time increases as the number of support locations increases.

While the general direction of these relationships is fixed, the specific details are not. The arrow on the graph indicates the effect of reengineering processes or implementing new technologies, such as developing lightweight munitions or support equipment. New technologies or processes can shift the timeline curve downward (light gray wedge). This allows positioning of resources more rearward than would otherwise be possible.[5]

AN ANALYTIC FRAMEWORK FOR STRATEGIC ACS PLANNING

How can Air Force leaders evaluate and choose among ACS options? To assist this effort, we are using an employment-driven modeling framework. The core of this framework is a series of models of critical support processes that can calculate equipment, supplies, and personnel needed to meet operational requirements.[6]

These models are "employment driven" because they start from the operational scenario—the employment requirements—to provide time-phased estimates of support resource requirements. Once support requirements are computed, the models can be used to evaluate options—such as prepositioning support resources or deploying from consolidated locations—for satisfying them. The evaluation includes such metrics as spin-up time, airlift capacity, investment and recurring costs, and political and military risks. Figure 10.2 depicts our modeling framework.

This framework addresses the uncertainties of expeditionary operations. The models can be run for a variety of mission requirements, including the support needed for different types of missions (humanitarian, evacuation, small-scale interdiction, etc.); the effects of different weapon mixes for the same mission on support system

---

[5]See Peltz et al. (1999) for a more-specific discussion of trade-offs regarding one part of the support process.

[6]Tripp et al. (1999a) discusses this model in more detail.

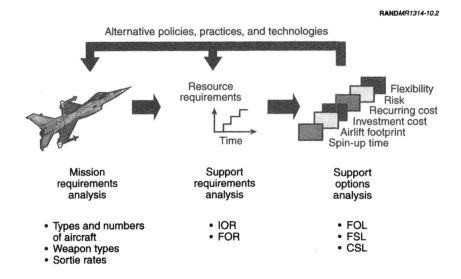

RANDMR1314-10.2

Figure 10.2—Employment-Driven Analytical Framework

requirements; the effects of different support policies, practices, and technologies; and other operation support needs.

The models estimate mission requirements at a level of detail appropriate for strategic decisions. This level includes the numbers of personnel and large pieces of equipment that account for most mission support airlift footprint, as well as enough additional detail to allow the model to reflect major changes in support processes and evaluate them against important metrics.

The final output of the modeling framework is an evaluation of the effects of each support option on metrics of spin-up time, airlift footprint, investment and recurring costs, risks, and flexibility. This shows the details of the trade-off between moving resources from centralized support locations or prepositioning them at FOLs.

ACS analyses may find that an option cannot be supported because of cost or process constraints. If so, senior leaders can consider options with less cost or risk that would still achieve their goals. Such a framework thus can be used not only for ACS system analysis, but also to support integrated analysis of operations, ACS, and mobility options.

KEY FINDINGS FROM ACS MODELING RESEARCH

Using this analytic framework has helped clarify the broad characteristics of the ACS system needed to support future expeditionary operations. An important finding of our research is that the "old" support processes cannot meet the Air Force goal of deploying to an unprepared base and sustaining a nominal expeditionary force (a 36-ship package capable of air defense suppression, air superiority, and ground attack aircraft) at a high operational tempo. Only judicious prepositioning would allow a 48-hour timeline and, even then, only under ideal conditions.

Table 10.1 illustrates this point. We input data from our commodity models for munitions, fuel, vehicles, shelter, F-15 avionics components, jet engines, and Low Altitude Navigation and Targeting Infrared for Night (LANTIRN) needs for the 36-ship force mentioned above into a preliminary integrating model. The objective was to minimize support costs and meet employment timeline goals while satisfying resource requirements for an employment scenario of surge levels for seven days. Table 10.1 shows the results for three different timelines.

A 48-hour timeline requires prepositioning substantial material at the FOL. A bare base can be used only if the deployment timeline is extended to 144 hours and substantial materiel is prepositioned at a regional FSL and if intra- and intertheater transportation is available to move resources to the FOL.

The reason for this conclusion is simple. Current support resources and processes are "heavy." They are not designed for quick deployments to FOLs having limited space for unloading strategic airlift. Significant numbers of vehicles and material handling equipment, such as forklifts and trailers, are required to meet EAF operational requirements. The airlift required to move this material, not including munitions, is enormous and may not always be available.

Shelter needs present another constraint on options for quick deployment. The current Harvest Falcon shelter package for bare bases requires approximately 100 C-141 loads to move and almost four days to erect with a 150-man crew. The construction time for the Harvest Falcon shelter package alone means shelter must be prepositioned to meet a 48-hour or even a 96-hour timeline.

Table 10.1
Cost Versus Timeline Resource Allocation Trade-Offs

Initiate and Sustain	Forward Operation Location	Forward Support Location	CONUS
At 48 hours	Bombs (IOR) Fuel Fuel mobility support equipment (FMSE) Shelter Vehicles	Missiles (IOR and FOR) Bombs (FOR) Repair: F-15 avionics, LANTIRN, engines, EW pods	Unit equipment Two-level repair
At 96 hours	Bombs (IOR) Fuel Shelter Vehicles	Bombs (FOR), FMSE Repair: selected avionics, LANTIRN, engines, EW pods	Unit equipment Two-level repair Missiles (IOR and FOR)
At 144 hours	Fuel	Bombs (IOR and FOR) Repair: F-15 avionics, LANTIRN, engines, EW pods Shelter Vehicles	Unit equipment Two-level repair Missiles (IOR and FOR) FMSE

These results do not mean expeditionary operations are infeasible. Technology and process changes may reduce the need to deploy heavy maintenance equipment. For now, however, these results do mean that setting up a strategic infrastructure to perform expeditionary operations involves a series of complicated trade-offs.

Expensive 48-hour bases may best be reserved for such areas as Europe or Southwest Asia (SWA) that are critical to U.S. interests or under serious threat. In other areas, a 144-hour response may be adequate. In still other areas, such as Central America, most operations may be humanitarian relief missions that could be deployed to a bare base within 48 hours because combat equipment would be unnecessary. For all these cases, the models and analytic framework that we have developed can help in negotiating the complex webs of decisions (Amouzegar, Galway, and Geller, 2002; see also Kumashiro, 2001).

One key parameter that affects ACS design is resupply time. If resupply time is cut, the IORs and initial deployment can also be cut. In addition to IORs, resupply time affects repair locations. If resupply time is long, more maintenance equipment and personnel must be deployed to keep units operating, and greater quantities of supplies will be needed to fill longer pipelines.

Short resupply times can help in dealing with uncertainties caused by inability to predict requirements or by changes in requirements resulting from enemy actions. A short resupply time provides the ability to react quickly to inevitable surprises, mitigating their effects.

The ACS system needs to be designed around expected wartime resupply times, not peacetime resupply possibilities. To examine its constraints, we analyzed resupply time as it varies by delivery process and assumptions. Some of these data were gathered from actual delivery times. Others were generated with models, using optimistic assumptions, which help show differences between possible and actual system performance.

The leftmost curve in Figure 10.3 (Air Mobility Express–Commercial [AMX-C]) shows the distribution of best expected resupply times for small items (less than 150 pounds) that could be shipped via express carriers to SWA from CONUS. This distribution includes the entire

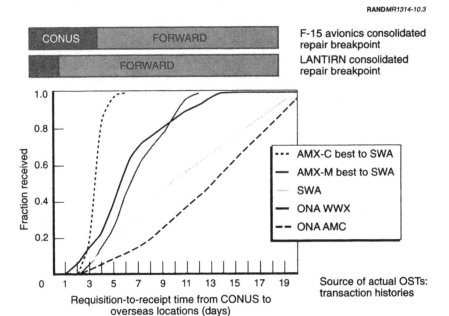

Figure 10.3—CONUS to SWA Resupply Times and
Support Breakpoint Solutions

resupply time, from requisition to receipt, and has a mean of about
four days, including weekends, holidays, and pickup days. This distribution was generated from a simulation model using very optimistic times for each part of the resupply process. The simulation
model assumes the processes are perfectly coordinated with no
delays due to weather, mechanical problems, or enemy actions. This
curve represents a "current process optimum" to SWA.

The third curve (Air Mobility Express–Military [AMX-M]) shows the
expected distribution of best resupply times to SWA for AMX-M, the
system used for large cargo in wartime, under optimistic assumptions. Median resupply time for this system is about seven days. The
fourth curve (SWA) shows the current actual delivery times for high-
priority cargo to SWA units. The data include delivery times for both
small and large cargo. Note that half these requisitions took more
than nine days to deliver.

Operation Noble Anvil (ONA), the air war over Serbia, provided extensive evidence of this challenge. The second curve from the left (ONA Worldwide Express [WWX]) shows the distribution of WWX deliveries during ONA. WWX is a DoD contract with commercial carriers to move small items within CONUS and from CONUS to the rest of the world. The contract specifies "in-transit" delivery times for shipments between specific locations. Most in-transit times to overseas theaters are about three days, excluding the day of pickup and weekends.

During ONA, resupply using WWX to Europe averaged about five days, while more than 10 percent of deliveries took more than ten days. As shown in Figure 10.3, deliveries of large items using military flights averaged more than 15 days (Air Force Materiel Command, 1999). Even in a highly developed theater for a benign conflict environment, resupply times are long and far from offering the benefits of short response times.

DoD recently established a resupply goal of five days to overseas locations and ordered reductions in inventory levels to reflect this new delivery goal. Our research, however, indicates that it may not be possible to deliver small items to overseas FOLs within five days in wartime environments. The goal is probably also not achievable for large items because the median expected delivery time for them, under optimistic assumptions, is seven days.

As mentioned above, resupply time affects decisions about repair locations. We have completed separate studies on maintenance support for key equipment in an expeditionary environment, including jet engines, F-15 avionics, and LANTIRN (see Amouzegar, Galway, and Geller, 2002). The top of Figure 10.3 shows the breakpoints for locating repair facilities in CONUS or forward locations for two cases in which the analysis is complete: F-15 avionics (Peltz et al., 2000) and LANTIRN pod repairs (Feinberg et al., 2001).

For F-15 avionics, consolidating repairs at regional or CONUS facilities sharply reduces personnel needs, as well as the need for some upgrades currently being considered for repair equipment. Unless the resupply time for any consolidated repair facility is less than six days, the longer pipeline will require substantial investments in new spare parts. Figure 10.3 shows that it may be difficult to achieve such delivery times from CONUS, although data from theater support of mission capable requisitions show that transportation times from regional FSLs can meet the six-day breakpoint.[7]

For LANTIRN targeting pods, for which no new acquisitions are planned, the breakpoint timeline is even shorter because of the lack of spares. Maintaining the availability of working pods in an MTW requires transportation times of less than two days from a consolidated repair facility. Figure 10.3 shows this is out of reach of CONUS and might even be difficult to achieve within theater. At the same time, however, deployment of LANTIRN repair to FOLs is not an attractive option. The test equipment is old, very heavy, and increasingly unreliable, so it may be necessary to consolidate repair locations to reduce the need to deploy test equipment.

Models of individual support processes yielded important insights on supporting processes for expeditionary operations. To plan an ACS system, we needed to integrate the outputs of models for different processes and consider mixes of options. This may include a mix of prepositioning some materiel, deploying other materiel from FSLs, and deploying still other materiel from CONUS. Our continuing research on this topic is exploring the use of optimization techniques to integrate options for several support processes.

These analyses led to the conclusion that performing expeditionary operations for the current force with old-style support processes and technologies would require judicious prepositioning of equipment and supplies at selected FOLs. This must be backed by a system of FSLs providing equipment and maintenance services. Such a system would require a transportation system linking FOLs and FSLs.

The Air Force already makes some use of FSLs, particularly for munitions and storage of war readiness materials (WRM). Consolidated regional repair centers have also been established to support recent conflicts. During Desert Storm, C-130 engine maintenance was consolidated at Rhein Main Air Base, Germany. During ONA, intermediate F-15 avionics repair capabilities were established at Royal Air Force Base Lakenheath, England.

---

[7]Data collected from the 4th Aerospace Expeditionary Wing deployment to Doha, Qatar, from May to August 1997. Mission capable requisitions that were processed at Prince Sultan Air Base in Saudi Arabia averaged less than five days. At that time, scheduled military resupply flights connected Prince Sultan Air Base and Doha.

OVERVIEW OF A GLOBAL ACS SYSTEM

Our results to date suggest a vision for an evolving ACS system to support expeditionary operations. The system would be global and would have several elements based at forward positions, or at least outside CONUS. Figure 10.4 gives a notional picture.

The system has five components:

1. FOLs. Some bases in critical areas under high threat should have substantial equipment prepositioned for rapid deployments of heavy combat forces. Other, more-austere, FOLs requiring more time to initiate operations might augment these bases. When conflict is not likely or humanitarian missions will be the norm, the FOLs might all be of the second, more-austere, form.

2. FSLs. The configurations and functions of these would depend on geographic locations, the presence of threats, and the costs and benefits of using current facilities. Western and Central Europe are presently stable and secure; it may be possible to support operations in such areas as SWA or the Balkans from European FSLs.

3. CSLs. CONUS depots are one type of CSL, as are contractor facilities. Other types of CSLs may be analogous to FSLs. Such support structures are needed to support CONUS forces, since some repair capability and other activities may be removed from units. These activities may be set up at major Air Force bases, convenient civilian transportation hubs, or Air Force or other defense repair depots.

4. A transportation network connecting the FOLs and FSLs with each other and with CONUS, including en route tanker support. This is essential; FSLs need assured transportation links to support expeditionary forces. The FSLs themselves could be transportation hubs.

5. A logistics command-and-control system to organize transport and support activities and for swift reaction to changing circumstances.

The actual configuration of these components depends on several elements, including local infrastructure and force protection, political aspects (e.g., access to bases and resources), and how site loca-

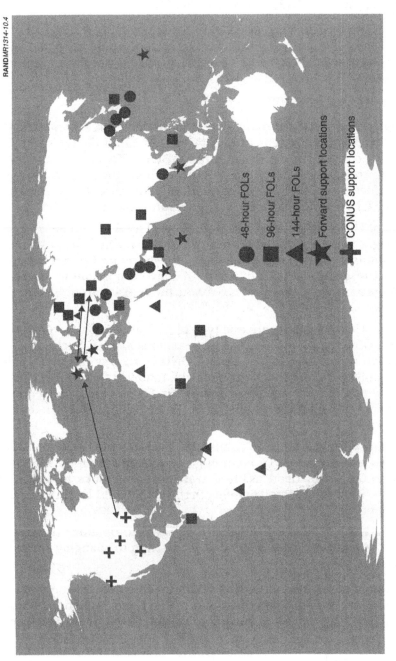

Figure 10.4—Potential Global ACS Network

tions may affect alliances. The analytical framework introduced here is being expanded and linked with methods for taking additional issues into account. The primary focus is on areas of vital U.S. interests that are under significant threat (Figure 10.4 shows clusters of FOLs in Korea, SWA, and the Balkans).

As new policies are developed and implemented, as the Air Force gains experience with expeditionary operations, and as new technologies for ground support, munitions, shelter, and other resources become available, the system will need adjustment to reflect new capabilities. Improvements in transport times, weight, and equipment reliability may favor greater CONUS support and shrinking the network of FSLs.

An advantage of using an analytic framework, such as the one described here, is that it helps focus research and attention on areas where footprint reductions could have big payoffs. Munitions is a key area where reductions in weight and assembly times could pay big dividends in deployment speed. For operations at bare bases, where shelter must be established, the development and deployment of more lightweight shelters (e.g., the Aeronautical Systems Center small-shelters program; "AEF Hotels," temporary shelters) can also pay dividends in deployment speed and footprint. Changes in these areas will not be made immediately, however, and the structure outlined above will enable expeditionary operations in the near term.

Peacetime cost is an important element of this and other analyses. The new support concept may help contain costs by consolidating assets; reducing deployments for technical personnel; using host-nation facilities; and, possibly, sharing costs with allies. Considerable infrastructure, including buildings and large stockpiles of WRM, may already be available in Europe.

Limited testing of our envisioned ACS occurred during ONA. Before the air war over Serbia, The U.S. Air Forces in Europe (USAFE) Director of Logistics (USAFE/LG) consolidated storage of WRM at Sanem, Luxembourg. During ONA, USAFE/LG established consolidated repair facilities at Royal Air Force Base Lakenheath and Spangdahlem Air Base, Germany, and created an intratheater distribution system to provide service between FSLs and FOLs. Munitions ships designated for use in another AOR were moved to support ONA munitions resupply. This transfer of assets between theaters raised several issues about how nonunit resources should be stored for use in multiple AORs.

ONA raised several general issues for those designing the future ACS system. Organizing the support for ONA took time that may not always be available in war. Heroic efforts were required to overcome shortfalls in system, training, and concepts of operation, which raised questions about what new efforts should be institutionalized in an ACS system. And some resources needed for ONA were tied to other AORs, which lead to questions about how logistics support can become more of a strategic rather than a tactical asset.

STRATEGIC AND LONG-TERM PLANNING FOR THE ACS SYSTEM

Building an ACS system requires many decisions about prepositioning and the location of support processes, including the categories of FOLs and FSLs. While the prototype models we used in this analysis deal with process characteristics and rough costs, support decisions must also account for threat situations and political considerations that change over time.

Strategic planning for an ACS system must be global and evolving. A global perspective is needed because the combination of cost constraints, political considerations, and support characteristics may dictate that some support for a particular theater or subregion be provided from facilities in another region.

This point is not theoretical. Much of SWA is politically volatile, and support there might better be provided from outside the region, as, indeed, some is now from Europe and Diego Garcia Island. The configuration of FOLs and FSLs is critical in sizing the aircraft fleet and in setting up its refueling infrastructure to support all theaters.

Strategic planning must evolve because the new security environment includes small, short-notice contingencies and continually changing threats. Geographic areas of critical interest will change over time, as will the specific threats within them. An expeditionary ACS system designed today would be oriented toward SWA and Korea, but within a decade, these regions could be at peace, and new threats could emerge elsewhere.

In addition to political changes, support processes and technologies may also change, as the Air Force continues to move to a more expeditionary footing and seeks to reduce support footprints while maintaining effectiveness. We expect to see many process and technology changes over the next ten years that will force reevaluations of the ACS system.

The need for global and evolving planning will require centralized planning in which trade-offs of cost, politics, and effectiveness are made for the system as a whole and ensure that each theater is appropriately protected and supported. This goes against the current practice of giving each theater commander control of all theater resources. Peacetime cost considerations alone require that facilities not be duplicated unnecessarily across theaters.

Changes in the force structure will also require changes to the support structure. The F-22, for example, is designed to have one-half the support footprint of the F-15. The Joint Strike Fighter is also designed to reduce support requirements. Air Force war games, particularly the Future Capabilities games, have experimented with radically different forces relying on standoff capabilities or space-based weapons. All these developments will lead to changes in both support requirements and in the options that are most attractive under peacetime cost constraints.

In assessing all these issues, it is advantageous to handle long-term changes in the same way as short-term modifications to policy and technology (see Tripp et al., 1999b). New technologies, political developments, and budget changes require continual reassessment of the support system configuration. New force structures will require different support resources, in turn requiring new support structures. For long-term decisions, the ability to perform quick-turnaround, exploratory analysis of different support structures becomes even more important.

REFERENCES

Air Force Materiel Command, briefing, Wright-Patterson AFB, Ohio, July 6, 1999.

Amouzegar, Mahyar, Lionel Galway, and Amanda Geller, *Supporting Expeditionary Aerospace Forces: Alternatives for Jet Engine Intermediate Maintenance*, Santa Monica, Calif.: RAND, MR-1431-AF, 2002.

Feinberg, Amatzia, Hyman L. Shulman, Louis W. Miller, and Robert S. Tripp, *Supporting Expeditionary Aerospace Forces: Expanded Analysis of LANTIRN Options*, Santa Monica, Calif.: RAND, MR-1225-AF, 2001.

Feinberg, Amatzia, Eric Peltz, James Leftwich, Robert S. Tripp, Mahyar Amouzegar, Russell Grunch, John Drew, Tom LaTourrette, and C. Robert Roll, *Supporting Expeditionary Aerospace Forces: Lessons from the Air War over Serbia*, Santa Monica, Calif.: RAND, MR-1263-AF, 2002.

Galway, Lionel A., Robert S. Tripp, Timothy L. Ramey, and John G. Drew, *Supporting Expeditionary Aerospace Forces: New Agile Combat Support Postures*, Santa Monica, Calif.: RAND, MR-1075-AF, 2000.

Kumashiro, Maj Patrick T., *USAF Centralized Intermediate Repair Facilities (CIRF) Test Plan, 1 September 2001–1 March 2002*, April 2001.

Menendez, Lt Col Michael, AF/IL, Logistics Transformation Team, electronic correspondence to Robert S. Tripp, October 5, 1999.

Peltz, Eric, et al., "Evaluation of F-15 Avionics Intermediate Maintenance Concepts for Meeting Expeditionary Aerospace Force Packages," *Air Force Journal of Logistics*, Vol. 23, No. 4, Winter 1999.

Peltz, Eric, Hyman L. Shulman, Robert S. Tripp, Timothy Ramey, Randy King, and John G. Drew, *Supporting Expeditionary Aerospace Forces: An Analysis of F-15 Avionics Options*, Santa Monica, Calif.: RAND, MR-1174-AF, 2000.

Tripp, Robert S., et al., "Strategic EAF Planning—Expeditionary Air Power, Part 2," *Air Force Journal of Logistics*, Vol. 23, No. 3, Fall 1999a, pp. 4–9.

Tripp, Robert S., Lionel A. Galway, Paul S. Killingsworth, Eric L. Peltz, Timothy L. Ramey, and John G. Drew, *Supporting Expeditionary*

Aerospace Forces: An Integrated Strategic Agile Combat Support Planning Framework, Santa Monica, Calif.: RAND, MR-1056-AF, 1999b.

Tripp, Robert S., Lionel A. Galway, Timothy L. Ramey, and Mahyar Amouzegar, *Supporting Expeditionary Aerospace Forces: A Concept for Evolving the Agile Combat Support/Mobility System of the Future*, Santa Monica, Calif.: RAND, MR-1179-AF, 2000.

Chapter Eleven

STRATEGIC SOURCING IN THE AIR FORCE
Frank Camm

Over the last 20 years, total quality management has penetrated the best-led organizations in the United States so thoroughly that it has disappeared into the day-to-day routines of their planning and operations. The result has been increasing adoption of a simple, customer-oriented perspective (Levine and Luck, 1994[1]), in which each organization strives to

1. identify who its customers are and what they want

2. identify the processes it uses to serve the customers and align the processes with customer demands as closely as possible

3. work continually to improve the quality of its knowledge about its customers and their needs and to improve the performance of the processes it uses to serve those needs.

This perspective has aligned whole organizations to central goals of continuing, customer-focused improvement. When looking to their own suppliers, these organizations naturally applied a similar perspective, one that aligns whole supply or value chains to central goals focused on the ultimate customers of these supply chains. As suppliers came to play a more and more integral part in efforts to align all relevant processes to customer needs, the strategic importance of the suppliers became increasingly obvious. These organizations evolved

---

[1]Compare Kaplan and Norton (1996). Many of these patterns emerged earlier in Japan. Since 1990 or so, they have been coevolving in high-quality organizations all over the world.

the concept of *strategic sourcing*, which they used to select sources and manage them in ways that explicitly served the broader strategic goals of the firm and, in particular, the needs of the firm's customers.

Strategic sourcing and broader innovative purchasing and supply management practices are increasingly helping the best organizations to *align each source they rely on—internal or external—with customer preferences and to coordinate these sources so that they work together toward common goals* (Pint and Baldwin, 1997[2]). What do these practices mean for the Air Force?

The Air Force already spends over half of its budget on purchased goods and services other than weapon systems.[3] It has an active program to expand such purchases through additional outsourcing, privatization, public-private partnerships, and other agreements with external providers. Under acquisition reform,[4] to help improve its relationships with its external sources and the goods and services that they provide, it is committed to import many services acquisition practices observed in a pure commercial setting into the Air Force.[5] As part of that effort, the Air Force contracting community refocused itself strategically to become the "business advisor to the Air Force," an advisor that can help all parts of the Air Force make better use of external sources for the services that they require (SAF/AQC, 1998). In the Quadrennial Defense Review (QDR) and the

---

[2]As Moore, Baldwin, et al. also noted in an unpublished 2001 RAND work.

[3]As Moore, Baldwin, et al. noted in an unpublished 2001 RAND work.

[4]The Air Force approach to acquisition reform is spelled out in a series of 18 specific initiatives called *Lightning Bolts*. The first 11, issued in 1995, focused primarily on issues relevant to weapon system acquisition. The next seven, issued in April 1999, gave increasing attention to issues relevant to acquiring services. Of particular importance to strategic sourcing is Lightning Bolt 99-7, "Product Support Partnerships," which commits the Air Force to "pursue public-private partnering to take advantage of the best government or commercial repair sources" for weapon systems. For information on the Lightning Bolts, see SAF/AQ (1999). The Air Force is also implementing other recent changes in DoD contracting policy that affect a wide range of policies relevant to services acquisition. Many of these changes are similar to those observed in the commercial sector. For information on these, see SAF/AQC (1998).

[5]Unless the context clearly indicates otherwise, this chapter always uses the word *commercial* to refer to activities in the private sector that government purchasing policies do not affect. The term does not refer to any activity found only in the defense industrial base. This definition emphasizes practices private firms use when they deal with one another and excludes practices of private defense contractors that may result from defense contracting policy of some kind.

Defense Reform Initiative, the Office of the Secretary of Defense (OSD) invoked the ongoing "revolution in business affairs" in the commercial sector and encouraged the Air Force and other parts of the DoD to import relevant parts of that revolution to improve the performance of the defense infrastructure (Cohen, 1997; DoD, 1997). Broadly speaking, then, the spirit of strategic sourcing and other forms of innovative purchasing and supply management is alive and active in the Air Force.

In practice, the Air Force is focusing its efforts to emulate best commercial sourcing practices on *competitive sourcing*—public-private competitions for support activities other than depot-level maintenance that create incentives for improved performance and, in the process, determine where external sources should replace internal sources. In fact, the entire Air Force program to improve the performance of its infrastructure in the 1997 QDR focused on expanded competitive sourcing. The Air Force has since discovered that competitive sourcing offers only limited opportunities and is turning its attention to other options. Some in the Air Force call this expanded perspective *strategic sourcing*, after a similar Navy effort. These Air Force and Navy strategic sourcing efforts are not the same as the prevailing commercial version but do open the door for broader efforts to align sources in DoD with appropriate warfighter and quality-of-life priorities of DoD's ultimate customers.

This chapter uses the commercial version of strategic sourcing as a frame through which to view the Air Force's ongoing efforts to improve its use of external sources of support services. It first provides more information about commercial-style strategic sourcing and its relationship to supply-chain alignment. It then asks where the current Air Force emphasis on competitive sourcing has come from. It compares and contrasts competitive sourcing with privatization, public-private partnerships and other forms of gain sharing, improved service-acquisition practices, and reengineering, all vehicles that the Air Force might employ in a broader approach to strategic sourcing. It asks how much of the benefit of commercial-style strategic sourcing the Air Force can realize via competitive sourcing alone. It then looks beyond competitive sourcing to suggest ways the Air Force could move its current, tentative view of strategic sourcing toward a broader commercial view more in keeping with a real revolution in business affairs.

STRATEGIC SOURCING AND SUPPLY-CHAIN ALIGNMENT

Supply-chain alignment typically begins with an end-to-end mapping of each process relevant to the supply or value chain. The supply chain starts with an ultimate customer and identifies each action that must occur to serve that customer. When all processes are identified and mapped, the map extends from the customer back to the source of each good and service used to provide what the customer wants.

When all processes lie within one organization, that organization can walk through the map and ask whether every action in the map really adds value to the final customer. The organization then identifies ways to eliminate actions that add no value. A typical outcome of this approach is that the organization tracks materiel more carefully and seldom allows it to sit waiting for something to happen to it. Another is that the organization replaces inspections after the fact with planning before the fact to engineer quality into a process in a way that reduces the need for after-the-fact inspection. The map also helps the organization identify ways to change the process that reduce resource consumption and increase the predictability of each element of the process, from forecasts of demand through each element of the production process to arrivals of materiel and other inputs to the process. A typical outcome is that the organization manages data on all parts of the process centrally and ensures that all parts of the process use the same data to serve the same final goals.

Of course, all processes in a supply chain rarely lie within one organization. But supply-chain alignment across organizations benefits from the same kind of process mapping and redesign described above. The key to such coordination is partnership among the relevant organizations. Such partnership supports sharing information and the gains from coordination in a way that makes each participant feel that it gains more from being in the partnership than from dropping out. Process mapping and redesign take time and often advance in blocks as mutual learning about relevant processes advances. So, supply-chain alignment across organizational boundaries benefits from long-term relationships in which all participants invest in their mutual interests and become increasingly tied to one another.

Observers of partnerships like those described above often use metaphors from marriage to describe the dynamics of establishing and sustaining supply chains across organizational boundaries. That should not lead us to believe that such partnerships are loving relationships full of mutual giving. More often than not, they involve constant tugging and hauling among talented participants, each seeking to gain as much as possible from these relationships without threatening them over the long term. The more-powerful partners can be intensely demanding; less-powerful partners must continually assess whether what they give up to remain in a partnership is worthwhile. Partnerships typically establish basic standards of fairness and trust to maintain the peace. But these standards do not prevent tough bargaining; rather, they seek to ensure that coordination occurs in a way that sustains a partnership that continues to benefit everyone involved.

Figure 11.1 provides a stylized version of two supply chains relevant to the Air Force. The first case provides component maintenance services to a warfighter, for which one important input is materiel management. So a single, unified supply chain flows from materiel management through component maintenance to the warfighter. The goal is to align all these activities to ensure that they serve the needs of the warfighter, not the parochial interests of the maintenance function that manages the details for the warfighter, of the actual providers of services, or of some other priority of less importance.[6] The second case provides family housing to a military family. One important input to family housing is pest control. So a single, unified supply chain flows from pest control through family housing to the military family. The goal is to align all these activities to ensure that they serve the needs of the military family, not the parochial interests of the civil engineer that manages the details for the military family or of the providers themselves.

In looking across all the customers that Air Force supply chains serve, what the customers want from providers can be thought of as

---

[6]In practice, of course, supply chains are far more complex. Even in this simple example, warfighters actually buy services from materiel management, which buys services from maintenance, which in turn buys other materiel management services, and so on. A complete analysis would address all elements of the process. Figure 11.1 offers a highly stylized simplification of the facts, which can become quite complex.

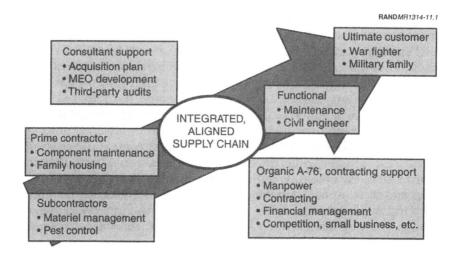

Figure 11.1—Players Relevant to an Integrated Air Force Supply Chain That Includes Contractors

high-level, Air Force–wide, strategic goals. As the Air Force serves its various customers, it will find itself attempting (1) to increase military capability, safety of flight, quality of life, and conformance with administrative law and socioeconomic goals and (2) to decrease the total ownership cost to the Air Force as a whole. These broad, strategic goals provide the basis for aligning all providers—contract and organic—with a single, coordinated set of strategic priorities.

Figure 11.1 depicts these supply chains in a setting in which the responsible Air Force functional is a buyer, not a provider, of services.[7] The services the prime and subcontractors in the figure provide could easily be—and often are—provided internally by the functional communities shown. Figure 11.1 uses a contracting setting to make the point that a supply-chain perspective is important,

---

[7]Within DoD, the term *functional* is shorthand for an organization with a specific functional responsibility or for the career field ("community") of personnel who maintain skills required to execute specialized activities associated with a functional responsibility. Examples of functionals are maintenance, civil engineering, personnel services, and financial management.

whether the services in question are provided inside or outside the Air Force.

When the providers are outside the Air Force, additional organizations become involved, as indicated in the boxes in the lower right and upper left. The manpower community coordinates and implements the Air Force competitive sourcing program. The contracting community designs and manages aspects of the sourcing process that involve external sources, from source selection, to definition and negotiation of contracts and quality assurance plans, to contract closeout. This community ensures that, at each stage in the process, the Air Force applies administrative law appropriately. The financial management community accounts for and disburses funds and thereby affects key aspects of sourcing design, particularly pricing. A variety of specialists ensure that the Air Force observes its socioeconomic commitments to full and open competition, support for small business, and so on. These specialists can in turn rely on the help of external sources, among other things, in developing acquisition plans, developing the "most efficient organizations" (MEOs) that define how internal sources make formal offers in competitive sourcing, and auditing contract and program compliance.

Observers tend to focus on these organizations in discussions of sourcing inside the Air Force. Despite currently having important responsibilities for various aspects of sourcing policy from the Air Force, these organizations are all ancillary to the supply chain itself. A clear focus on supply-chain alignment would ensure that they manage any relationship with an external source to link the source as closely as possible to the ultimate Air Force customer, not to pursue parochial, procedural concerns of their own.

Strategic sourcing allows a buying organization to think about what kinds of relationships it wants, with the sources of each good and service it buys, to link these goods and services to its customers' needs.[8] Closer links typically require more effort; they are appropriate only where they generate benefits large enough to justify the effort required.

---

[8]This discussion draws heavily on an unpublished 2001 RAND work by Moore, Baldwin, et al.

Where tight links are critical to success, a close partnership is required. The link required may be so close that the organization decides to keep the source in house. That is how the Air Force views activities directly linked to combat, for example. When the goods and services are available from many external providers and are easy to define in generic terms, partnerships are less important. More traditional, arm's-length arrangements may be adequate.

Organizations using strategic sourcing define a continuum that extends from (1) in-house provision to (2) tightly linked, long-term partnerships with sole, best-in-class external sources to (3) medium-term relations with a few partners who all work closely with the buyer to (4) short-term relations based on simple source selections to (5) simplified acquisition that uses contractual vehicles no more complicated than those associated with a credit-card transaction. As a service grows in importance to the buyer's customers, grows in dollar value, requires more investment in assets specific to the service, or is available from fewer qualified providers, these buying organizations seek a more strategic relationship, one that lies toward the first part of this continuum.[9]

Strategic sourcing then helps the buyer to structure and maintain each type of relationship, where appropriate, and identifies metrics that can be used to align a supply chain in any type of relationship considered. Strategic sourcing uses these metrics to support benchmarking that measures the performance of the supply chain relative to similar supply chains outside the organization. Also drawing on these metrics, this technique supports formal incentives, such as gain-sharing programs and programs to enhance a supplier's importance to the buyer, to encourage each participant in the supply chain to give the buyer's goals more attention than its local concerns. Looking across supply chains, strategic sourcing coordinates the buyer's demands from each source to get as much leverage from its size and strategic vision as possible. An organization using strategic sourcing can apply this approach to both internal and external sources.

---

[9]See Camm (1996) and Pint and Baldwin (1997); this is also addressed in an unpublished 2001 RAND work by Moore, Baldwin, et al.

Such an effort succeeds only if the buying organization elevates the importance of purchasing by giving it more senior-leadership visibility, integrating it effectively with the internal functional communities whose performance depends on effective sourcing actions, and developing the in-house skills required to integrate purchasing with strategic concerns. When strategic sourcing calls for a partnership with an external source, the partners must often shadow one another, maintaining many links at many levels, to sustain coordination in long-term planning and near-term execution. In practice, the interactions required to sustain shadowing between two organizations tend to blur the formal organizational boundary between them.

This approach to strategic sourcing immediately reveals three things about outsourcing. First, it is not always a good idea. Almost every organization using strategic sourcing determines that it should maintain in-house sources for some of the goods and services on which it relies. Second, outsourcing can take many forms. When outsourcing does make sense, it should be tailored to the kind of relationship the buyer wants with the source in question. Third, outsourcing is more attractive when it is managed systematically to ensure that it proceeds in the context of a relationship that promotes the buyer's strategic goals. Learning more about strategic sourcing teaches a buyer to assess risk better and to manage it in the form of the relationship of choice; as a result, the buyer associates less risk with outsourcing. This third point helps explain why outsourcing continues to expand in the commercial sector; the buyers and sellers involved are learning to use relationships that make outsourcing increasingly attractive relative to organic provision.

WHY IS THE AIR FORCE INTERESTED IN OUTSOURCING?

Outsourcing was an integral part of every proposal to improve DoD's infrastructure during the 1990s.[10] These proposals often relied, implicitly or explicitly, on an almost self-evident argument that the private sector is inherently better than the government at providing a wide variety of support services that private firms commonly sell to

---

[10]DoD has had programs in place to support outsourcing since at least the 1950s. But the defense reform efforts of the 1990s have given outsourcing a new emphasis. See, for example, Commission on Roles and Missions of the Armed Forces (1995); Defense Science Board (1996a, 1996b); Cohen (1997); and DoD (1997).

other private firms. Examples include logistics, facility management, information, personal, and other generic business services. Advocates have offered such an argument repeatedly, almost as an article of faith in the capitalist way of life, which proved its superiority to anyone who still doubted it with the dissolution of the Soviet Union.[11] The obvious implication: DoD should get out of the business of providing such services for itself and turn to providers in the private sector to get the supposedly inherent higher performance at lower cost.

More thoughtful observers have cited empirical data to support increased outsourcing. Through a series of influential studies, the Center for Naval Analyses (CNA) demonstrated that DoD has used public-private competitions for support services to reduce their expected costs to DoD.[12] Because these competitions have been DoD's principal method of outsourcing support activities, many have interpreted these results to say that outsourcing has increased DoD's savings. In fact, *competition* has created these savings, whether the public or private offeror won. An equally important finding in the CNA studies was that the public offeror often wins. That is, at least subject to the rules applied in these competitions, the private sector is not inherently better at providing support services than the government. The Air Force has applied the CNA methodology to look at public-private competitions during different periods, for different support activities, and for different parts of the Air Force. The qualitative results are always the same: (1) These competitions reduce costs to the Air Force, and (2) Air Force providers win many of the competitions. These results have been so influential in

---

[11]Ann Markusen's (2001, p. 14) review of relevant studies concludes that "there may indeed be savings and/or higher productivity to be gained from further Pentagon privatization. But advocates have not buttressed their case with hard evidence, especially given the complexity of the national security mission."

[12]See, for example, Marcus (1993); Tighe et al. (1996a, 1996b). The studies in footnote 10 also cited results from these studies without clarifying the caveats relevant to interpreting these results. These studies have shown that the public and private offers generated by one particular kind of public-private competition have systematically called for fewer people to perform services than DoD had used to provide these services internally before conducting the competition. Analysts have raised questions about this measure, but the estimated savings are so large and consistent over time, activity, and location that most observers accept that this kind of competition leads to significant savings over the long term. Compare Gates and Robbert (2000).

large part because they rely so directly on quantitative, empirical evidence from DoD's experience with public-private competitions.

Others noticed that outsourcing was expanding in the private sector. Private firms were relying more heavily on other private firms to provide support services that private firms had traditionally provided for themselves. This trend was even more pronounced among the "best commercial firms"—firms recognized by their peers as being the best in their class of business. This trend suggested that, even if the private sector was not inherently better than the government at providing support services, new ways of providing these services were making an external source more attractive than an internal source.[13] Alternatively, as firms focused their management attention on their own "core competencies"—the capabilities most important to their future success and survival—they were increasingly willing to rely on other firms—those with core competencies in support services—to provide the services they needed but did not want to focus management attention on.[14]

Why should this not be as true for government as it has been proving to be for commercial firms? In particular, why should DoD not focus on its military-unique mission and rely on others more expert at providing support services that were not unique to the military? John P. White, then–Deputy Secretary of Defense, has said that

> outsourcing allows the department to focus on its core competencies—that is, conducting military operations in the interests of the United States. That is where our focus ought to be: not on the ancillary services, but on the mission. (White, 1996.)

Gen Ronald Fogelman, Chief of Staff of the Air Force at the time, stated this perspective clearly when he said that

> Our warfighting activities will be designed for effectiveness and our support will be designed for efficiency . . . . support activities not

---

[13]For more information, see Camm (1996), Pint and Baldwin (1997), Laseter (1998), Lewis (1995), and Gattorna (1998).

[14]These arguments are flip sides of the same coin. As better methods emerge for managing a relationship between two organizations, they are both better able to focus on their own expertise and are more comfortable doing so.

deployed for combat will be performed by a robust civilian and competitive private sector.[15]

Such arguments built the case for increased outsourcing and privatization in the Air Force and the rest of DoD. They help explain why the Air Force has given outsourcing such close management attention and has developed a variety of programs to identify opportunities to make greater use of external sources to provide the support services that it needs. Historical evidence on DoD's experience to date with public-private competitions, in particular, helps explain the heavy emphasis that the Air Force gave it in its 1997 QDR plan. Arguments about core competencies and management focus begin to touch on the issues raised in commercial-style strategic sourcing. But for the most part, to date, the Air Force has shown little awareness of how important the details of strategic sourcing are likely to be to a successful outsourcing program.

POLICY ALTERNATIVES RELEVANT TO AN AIR FORCE STRATEGIC SOURCING PROGRAM

Policy debates about outsourcing, privatization, competitive sourcing, reengineering, and so on, often lose sight of a simple but important point. They are all simply alternative ways to do one thing: improve Air Force support services by improving their performance and/or reducing their total ownership cost to the Air Force as a whole. Because each dominates the others in particular places, the Air Force management portfolio should retain all of them. A formal strategic sourcing program could provide an effective way to organize and manage such a portfolio. Although the focus here is on competitive sourcing, it is worth taking a moment to define and contrast these alternatives, to make them clear in the discussion ahead.

Outsourcing

Outsourcing involves a decision to continue using a service but to stop providing it internally and buy it from a new external source. That means the Air Force buyer must remain engaged and must find

---

[15]Quoted in Northington (1997).

new, contract-based methods to ensure that it gets at least as much from an external source as it has traditionally gotten from an internal source. For practical purposes, with a few significant exceptions, an Air Force buyer can do this only by holding a *public-private competition* to determine whether an external source is really better than an internal source. The exceptions involve small activities, for which the costs of a public-private competition overwhelm any potential benefits, and activities for which no government civilian jobs are at risk.

Private sources of services have consistently argued that inherent differences between the public and private sectors, inherent biases in favor of government sources on the part of government buyers, and persistent differences in cost accounts make fair public-private competitions impossible.[16] Such concerns have led to many adjustments in the rules of engagement for such competitions, but congressional and administration support for public-private competitions remains strong. In effect, these competitions help protect government jobs and give government workers a chance to fight for their jobs by improving their own performance. So public-private competition is likely to remain the dominant road to any outsourcing in the Air Force for the foreseeable future.

Office of Management and Budget (OMB) Circular A-76 has jurisdiction over all public-private competitions other than those for depot-level maintenance activities (OMB, 1996). This circular prescribes very specific guidelines and accounting systems for these competitions; depot-level maintenance activities, when they occur, follow similar guidelines and use accounts that attempt to level the playing field.[17]

Privatization

In the Air Force, privatization can take its common meaning—government sale of a physical asset and the service activities associated

---

[16]As discussed in an unpublished 1993 work by Camm, Dreyfuss, and Hoffmeyer. For a recent variation on this theme, see Conklin (1999).

[17]Depot-level competitions are currently quiescent, because Congress requires each armed service to keep at least 50 percent of its depot-level maintenance workload in house. This rule currently prevents any further depot outsourcing and hence stops depot-level competitions.

with it to a private-sector source—but such sales are rare for two reasons. First, the Air Force produces very few services that it does not also consume itself. Hence, after transferring an in-house activity and its associated assets to an external source, the Air Force almost always continues to buy this service from the new source. Such a change is typically called outsourcing and is governed by the guidelines for outsourcing described above. In contrast, simply ending the demand for a service and selling the assets associated with it to a private firm is typically thought of as property disposal, not privatization. So, the kind of privatization most often reported in the press—the sale of a steel mill, say, or an airline—rarely occurs in the Air Force. Second, even if such a sale did occur, the Air Force would have only limited rights to retain any revenues the sale generated. With a few, effectively experimental, exceptions, the Air Force must remit such revenues to the Treasury and so has little incentive to sell assets.

Privatization in the Air Force more typically refers to an agreement in which a private firm (1) gets the use of an Air Force asset in exchange for providing some, sometimes unrelated, service in kind or (2) finances a new asset and gives the Air Force first right of refusal to use that asset in exchange for payments for that use. An example of case 1 would be an air base transferring its electricity distribution grid to a local private utility in exchange for some amount of delivered power at discounted rates over a specified period. An example of case 2 would be a private firm building family housing on or near a base and offering to rent it to Air Force personnel in exchange for their housing allowances. If no one is interested, the firm is free to rent the housing to nonmilitary families, with some restrictions. Seen this way, privatization serves primarily to provide additional (private) investment funding for support services, thereby allowing the Air Force to free the budgetary funds Congress provides for use elsewhere.

Gain Sharing

Gain sharing works in a similar way. For example, the Air Force can potentially give a private firm access to land it controls in exchange for receiving buildings and/or facility and property management services from the new building assets that the firm creates on the land. Alternatively, a private firm could invest in energy conservation measures at an Air Force facility. The firm would measure the change in energy consumption at the facility relative to a mutually negotiated baseline, then take a share of the resulting financial saving as payment. In each case, a private source applies private financial capital to an Air Force asset to generate new value and shares that value with the Air Force.[18]

Innovative Contracting

Innovative contracting allows the Air Force to revisit its approach to acquiring the services that it already buys from external sources in ways that improve the quality or reduce the cost of the services. Acquisition reform offers the Air Force a wide variety of mechanisms for doing this,[19] such as the following:

- improving communication with potential sources during source selection

- bundling services in ways that improve coordination within a bundle and that attract higher-quality providers

- writing statements of objectives that more clearly identify the Air Force's true needs

- designing source selections to attract and select higher-quality providers

- writing contracts that better link providers' capabilities to the Air Force's true needs

- managing contracts to encourage improvement over time.[20]

Such improvements offer large benefits to organizations that, like the Air Force, spend a large portion of their total budgets on purchased goods and services.

---

[18]For more details, see U.S. General Accounting Office (1999), Chang et al. (1999), and Koegh (1999).

[19]As will be discussed below, these techniques can be applied as part of competitive sourcing or privatization as well. But they are available even in the absence of any new competitive sourcing or privatization; they work equally well (and probably better, because of fewer constraints) for services already acquired from external sources.

[20]As discussed in an unpublished 1999 work by Camm and Moore.

Reengineering

All the alternatives above look to the private sector to help the Air Force improve its support services. The Air Force can also use *reengineering* to improve the processes of its internal sources. Reengineering means different things to different people. In the commercial sector, it typically refers to quantum process changes that break down traditional functional and organizational boundaries, often relying on new information technology to do so, and offer qualitatively new processes that dramatically improve performance and reduce cost. Although nothing prevents the Air Force from seeking such radical changes in its support services, such changes are unusual. More-gradual process changes do occur, especially within functional communities. For parts of the Air Force in which private sources play little or no role, this may be the only alternative available to improve support services (see, for example, Girardini et al., 1995).

Summary

The Air Force is currently pursuing all these alternatives in one way or another. As noted above, however, it currently clearly favors competitive sourcing over the others. We next turn to a closer examination of how the Air Force can pursue strategic sourcing and supply-chain alignment in the context of its competitive sourcing program.

PURSUING STRATEGIC SOURCING AND SUPPLY-CHAIN ALIGNMENT IN COMPETITIVE SOURCING

Figure 11.2 presents a stylized end-to-end map of the process that the Air Force could use to identify candidates for competitive sourcing, hold competitions, and develop and manage relationships with the sources that win these competitions.[21] Very briefly, the process works as follows:

- Determine whether the activity is eligible for competitive sourcing. It is eligible if it is not inherently governmental and does not present significant risks if provided by an external source.

---

[21]This map is a variation on that explained in an unpublished RAND 1997 work by Camm and Moore.

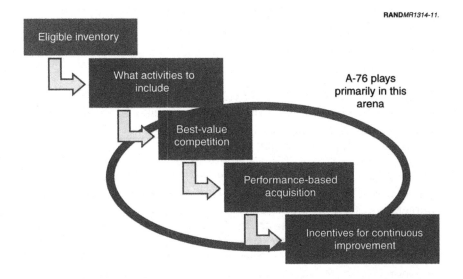

Figure 11.2—Process Steps for Choosing Competitive Sourcing Candidates, Running Competitions, and Managing Relationships with the Winners

- Bundle the activity to achieve the best economies available from commercial sources.

- Hold a competition among private sources that allows the Air Force to choose the source that provides the best value to the Air Force. The Air Force will then compare the cost of the source chosen with that of providing a similar level of service in house.

- Define the acquisition in terms of measures of performance relevant to the ultimate customer for the service. Use these measures to define the scope of the work and to define the quality assurance plan.

- Translate the performance measures identified above into incentives that induce the source to work with the Air Force, over time, to improve continually the level of service its provides.

OMB Circular A-76, which so heavily shapes current Air Force policy on competitive sourcing, operates mainly in the area indicated in the figure. But the Air Force must ensure that all steps of any competitive sourcing program are compatible with the elements of the circular that operate in this region. Taking this simple map as given, how can the Air Force exploit opportunities from strategic sourcing and supply-chain alignment in each step of the map?

Eligible Inventory

In 1998, Defense Reform Initiative Directive (DRID) 20 required the Air Force and other DoD components to review every position to identify which were eligible for competitive sourcing (Hamre, 1998). An extensive internal review that Headquarters Air Force and the major commands (MAJCOMs) conducted determined that only about 61,000 positions (9 percent of the 673,000 positions in the Air Force) were eligible for competitive sourcing review.[22] This restrictive percentage is the result of very conservative assumptions about the risks associated with outsourcing. The guidelines the Air Force used to execute the DRID 20 assessment instructed reviewers to err on the conservative side wherever they had doubts about the advisability of outsourcing.

Under current guidance, the Air Force will conduct such reviews repeatedly in the future. An approach based on strategic sourcing can help identify additional opportunities for competitive sourcing by helping the Air Force (1) to revise the process and guidelines for the review so that it properly reflects a strategic view of outsourcing and how it fits in the Air Force's broader plans and (2) to devise better relationships with key sources of services, thereby reducing the risks of outsourcing and increasing the relative desirability of using external sources. First, this section will consider how to improve the process and guidelines the Air Force uses to choose competitive sourcing candidates.[23]

The Air Force relied heavily on functional channels to execute the DRID 20 review.[24] Such functions as maintenance, civil engineering,

---

[22]The Headquarters Air Force Directorate of Manpower and Organization (AF/XPM) reports that the Air Force reviewed 721,000 positions (U.S. Air Force, 1998). The number in the text is the result of a recount at RAND, by Gary Massey, Craig Moore, and Joe Bolten, designed to eliminate double counting of positions with multiple identifiers.

[23]This discussion draws on material in an unpublished RAND 1999 work by Camm, Moore, and Roll.

[24]It used a very similar approach in a 1996–1997 review of competitive sourcing candidates, called Project Jump Start, and identified almost exactly the same number of candidates.

and financial management decided which of their positions would be risky to outsource. The best commercial firms have learned that such functionals have strong incentives to be conservative to protect their own career fields from outsourcing. These firms conduct comparable strategic reviews of candidates for outsourcing at a higher level, asking the customers of the services in question what risks an external source would present. The Air Force would find it difficult to go that far because it relies heavily on a functional structure for decisionmaking. But it could empower nonfunctional players, such as the manpower community or the command staffs of the MAJCOMs, by providing them the resources and the authority to review the functional inputs and to refine them over time.

DRID 20 effectively imposed a series of screens for determining what risks each position might pose if outsourced. A position had to pass every screen to become a candidate for competitive sourcing. Table 11.1 summarizes the most important screens. Any future review will likely use a similar approach. The challenge is to ensure that the screens embody the true strategic values of the Air Force: That is the first step toward ensuring that competitive sourcing serves the interests of the ultimate customers—the warfighter and the military family.

For example, the term *inherently governmental* is precisely defined in the Office of Federal Procurement Policy's Policy Letter 92-1 (1992)

Table 11.1

Expanding the Inventory of Candidates Eligible for
Competitive Sourcing

Screen	Adjustment
Inherently governmental	Move line between real policy discretion, administration
Deployed in war plans	Consider positions that deploy outside the area of responsibility
Required in CONUS to support rotation base	Return forces to CONUS; adjust rotation base policy
Required to sustain career field	Refine career field; expand use of lateral entry
Traditionally military or organic	Change policy (e.g., military bands)

and OMB Circular A-76. But the definition allows a great deal of subjective judgment about the distinction between activities that require policy discretion, and therefore must remain in house, and activities that simply administer, support, or execute policy decisions made elsewhere. Policy can be adjusted to reduce the number of positions that must remain in house. Appropriate experience and feedback can help those making subjective judgments in the field while implementing this screen refine their judgment over time and more accurately reflect the true intent of the policy letter.

The definition of *deployment* also potentially allows discretion. Today, no manpower position that moves geographically during a contingency can be competitively sourced; all such positions must remain in house. The Air Force could potentially open to competitive sourcing review some or all positions (1) deployed within the continental United States (CONUS) or (2) deployed from CONUS to destinations outside CONUS but still outside the area of responsibility during a contingency.

The same could be said of each screen shown on the chart. The Air Force can potentially adjust its policy on the rotation base or the locations of forces during routine peacetime periods. It can redefine career fields and the activities used to build skills in these career fields. As part of such a consideration, the Air Force can consider the desirability of lateral entry, in areas where the Air Force does not currently use it, as a way of reducing the need for the less-skilled positions that commercial firms often consider for outsourcing. And the Air Force can ask how many traditionally military activities, such as bands, it really needs.

All these decisions have substantive implications; they are not simply statements on paper. The Air Force cannot make these decisions lightly and should not allow a desire to expand the competitive sourcing pool to dominate such strategic decisions if that is inappropriate. That said, the Air Force has the discretion to revisit its past decisions in each of these areas.

Greater strategic interest in competitive sourcing could suggest a less conservative interpretation of each of these screens. As the Air Force continues to monitor these screens, as well as their effects on performance and cost, over time, it can continue adjusting the screens. Such adjustments could easily lead to a progressively less conservative approach over time, one that opens significantly more than 9 percent of the Air Force to review.

What Activities to Include

As part of their strategic sourcing programs, the best commercial firms are progressively expanding the bundles of services they buy from external sources.[25] Larger bundles are not always better, but expanding bundles from their current size can offer a number of advantages:

- greater access to economies of scale and scope allowed by new information systems, benchmarking networks, and better coordination or alignment of related activities in a supply chain

- better ability to hand off noncore activities in a controlled way that supports efforts to maintain management focus in house on the strategic core

- better access to the best providers, which seek the relationships that can add the most value through partnership arrangements; the cost of creating and sustaining such relationships is easier to justify in larger bundles, which also allow more opportunities for mutual gains in a partnership.

The Air Force can benefit from these advantages of bundling but must always keep in mind that it has socioeconomic goals that may conflict with bundling. In particular, the Air Force has a legislatively mandated commitment to support and promote the interests of small businesses as sources for Air Force services. As the Air Force examines the appropriate approach to bundling, it must balance its strategic concerns for the warfighter and military family with its strategic socioeconomic concerns.

Such concerns are only tangential to competitive sourcing. The Air Force does not define the screens above in terms of the bundles that will be reviewed; it defines them more in terms of individual posi-

---

[25]Larger bundles cover, for example, more separate but related services, more sites, or more-devolved management responsibility. This discussion draws heavily on an unpublished 2001 RAND work by Moore, Camm, and Baldwin; see Baldwin, Camm, and Moore (2001).

tions required to deal with a variety of manpower concerns in the force as a whole. Circular A-76 says nothing about bundles. It works with whatever bundles the Air Force chooses for the specific competitions and cost comparisons required under A-76. So, although the Air Force must define bundles as part of competitive sourcing, competitive sourcing per se places no limits on those bundles. The Air Force has full latitude to structure whatever bundles match its broader strategic goals.

The best way to learn what bundles will meet its strategic goals is to conduct market research. Ask the best providers what they can offer; how they structure these bundles to give the buyer effective control, over direction if not day-to-day operations; and what roles small business can play in these bundles as subcontractors or primes working in close coordination with partners. The Air Force will have to improve its in-house cross-functional coordination to conduct such market research. Although the contracting community is committed to expanding the use of market research in the Air Force, the functionals are likely to have the skills needed to understand how best to structure bundles.

For example, the Air Force might determine that regional bundles offer greater benefits than base-specific bundles of equipment maintenance or business services, such as payroll. Pursuing such bundles may require the Air Force to execute reengineering before it turns to a comparison of internal and external sources. The functionals are probably better able than contracting to execute the market research that might help the Air Force coordinate reengineering and competitive sourcing efforts in a particular area.

Best-Value Competition

The success of a long-term partnership depends heavily on choosing the right partner. An effective buyer seeks a partner to grow with over time, to sustain the kind of mutual learning likely to yield effective supply-chain alignment over time. Whatever the source-selection method, it must help the buyer identify a partner with matching strategic needs and values today and in the future, whatever those needs and values might turn out to be. The best commercial firms use "best-value" competitions or comparable selection processes to winnow out competing sources until they find ones they are willing to invest in over the long term. Best-value competitions compare alternatives using not only current or historical costs but also a variety of current and historical quality measures, as well as indicators of such future capabilities as financial strength, investment patterns, training programs, and strategic goals.[26]

Providers seeking partners similarly review the buyers to whom they offer services. Providers clearly do not control competitions as buyers do but can choose which competitions to participate in. Properly structured and conducted best-value competitions convey signals to providers that the buyers understand partnership and will be able to work with the providers in a way that both can add value and prosper over the long term. Given an expanding pool of commercial opportunities to participate in, the best providers can be quite selective about how much to invest in offers to the Air Force as a buyer.

The Air Force is constrained in the kinds of best-value procedures it can apply to competitive sourcing. Most directly, Circular A-76 does not allow the Air Force (or any other federal buyer) to compare public and private offers on any grounds other than monetary cost. But the Air Force can use best-value criteria to compare private offerors with one another before choosing one to compare with the government offeror. Because its competitions are subject to broader and more formal oversight than most commercial competitions, the Air Force must be more careful about how it structures and executes a best-value competition to avoid protests. That said, commercial buyers know that they can attract good offers only if they conduct fair competitions themselves. Formal federal acquisition procedures typically lead to source selections that are more costly and take longer for offerors than do their commercial counterparts; best-value competitions simply add to this distinction. And relative to their commercial counterparts, formal federal acquisition procedures require greater reliance on formal competitions and less on a buyer's administrative decision, based on performance to date, to extend existing contracts to additional periods or to new activities.

Still, the Air Force can do a great deal more to incorporate best-value procedures into its competitive sourcing program than it does today. To do this, the Air Force must move beyond simple compliance with

---

[26]As discussed in an unpublished 1997 RAND work by Moore et al. (1997).

OMB Circular A-76 and become more proactive. It is true that A-76 is a burdensome regulation that the Air Force is legally required to implement. The A-76 process requires the Air Force to announce any cost comparison to Congress early in the sourcing process and then limits the time available to complete a cost comparison.[27] Given the Air Force's current skills, resources, and organizational structure, simply complying is a challenge.

But the competitions Circular A-76 requires also offer an opportunity to identify new partners who can support improved supply-chain alignment. These competitions offer literally hundreds of potential openings to begin bringing commercial-style strategic sourcing into the Air Force. Put another way, a strategic-sourcing perspective would allow the Air Force to get significantly greater increases in performance and reductions in costs than simple compliance offers. It would help the Air Force see the future value of attracting and choosing good sources and hence the value of investing today to provide the additional skills and resources and the organizational structure that could give the Air Force access to these sources.

Until the Air Force uses best-value procedures more effectively, it will not be able to identify sources that could allow it to develop the forms of productive long-term partnerships that support expanding outsourcing in the commercial sector. Perhaps even worse, it will not even be able to attract the interest of providers who could become the best partners to participate in Air Force competitions.

Performance-Based Acquisition

Once a best-value competition allows access to the best service providers, performance-based acquisition brings in the tools required to align the providers with the needs of the ultimate customers.[28] This is important in three ways:

---

[27]The A-76 process allows two years for single-activity and four years for multiple-activity studies. Such times are very long relative to commercial competitions but short relative to DoD's historical experience with A-76. Federal Acquisition Regulations slow these studies, but critical-path analyses make it clear that much shorter completion times are possible if the Air Force applies the will and skills required to complete them expeditiously.

[28]This discussion draws on Baldwin, Camm, and Moore (2000).

- A performance-based approach is a signal, like an effective best-value competition, to the best providers that a buyer understands how to work with them to their mutual advantage. By proposing to use a sophisticated best-business practice, a buyer demonstrates that it is prepared to act as a worthy partner.

- A performance-based approach allocates roles effectively to the buyer and provider. The buyer focuses on translating strategic goals into specific measures that link any support service to those goals, an inherently demand- and value-oriented approach. The provider focuses on translating strategic capabilities into specific actions that advance the buyer's goals, an inherently supply- and capability-oriented approach. Each party stays focused on its core interests.

- A performance-based approach provides an objective way for the buyer and provider to continue talking over time to update priorities, change practices that are not working, identify additional areas for cooperation, and refine the relationship in general. Performance measures provide a specific vocabulary for communication, even if that communication reveals that they need to be adjusted over time to remain effective. Such measures allow buyer and seller to align their core interests to their mutual benefit.

Circular A-76 does not say anything directly about what role performance measures might play in a cost comparison. Cost must remain a significant factor in the source selection to allow a public-private cost comparison. But in principle, it should be possible to integrate an aggressive approach to performance measurement with the basic elements A-76 requires. Once a source is chosen, A-76 has little more to say; performance-based management comes to the fore.

Three issues seem to discourage more-aggressive consideration of a performance-based approach in the Air Force. First, like best-value competition, performance-based acquisition requires an incremental initial investment at the very time when an Air Force organization conducting an A-76–based cost comparison is struggling to comply with A-76 itself. Time limits on the execution of A-76 studies appear to aggravate this concern. A more-strategic view of A-76 would allow the Air Force to recognize and perhaps even quantify the benefits of a performance-based approach. Such a view would make it easier to invest in the skills, resources, and organizational changes required to support specific performance-based service acquisitions.

Second, performance-based acquisition presents a direct challenge to the Air Force's traditional reliance on functionals to specify not just *what* the Air Force wants from a provider but *how* it will be provided. Performance-based acquisition explicitly redirects the buyer's attention away from how to provide a service to what the ultimate customer wants and how to express that in terms of its value to the customer not in terms of the means the provider uses to produce it. As traditional providers of most support services, Air Force functionals have typically developed detailed instructions and other documents on how to provide these services. A performance-based approach explicitly asks them to de-emphasize this "expertise" and give more attention to reassessing what the warfighter and military family really want. Such an approach requires (1) a new role for the functional and (2) greater cross-functional communication among the functional; the ultimate Air Force consumers; and others, such as manpower and contracting, who are more directly involved in the details of the A-76-shaped acquisition.

Third, a performance-based approach affects not only (1) how the Air Force, as a buyer, communicates with an external source over the course of a relationship but also (2) how the Air Force, as a buyer, communicates with an internal source that wins a cost comparison. In effect, when an internal source wins, performance-based acquisition points to a formal contract between the internal buyer and the internal provider and holds the internal provider accountable in a way that the Air Force rarely has in the past. Just as DoD has discovered over time that the benefits of competitive sourcing come less from outsourcing than from competition, the best commercial firms have discovered that the benefits of performance-based management of a source often come less from outsourcing than from establishing effective accountability for any source, internal or external. This view is quite new to the Air Force.

Many external sources argue that, because DoD cannot establish the same kind of specific cost and performance controls over an internal source that it routinely seeks over an external source, the execution of any agreement prevents the playing field from being level. For example, so the argument goes, the Air Force will not hold its internal sources to the same standards that it expects from its external sources. Performance-based acquisition brings this concern to center stage. Unless the Air Force can maintain real performance-based management of all its sources, internal and external, this is a fair criticism. Looked at another way, Circular A-76 offers a strategic opportunity to introduce performance-based management throughout the Air Force, for internal and external sources. Such an approach will succeed only if the Air Force leadership displays the will to take advantage of this strategic opportunity.

As with best-value competition, the Air Force must use performance-based acquisition more effectively (1) before it can identify sources that will allow the Air Force to develop productive longer-term partnerships and, (2) perhaps even worse, before it can attract the interest of providers who could become the best partners to participate in Air Force competitions.

Incentives for Continuous Improvement

Once performance-based acquisition identifies measures that the Air Force can use to align the capabilities of its sources with the needs of its ultimate customers, incentives for continuous improvement can induce these sources to act on that alignment, every day and over the course of a continuing relationship. Each day, incentives induce each source to apply its resources to the needs of the moment. Over time, incentives induce each source to seek process improvements that reduce its costs and/or increase its performance.

"Incentives" can be direct cash payments for performance. For example, an award fee gives the buyer great latitude periodically (say, quarterly) to review performance and translate it into an overall monetary judgment of the value of that performance. Formal incentive contract terms give the buyer the ability to highlight specific aspects of performance, specify expectations with an explicit formula, and reward performance in targeted areas. Periodic gainsharing agreements offer the provider the opportunity to propose investments that will benefit the buyer and that will recover the provider's investment with an additional payback only if the benefits promised materialize.

But in best commercial practice, incentives can just as often be less direct and explicit than cash. Benchmarking tracks performance improvements elsewhere and gives the buyer information it can use to "encourage" the provider to achieve similar performance improvements if it wishes to retain the contract. In a more positive vein, a buyer can extend the term or scope of a contract as a reward for good performance. A buyer can reduce oversight when performance is good, reducing administrative costs for both buyer and seller. And contracts are typically written so that either partner can terminate without cause, on 60 to 90 days' notice, if it concludes that it no longer benefits from the relationship. Such language creates incentives for both partners to deal with one another in a way that benefits them both as long as the relationship provides room for mutual benefits.

As in so many things, the Air Force cannot pursue such practices as aggressively as the best commercial firms. For example, formalized incentives typically require greater specificity in a DoD setting than in a purely commercial setting. As noted above, regulations limit the Air Force's ability to extend a contract without further formal competition. And the time required to execute a full source selection limits the plausibility of the government's threat to withdraw from a relationship on, say, 90 days' notice, thereby limiting its ability to demand performance from an internal or external source.

Furthermore, competitive sourcing limits the Air Force's ability to create such incentives to the extent that (1) the cost comparison itself affects the definition of such incentives or (2) the Air Force cannot apply incentives equally to public and private sources. The two problems are related. The Air Force has more freedom to use formal incentives with external sources than with internal. This immediately presents difficulties associated with a level playing field. And Circular A-76 presents formal accounting rules for evaluating formal incentives that differ for potential internal and external sources. Even though OMB and DoD seek rules that level the playing field, the accounting rules can themselves add to concerns about comparing internal and external sources fairly. These problems will always be present.

But as in so many things, the Air Force can make greater use of incentives to induce continuous improvement in competitive sourcing than it does today. Use of incentives presents benefits and challenges similar to those for best-value competition and performance-based acquisition. On the benefit side, it supports the creation and sustainment of the long-term partnerships so important to effective supply-chain alignment. And because of this, the Air Force's use of incentives demonstrates to potential providers that it is serious about being a good partner. On the challenge side, the use of incentives requires an initial investment to define incentives, yet time is short and the Air Force is still grappling with the basics of OMB Circular A-76 itself. The Air Force can use the benefits that effective incentives offer to justify the investments in skills, resources, and organizational change necessary for introducing such incentives into competitive sourcing actions.

Discussion

This brief review has shown the following:

1. A strategic perspective, which seeks to use supply-chain alignment to link the Air Force's ultimate customers with each of its sources, can potentially help the Air Force improve each step in the end-to-end planning and management process relevant to its competitive sourcing program. This perspective offers value by aligning the steps with one another and with what the Air Force ultimately really cares about. It does this by emphasizing that, although different internal organizations often execute different steps in the process, a coordinated, strategic view can help ensure that these organizations work together toward similar goals.

2. At each step in the process, the Air Force can learn from best commercial practice but must also keep its own goals and circumstances in mind. Development of the "eligible inventory," for example, emulates commercial practice by formally linking sourcing decisions to the strategic priorities of the Air Force. But it alters commercial practice to reflect the importance to the Air Force of administrative law on inherently government activities, war plans, a peacetime rotation base, internal labor markets within career fields, and so on. As it decides "what activities to include" in a bundle, the Air Force can examine commercial bundling trends but must adapt them to reflect its own socioeconomic goals. On best-value competition, performance-based acquisition, and incentives for continuous improvement, administrative law prevents the Air Force from doing everything it observes in the commercial sector, but acquisition reform and

other recent policy changes in DoD allow the Air Force to do a great deal more than it has in the past. It must shape its specific approach in each step to the nature of the specific opportunities now available.

3. The competitive sourcing program presents special challenges to the application of a strategic approach to sourcing. The "eligible inventory" today limits the competitive sourcing program itself to only 9 percent of the force. The Air Force cannot effectively use best-value competitions to compare internal and external sources. Performance-based acquisition and incentives for continuous improvements raise a variety of questions about the equal treatment of internal and external sources in competitive sourcing. And the challenge of executing the basic elements of OMB Circular A-76 alone is testing the will and capabilities of many parts of the Air Force. Doing more without some change in policy and leadership direction may be asking too much. That said, none of these challenges is great enough to prevent the application of a strategic perspective to each step of the planning and management processes relevant to competitive sourcing. If the senior leadership of the Air Force shows the will to pursue a strategic approach in competitive sourcing, the Air Force can potentially benefit in terms of reduced costs and improved performance.

In sum, a more-strategic approach to competitive sourcing could enhance supply-chain alignment for the activities affected and hence help the Air Force pursue its broader strategic goals. But a more-strategic approach will require a more-strategic commitment.

LOOKING BEYOND COMPETITIVE SOURCING

As noted above, only 9 percent of the Air Force is currently eligible for competitive sourcing. Even if the eligible inventory were doubled to 18 percent, which would take considerable time if it is possible at all, the scope of competitive sourcing is narrow. A truly strategic approach to sourcing would consider the sources of all support services and seek ways to align these more closely with the needs of the Air Force's ultimate customers. Such a strategic approach would set the stage for complete review and improvement of the processes the Air Force uses to provide or acquire all support services.

What would such a program look like? It would recognize the value of integrating all the options the Air Force has available to align its supply chains for support services and to improve the processes in them, including privatization, gain sharing, innovative contracting, reengineering, and competitive sourcing. This integration would ensure that the Air Force (1) uses these options in a coordinated fashion to pursue similar goals and (2) allocates its leadership resources and other scarce resources among these alternatives in a way that maximizes their joint effects on broad Air Force goals. How can that be done?

First, the Air Force would use similar metrics to evaluate all the alternatives. The metrics would reflect specific expected effects on military capability, quality of life, and total ownership cost. Such metrics would not reflect numbers of positions reviewed, numbers of contracts written using a simplified format, and so on. True outcome-oriented metrics provide a common basis for viewing these alternatives together. It is at least as important that such metrics help the Air Force leadership understand the strategic importance of these improvement programs. A leader unimpressed by the number of reengineering candidates identified can be very impressed by a reduction in deployment footprint or an increase in job satisfaction among senior enlisted maintenance personnel.

Second, the Air Force would use these metrics to track the real effects of expanded privatization, gain sharing, and so on. Support for competitive sourcing exists in DoD today primarily because of the evidence that it is likely to reduce operating costs; no comparable evidence exists for any of the other programs mentioned here. The Air Force will have to build the case for these as it goes along. What does that mean? In each case, the Air Force needs a reliable baseline, a way to measure departures from it over time, and third-party verification of the measurements of the baseline and departures from it. In addition, the measures should ultimately provide clear evidence of improvements in outcomes relevant to the Air Force as a whole.

Third, building on a common set of metrics and growing evidence of real success against these metrics, the Air Force would elevate the importance of these programs within the organization by assigning their planning and execution to higher-echelon, higher-quality personnel and placing greater weight in personnel reviews and promotion decisions on successful program execution. The Air Force would need to create new multifunctional organizations to design and coordinate such programs. Common metrics would give the Air Force leadership a basis for understanding why each of these changes was worthwhile. Greater leadership emphasis would in turn encourage Air Force personnel to take the metrics seriously. The metrics would also simplify efforts to justify investments in training and in the new structures required to make these programs succeed. Without such metrics, the investments become unjustifiable cost increases precisely when money is dangerously short.

Fourth, the Air Force would invest in training to help its personnel implement each specific change undertaken and to understand how to continue aligning its supply chains for support services and for improving its support processes in a more global sense. This training would include generic skills relevant to working effectively in multifunctional teams and organizations, identifying new opportunities for improvement, and structuring improvements in effective ways, as well as skills specific to each new change made. The training would feature both ongoing classroom training and experience on the job actually doing all these things in increasingly demanding settings over a career. A career may well include time in nontraditional functional settings and perhaps even outside the government with an external source or a best-in-class provider.

Outcome-oriented metrics relevant to the Air Force leadership, a closed-loop system for measuring them when the Air Force implements a change, leadership support to ensure that personnel take the metrics and measurements seriously, and training that allows the personnel to use them effectively will together provide the ingredients for a different approach to managing support services in the Air Force. Each is required for the others to contribute to success. To be successful, the Air Force cannot drift tentatively in a new direction. It must prepare a systematic effort to change its management of support processes in a decisive, coordinated fashion. Such change is never easy for a large organization. But large, successful organizations, such as General Electric, have shown that it can be done in a way that yields remarkable benefits for the organization as a whole.[29]

---

[29]General Electric is particularly interesting, because it is larger than the Air Force and at least as global and diverse. An unpublished 2001 RAND work by Moore, Baldwin, et

The changes discussed here are process changes; they do not target specific programs, such as privatization of family housing or reengineering of base-level logistics. Particular privatizations and reengineering efforts are examples of specific programs the Air Force would pursue under this broad strategic-sourcing approach. It would pursue each program in a way that links it more closely to the Air Force's basic strategic interests and, in so doing, probably coordinates it more closely with other specific programs under way.

For example, it would be easier for the Air Force to reengineer base-level supply or payroll to develop regional or MAJCOM-level approaches before entering a competitive sourcing study. At any particular base, it would be easier to encourage a public-private partnership that provides new base office space, privatization of the utilities that support this space, and competitive sourcing of other traditional civil-engineering activities that affect the use of the space in a way that all these initiatives work together to cut total ownership costs and increase the quality of office space at the base. It would be easier to justify the investments in training and information systems required to make such efforts succeed. And it would be easier to compare efforts across activities and bases to decide where to invest the Air Force's scarce resources.

If recent experience in the commercial sector is any indicator, such an approach should yield the following kinds of changes in the Air Force:[30]

- increasing reliance on (1) the services of external parties, which will finance, own, and provide full support for such assets as computers, vehicles, chillers, engines, generators, and call centers, rather than (2) buying the assets itself and supporting them with internal or external resources

- increasing reliance on larger, cost-plus-award-fee service contracts with long-term partners that delegate responsibility for managing and coordinating broad categories of support activities but preserve flexibility so that the Air Force can ensure that

---

al. also discussed recommendations on how to implement large-scale organizational changes of this kind.

[30]As discussed in two unpublished 1999 RAND works, one by Camm and Moore and another by Camm, Moore, and Roll.

its sources of support activities respond quickly and effectively as priorities change over time

* increasing reliance on quantum process changes that cut across functional and organizational boundaries and use common databases, shared data on performance metrics, and outcome-oriented incentives to coordinate these processes in ways that yield large and continuing improvements in cost and performance

* improving the results competitive sourcing realizes but increasing reliance on noncompetitive sourcing improvement programs to improve support activities ineligible for competitive sourcing.

The Air Force will do these things not because they are judged to be desirable in their own right but because they yield demonstrable, outcome-oriented benefits that the Air Force leadership values.

The Air Force does not need any changes in law or regulations to pursue such initiatives. It has the full authority and resources that it needs to initiate such a strategic approach today. But selected changes could make such initiatives easier to pursue. As the Air Force learns more about which specific policy initiatives interest its leadership the most, it can take proposals for change forward to OSD and Congress. And the Air Force can do this with the benefit of new, credible information about how such policy changes can improve the Air Force's military capability, quality of life, and total ownership cost.

SUMMARY

Air Force efforts to improve its sourcing program have given first priority to competitive sourcing. The Air Force is interested in competitive sourcing primarily because, historically, it has clearly helped cut the costs of Air Force support services. Growing commercial interest in outsourcing stems from a different source—a growing understanding of how to align supply chains that cross organizational boundaries in ways that improve the performance and cut the costs associated with these supply chains on a continuing basis. The Air Force could get more from its competitive sourcing program if it took a more strategic, commercial-style approach to sourcing. A commercial-style strategic sourcing program would also motivate the Air Force to look beyond competitive sourcing to a variety of other approaches to improving support services that could help the Air Force more than competitive sourcing ever will.

Commercial-style strategic sourcing helps participating organizations align each part of a supply chain with the goals of the ultimate customers it serves. Differences between the goals of the Air Force and the best commercial firms mean that the Air Force should not simply emulate commercial practice and push strategic sourcing as far as these organizations have. But commercial-style strategic sourcing offers opportunities that, properly adapted, could improve Air Force support services. Such an approach would align each supply chain with the relevant concerns of warfighters and military families in the context of the Air Force competitive sourcing, privatization, gain sharing, innovative contracting, and reengineering programs. This approach would allocate resources and leadership attention among the programs in ways that would help warfighters and military families the most. And it would coordinate these programs to increase their abilities to work together in pursuit of goals relevant to warfighters and military families.

Even focusing Air Force strategic sourcing efforts on competitive sourcing alone could significantly improve the performance levels and costs of support services addressed in this program. The Air Force could do this without any significant changes in the laws, regulations, or OSD guidance that affect its current sourcing programs. A broader approach is desirable because competitive sourcing currently covers only 9 percent of Air Force manpower slots, a percentage that is unlikely to increase in the near future. A broader approach would give the Air Force access to opportunities that yield much greater benefits per dollar of resource or hour of leadership time committed to improving support services.

REFERENCES

Assistant Secretary of the Air Force (Acquisition), Air Force Acquisition Reform site, Lightning Bolts '99 page, April 1999. Online at http://www.safaq.hq.af.mil/acq_ref/bolts99/ (as of February 11, 2002).

_____, Contracting (SAF/AQC), Contracting 21: Air Force Contracting Business Plan, Washington, D.C., November 1998. Online at http://www.safaq.hq.af.mil/contracting/con21busplan/cont21w. pdf (as of December 16, 1999).

Baldwin, Laura H., Frank Camm, and Nancy Y. Moore, *Strategic Sourcing: Measuring and Managing Performance*, Santa Monica, Calif.: RAND, DB-287-AF, 2000.

_____, *Federal Contract Bundling: A Framework for Making and Justifying Decisions for Purchased Services*, Santa Monica, Calif.: RAND, MR-1224-AF, 2001.

Camm, Frank, *Expanding Private Production of Defense Services*, Santa Monica, Calif.: RAND, MR-734-CRMAF, 1996.

Camm, Frank, David Dreyfuss, and Karl Hoffmeyer, "The Role of the DoD Cost Comparability Handbook in Air Force Public/Private Competitions," unpublished RAND research, 1993.

Camm, Frank, and Nancy Y. Moore, "Strategic Sourcing: A Key to the Revolution in Business Affairs," unpublished RAND research, 1997.

_____, "Acquisition of Services in 2010: Ideas for Thinking About the Future," unpublished RAND research, 1999.

Camm, Frank, S. Craig Moore, and C. Robert Roll, "RAND's Quick Review of USAF Manpower Requirements," unpublished RAND research, December 1999.

Chang, Ike Y., Steven Galing, Carolyn Wong, Howell Yee, Elliot I. Axelband, Mark Onesi, and Kenneth P. Horn, *Use of Public-Private Partnerships to Meet Future Army Needs*, Santa Monica, Calif.: RAND, MR-997-A, 1999.

Cohen, William S., *Report of the Quadrennial Defense Review*, Washington, D.C.: U.S. Department of Defense, May 1997.

Commission on Roles and Missions of the Armed Forces, *Directions for Defense*, Washington, D.C.: U.S. Government Printing Office, May 24, 1995.

Conklin, B., "A New Model for Public-Private Competitions: Leveling the Playing Field," Briefing at Annual A-76 Conference: Improving Cost Effectiveness and Enhancing Performance, Arlington, Va.,

December 6–8, 1999. Presented by ESI International and Center for Public-Private Enterprise.

Defense Science Board, *Task Force on Privatization and Outsourcing, Final Report*, Washington, D.C.: Department of Defense, April 15, 1996a.

_____, *1996 Summer Study, Final Report: Achieving an Innovative Support Structure for 21st Century Military Superiority—Higher Performance at Lower Costs*, Washington, D.C.: Department of Defense, November 1996b.

DoD—*See* U.S. Department of Defense.

Executive Office of the President, Office of Management and Budget, "Performance of Commercial Activities: Circular A-76 Revised Supplemental Handbook," Washington, D.C., March 1996.

Gates, Susan M., and Albert A. Robbert, *Personnel Savings in Competitively Sourced DoD Activities: Are They Real? Will They Last?* Santa Monica, Calif.: RAND, MR-1117-OSD, 2000.

Gattorna, John, ed., *Strategic Supply Chain Alignment: Best Practice in Supply Chain Management*, Aldershot, England: Gower Publishing Limited, 1998.

Girardini, Kenneth J., Nancy Y. Moore, Rick Eden, Carl Dahlman, and David Oaks, *Improving DoD Logistics: Perspectives from RAND Research*, Santa Monica, Calif.: RAND, DB-148-CRMAF, 1995.

Hamre, John J., Department of Defense Reform Initiative Directive No. 20: Review of Inherently Governmental Functions, Memorandum, Washington, D.C.: Department of Defense, January 16, 1998.

Kaplan, Robert S., and David P. Norton, *The Balanced Scorecard: Translating Strategy into Action*, Boston: Harvard Business School Press, 1996.

Koegh, P. J., "New Procurement Approaches to Create Public/Private Partnerships: A Request for Qualification (RFQ)," Third Privatizing and Competitive Sourcing of Military Services Forum, San Diego, Calif.: World Research Group, February 23, 1999.

Laseter, Timothy M., *Balanced Sourcing: Cooperation and Competition in Supplier Relationships*, San Francisco: Jossey-Bass for Booz-Allen & Hamilton, 1998.

Levine, Arnold, and Jeffrey Luck, *The New Management Paradigm: A Review of Principles and Practices*, Santa Monica, Calif.: RAND, MR-458-AF, 1994.

Lewis, Jordan D., *The Connected Corporation: How Leading Companies Win Through Customer-Supplier Alliances*, New York: The Free Press, 1995.

Marcus, Alan J., *Analysis of the Navy's Commercial Activities Program*, Alexandria, Va.: Center for Naval Analyses, CRM 92-226, 1993.

Markusen, Ann, *The Case Against Privatizing National Security*, New York: Council on Foreign Relations, 2001.

Moore, Nancy Y., et al., "Commercial Sourcing: Patterns and Practices in Facility Management," unpublished RAND research, 1997.

Moore, Nancy Y., Frank Camm, and Laura H. Baldwin, "Strategic Sourcing: Bundling Policies and Practices of Leading Firms," unpublished RAND research, 2001.

Moore, Nancy Y., Laura H. Baldwin, Frank Camm, and Cynthia Cook, "Implementing Best Purchasing and Supply Management Practices: Lessons from Innovative Commercial Firms," unpublished RAND research, 2001.

Northington, BGen Larry, "Outsourcing Challenges," Briefing, AF/XPM, Washington, D.C., July 22, 1997.

Office of Federal Procurement Policy, "Inherently Governmental Functions," Policy Letter 92-1, Washington, D.C., September 23, 1992.

OMB—*See* Executive Office of the President, Office of Management and Budget.

Pint, Ellen M., and Laura H. Baldwin, *Strategic Sourcing: Theory and Evidence from Economics and Business Management*, Santa Monica, Calif.: RAND, MR-865-AF, 1997.

SAF/AQ—*See* Assistant Secretary of the Air Force (Acquisition).

SAF/AQC—*See* Assistant Secretary of the Air Force (SAF/AQ), Contracting.

Tighe, Carla E., Derek Trunkey, and Samuel Kleinman, *Implementing A-76 Competitions*, Alexandria, Va.: Center for Naval Analyses, CAB 96-24, 1996a.

Tighe, Carla E., James Jondrow, Samuel D. Kleinman, Martha Koopman, and Carol Moore, *Outsourcing Opportunities for the Navy*, Alexandria, Va.: Center for Naval Analyses, CRM 95-224, 1996b.

U.S. Air Force, "Defense Reform Initiative Number 20," Briefing, Washington, D.C., 1998.

U.S. Department of Defense, *Defense Reform Initiative Report*, Washington, D.C., November 1997.

U.S. General Accounting Office, *Public-Private Partnerships: Key Elements of Federal Building and Facility Partnerships*, Washington, D.C., GAO/GGD-99-23, February 1999.

White, John P., "Outsourcing: DoD's Strategy for Better Management," *Roll Call*, April 29, 1996, p. 8.

Chapter Twelve

READY FOR WAR BUT NOT FOR PEACE: THE APPARENT PARADOX OF MILITARY PREPAREDNESS

Carl Dahlman and David Thaler

INTRODUCTION: THE CURRENT PARADOX OF READINESS

The purpose of a military organization is to enable the nation's political leadership to apply or, when useful as a negotiation strategy, threaten to apply force in the pursuit of national security objectives.[1] The U.S. experience since the end of the Cold War is clear on one thing: A great nation with worldwide strategic, political, and economic relations cannot avoid situations in which challenges to its interests require the application of force. A strong—and ready—military capability has proven essential for providing national leadership a wide range of options for supporting national objectives.

There is no discernible political disagreement on this fundamental starting point for a discussion of readiness. Whatever problems may occur relating to readiness—and real and serious problems exist in the current environment—it seems safe to state that they are not primarily political. Most of the polity supports the notion of a ready military (although this means different things to different people), and the only exceptions are a minority of isolationists in both major political parties. But in periods of prolonged peace without a clear

---

[1]The research underlying this chapter was sponsored by the Air Force Director of Air and Space Operations, Lt Gen Marvin Esmond, Air Force Deputy Chief of Staff, Air and Space Operations, and was carried out in the Resource Management program in RAND's Project AIR FORCE. We are grateful for the strong support of our research agenda we have received from the officers in the Air Force Readiness Center (AF/XOOA) and from our program director, C. Robert Roll. Several friends and colleagues have helped us with comments on this chapter, and we especially wish to acknowledge Laura Junor of the Center for Naval Analyses, and RAND colleagues Gerald Kauvar, Bernard Rostker, Leslie Lewis, and Jeremy Shapiro.

and immediate threat, there will always be a discussion of exactly how ready the U.S. military needs to be and how much the American people should be willing to pay for that readiness. Therein lies the seed corn of readiness problems: Readiness is a very worthy goal, but does it really have to be at such a high level at all times?

In the post–Cold War era, the United States has fielded capabilities designed to respond to the outbreak of two geographically separated major theater wars (MTWs), each of a magnitude roughly comparable to Operation Desert Storm, which led to the liberation of Kuwait in 1991 after Iraq's invasion. These basic planning scenarios have been in place since the first Bush administration, with some important changes in assumptions about their relative timing, the military capabilities and strategies enemy forces may be able to bring to the fight, and the strategies the United States would employ. These changes have helped bring about reductions in force structure (the number of units fielded), which have been accompanied by cutbacks in personnel, equipment, and training.

At the same time, the nation became embroiled in various contingencies, found it difficult to say no when called upon, and found it even harder to extract itself from the seemingly never-ending commitments that often resulted. The military now has to operate in a new era of peacetime preparation for major wars *and* continuous engagement in various smaller operations. In this era of "Boiling Peace," a new set of pressures on the military emerged. One writer on military affairs (Peters, 1999) expressed the pressure as follows: "The mathematics of readiness have altered radically, but we have not deciphered the new formula. The deployment of a reinforced brigade cripples multiple corps." One side of the new equation was that it was no longer sufficient to plan only for MTWs; it was also necessary to plan for many types of operations, some of very long duration.

But the new equation turned out to be an inequality instead, and its other side was that budget resources were not sufficient to pay for both activities. The United States could fight and win the major wars, but it had trouble *preparing* for the big ones *while* executing a variety of small operations. In the mathematical terms of set theory, the set of small contingencies was no longer contained inside the set of larger ones; they were complementary or additive. The military thus executed a combined peacetime mission: training for major war and simultaneously performing peacekeeping and peace enforcement in multiple locations around the world for extended periods. This would imply new resource requirements, more force structure in certain areas, and a set of new management challenges within the services, in the Office of the Secretary of Defense (OSD), and in Congress. Yet the resources were not available, and the nation was slow to tackle the management challenges.[2]

There is broad agreement that readiness has been declining over the last several years and that the cause is a mismatch between resources and new taskings, as suggested above. But this is only the beginning of a complex and subtle story. Readiness is not a simple concept that can be measured by some index that is reported day by day.[3] Some of what is at the heart of readiness *cannot* be adequately quantified, although it can be assessed in various qualitative ways. Also, readiness has many dimensions, and a central difficulty is that all these dimensions interact and change over time.

Since readiness is at the very heart of military activities in preparing to support national security objectives, the difficulty of expressing and measuring the readiness of the armed services objectively has been a source of unceasing frustration for all involved. Over the last few years, the desire to define and build appropriate readiness assessment and management systems has grown significantly inside and outside the military services. The concern over how to allocate very scarce resources is only one of the reasons. At the heart of the issue lies the problem of determining when readiness has sunk below an acceptable standard, and there is increasing suspicion that much of the U.S. military recently crossed that threshold.

---

[2]At the time of writing, the DoD was engaged in a top-to-bottom review of the national military strategy under Secretary of Defense Donald Rumsfeld. The outcome of the review was not yet clear.

[3]There have been calls for a readiness index that would measure short-term changes in the status of the force, a "Dow Jones for readiness." Apparently, there is a belief that such an index would represent a more objective statement than current readiness reporting. Technically, it would of course be feasible to compute an index for readiness. All that is required is to select a suitable number of quantifiable indicators and assign them appropriate weights in a monthly or quarterly index. However, that does not remove the subjectivity. The Dow Jones is likewise not an objective number but a subjective selection of stock prices; the daily weights and prices reflect the subjective evaluations of buyers and sellers in the stock market. The same would have to be true for an index of readiness—it cannot be "objective."

Surprisingly, assessments of whether the four services would be capable of fulfilling the objectives of the commanders in chief (CINCs) in the two MTW scenarios have not indicated significant shortfalls. That is, readiness, as DoD defines it, has not fallen so much that the service secretaries and the Joint Chiefs of Staff have had to inform the President and the Secretary of Defense that it is no longer possible to execute the war plans. Yet, as suggested above, there are still significant readiness problems around the force. Thus, the DoD is conveying a mixed message: Yes, we are prepared and capable of fighting the nation's wars; still, we have readiness problems.

A very important task for this chapter is to explain why this seeming paradox still conveys a coherent message: How can there be readiness problems when the armed forces are prepared to meet the objectives of the war plans? Recent RAND research indicates that there are significant readiness problems in several areas—which implies that we need a better definition for the term *readiness* and precise explanations of how and why readiness problems are currently manifesting themselves. Such diagnoses then permit discussion of what is to be done about the problems. This chapter begins by explaining how current operational readiness assessments are made, then discusses why this notion of readiness is too limited. We then describe why focusing on operational readiness is actually misleading because it misses how military leaders react to budget cuts, and it misses why the result has been a severe mismatch between requirements and resources in some major areas. The chapter concludes with a discussion of why this mismatch persists and of how the remedy lies in setting better operational and functional standards.

A word of caution is appropriate. We have focused much of our readiness-related research on Air Force readiness within RAND's Project AIR FORCE. Many of the specific examples of readiness issues in this chapter thus have an Air Force flavor. However, we believe that the themes in this chapter are common to all the military services, with specific manifestations that reflect differences in service roles and organizations. This belief has been borne out in many discussions with personnel from all the services. The intention of this chapter is to highlight certain very difficult, general themes in managing readiness, not to criticize one service unduly for causing problems that could have been avoided. Readiness may be a simple concept, but remains difficult to manage in practice. Problems are bound to emerge, and diagnosing them can help the DoD identify better management tools.

OPERATIONAL READINESS AND HOW IT IS CURRENTLY ASSESSED

Traditionally, readiness measures the ability to go to war and carry out certain assignments in a timely manner. The standard DoD definition of readiness is as follows:

> The ability of US military forces to fight and meet the demands of the national military strategy. Readiness is the synthesis of two distinct but interrelated levels: a. unit readiness—The ability to provide capabilities required by the combatant commanders to execute their assigned missions. This is derived from the ability of each unit to deliver the outputs for which it was designed. b. joint readiness—The combatant commander's ability to integrate and synchronize ready combat and support forces to execute his or her assigned missions. (DoD, 2001, p. 360.)

The requirements are set by combatant commanders who expect certain standards of performance so that their assigned missions can be met. Readiness is then an estimate of whether those who are to perform relevant tasks are capable of successfully completing them.

Since warfighting tasks exist at different echelons of responsibility, readiness metrics and assessments must also apply at different levels (see, for example Betts, 1995, Ch. 1). The nation may be ready to meet any military challenge even if one of the military services is less than ready, and each service may be ready to fulfill its roles in any one immediate operation even if some of its units or capabilities are not ready to carry out assignments in other operations. This discussion is limited to the question of how to assess the readiness status of a particular military service and does not address whether the war plans for a particular region are sufficient to defeat an aggressor and whether the military services jointly can carry out both planned and currently unforeseen contingencies. Clearly, readiness at the national or joint level is not simply additive; the services bring both synergistic and overlapping capabilities to the fight.

Nevertheless, it is necessary to begin with separate assessments of each service, because these will have profound implications for the national and joint assessments. The services base their readiness assessments on reports from their units that are identified in existing war plans.[4] For Army and Marine Corps land forces, these are battalions; for sea borne Navy and Marine units, they are ships; and for naval aviation and most of the Air Force, they are squadrons. The answers to two basic questions help determine the readiness status of units.

The first question is: When is the unit required to be ready? In deliberate war plans, units are ordered into a deployment schedule called Time Phased Force Deployment List (TPFDL). The TPFDL gives the planned mobilization day for when a unit is to be ready to deploy to a particular theater, that is, it sets a specific time at which the unit must be available for deployment. Thus, the unit has to be ready when the plans require it to be prepared to deploy to a theater of operations. For example, some Army units must be ready to go immediately after a deployment decision has been made, but many reinforcing and support units may not be required for several weeks or months after the initial combat units have deployed. The Army therefore organizes its units into a readiness hierarchy, with the units in the XVIII Airborne Corps at the top, through the remainder of the active units, and into the combat and support units in the Guard and Reserve. The Navy keeps all deployed units at peak readiness, but generally has a cycle of approximately 18 months with six months each in maintenance, preparation and training, and then deployment. The Marine Corps strives to keep all units ready for deployment on short notice, but keeps some at heightened alert and allows others a period of somewhat lower readiness during periods of recuperation after deployments. The Air Force keeps all combat squadrons and combat support assets at sufficiently high readiness to deploy them very early in any contingency.

The second question is: What is the unit supposed to be able to do when called on by a theater commander? Each unit has a written mission statement that specifies the capabilities it is supposed to

---

[4]Many units in the services perform support roles and are not in the war plans, so they do not report their readiness status.

bring to the fight. The amount of discretion left to the unit commander in designing a training schedule differs between the services, as
does the specificity of the mission assignment. Since the ground services may face very different conditions on various deployments,
their mission statements usually enumerate more very-specific tasks
and capabilities than a unit ever could become proficient at in
peacetime. On the other hand, an Air Force statement typically gives
a very general mission (provide precision strike, conduct air superiority missions, transport required materiel, etc.) while also being
very specific about how many pilots within the unit must have certain special qualifications. Suffice it to say that the written mission
statements for all units are designed to leave considerable leeway for
unit commanders (and the major service commands that provide
them resources) to manage their own training schedules given the
resources available to them.

The combination of planned wartime deployment schedules (the
"when") and unit mission statements (the "what") is the basis of the
current standard readiness reporting system, the Status of Resources
and Training System (SORTS). Each unit prepares a monthly report
that is a snapshot of the status and availability of equipment, personnel, and training. Using this snapshot and his own personal
assessment, the unit commander assigns a "C" rating to the unit, a
summary measure that states whether a unit has all the resources
required to go to war. All flying units in the Air Force are programmed to be C-1—that is, they are assigned resources in personnel, equipment, and training funds to reach at least 90 percent of
wartime requirements. If a unit falls below 90 percent for any one
resource—personnel, equipment, or training—the commander rates
the unit as less than C-1 (for example, a C-2 rating indicates that a
resource is between 80 and 90 percent) until the shortfalls have been
corrected. But it is possible for a unit to have all the resources available to it—all the equipment, all the personnel, and all training may
be current—and still receive a rating lower than C-1.[5] While this

---

[5]An additional complexity is that not all units are supposed to be C-1 at all times. The
Marine Corps and the Air Force resource all their units at C-1 status because they are
expected to deploy on very short notice. The Army designates certain units as first-to-
go and resources them at C-1, but later-deploying units, especially support units, may
be resourced at C-2, C-3, or in some exceptional cases as low as C-4 and would still be
expected to deploy. SORTS reports are based on whatever resources a unit is assigned,

could be because the commander knows of some specific shortfall that may be upcoming (aircraft going into depot, foreseen personnel rotations, training plans that cannot be executed), it is also his or her prerogative to make the ultimate judgment of whether the unit is ready. In principle, a commander could have all the resources required and still state that the unit is not ready to go to war. This is highly unlikely, but the system allows for it to happen.

Many of the questions concerning the use of SORTS as a readiness assessment system lie beyond our present purposes. Suffice it to note that SORTS is a standard input into an overall DoD readiness assessment system, the Joint Monthly Readiness Review (JMRR), which the Joint Staff, OSD, and the four military services execute together. JMRR is a procedure for determining whether the current war plans can be carried out. This review looks at all the units in the deployment schedules for each war scenario and checks whether the units are reporting a sufficient readiness status in SORTS to be available to fight the war. The assessment looks at each of the warfighting scenarios separately. It also examines them sequentially because some units are designated to "swing," that is, deploy to the first major war and, then, after a suitable period, have to be free to join the fight in the second. Note that this high-level readiness assessment is an attempt at reconciling a warfighting CINC's demand for forces with the supply of forces the military services can provide. However, the assessment does not actually fight the war under different assumptions about enemy capabilities and tactics; this operational analysis is the responsibility of each theater commander. All JMRR really does is determine whether deployable units report their readiness status as deployable. If they do, the CINC is assumed to be able to win the fight; if they do not, the affected service has to explain why the units are not reporting as ready, what remedial actions are being taken, and when restoration will be complete.

It is worth emphasizing how MTW-centric current operational readiness assessments are. In spite of the fact that the military services have become increasingly engaged in smaller-scale contingencies, readiness is still measured against the major wars in the Defense Planning Guidance. The reason is that there are no deliberate war

---

which in most cases—even for units designated to stay at C-1—is less than 100 percent of the resources required for high C-1 status.

plans for anything but two MTWs. A decision to alter the planning guidance to base force structures and operating budgets on new strategic concepts, including a steady-state level of smaller-scale contingencies, would undoubtedly affect the processes for assessing operational readiness. But the substance is likely to be very similar: Readiness will only be measured against deliberate plans not against the unforeseeable contingencies for which war plans have not been developed or analyzed.

The following sections discuss how the military services base their programming for readiness-related activities on a separate set of criteria altogether. Under Title 10 of the United States Code, the services are responsible for manning, equipping, and training their forces. This will raise issues that, of necessity, go beyond immediate operational warfighting assessments.

TOWARD A MORE-ENCOMPASSING NOTION OF READINESS

At present, at least two of the military services (the Air Force and the Army) face the seeming paradox mentioned in the introduction: Using the above approach to readiness (i.e., JMRR based on TPFDLs and SORTS), they report that they are able to carry out their assigned wartime tasks—yet they repeatedly state in public speeches and congressional testimony that they have significant readiness problems. Thus, there must be more to readiness assessments than simply whether warfighting tasks can be executed satisfactorily. The implication is unavoidable: Current readiness assessments do not include all service responsibilities that should be considered part of readiness.

Recent RAND research for the Air Force has focused on this particular question in considerable detail (see Dahlman and Thaler, 2000). Fundamentally, *readiness is the relationship between the tasks assigned to a unit and its ability to perform those tasks*; this is the heart of SORTS. The idea is then that units are assigned tasks and the resources to perform them to certain standards. Readiness is achieved when the resources—equipment, personnel, supporting processes, and funding—are sufficient for the unit to produce to the standards. If there is more to readiness than producing operational capabilities for a warfighting CINC, one must look at the tasks assigned to units and ask what more they are supposed to do. Figure 12.1 conceptually represents two tasks that *in practice* are expected of all military units: maintaining operational readiness *and rejuvenating the human and physical capital of the force.*

The left side of the figure represents the traditional focus of readiness assessments, as just discussed. The right side represents the other essential function of unit activities: ensuring the continuous recreation of the necessary knowledge and tools to produce and execute military operations. All organizations—commercial or governmental, service or goods producing—perform these two functions. That is, they produce some current output and simultaneously train their people and maintain their equipment to sustain the organization over time. These tasks are linked temporally: Today's rejuvenation efforts affect tomorrow's levels of production and output. What sets military organizations apart is that they are characterized by an extraordinarily high rate of flow-through of personnel. Military careers are short, averaging five to six years in the enlisted force and ten to eleven in the officer corps, and assignments to various duties are much shorter still. By design, this demands a level of on-the-job training (OJT) that simply has no parallel in the private sector or in other governmental institutions. Because this is such an integral element of military activities, it tends to be taken as a given, a cost of doing business. It therefore receives less attention to it than it deserves.

Continuous OJT is a unit task separate from producing current warfighting capabilities. Assume, for a moment, that the personnel flow-through were suddenly to stop. If all currently assigned personnel were to stay for a very long time in their current jobs, experience would allow them to perform their individual tasks at a peak level of proficiency. Since very few junior personnel would be needed to replenish the work force, continuous OJT would not be required. The unit could then focus on its ability to deliver the warfighting capability for which it was designed. Now, let the personnel flow-through restart. There is rapid turnover, with experienced personnel vacating their jobs and junior personnel entering the unit to take the place of departing people. OJT then becomes critical to passing the knowledge of senior personnel to the junior personnel who would replace them.

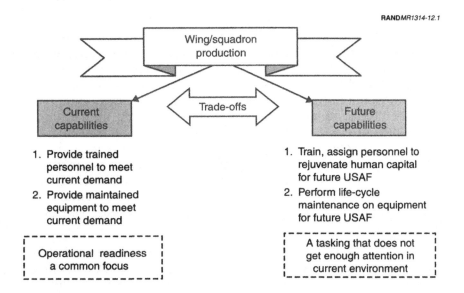

RANDMR1314-12.1

Figure 12.1—Unit-Level Readiness Related Taskings and Outputs

This task, in essence, is "relay training." Over time, the military services have all established their own professional expertise in their particular branches of warfighting. The only way an organization can preserve this knowledge—the baton, to use the relay parallel—is to ensure that it is handed over from one generation to the next, in a continuous training cycle that lasts the entire professional career of every member of the military. In the Air Force, the knowledge of how to conduct successful air and space operations must be recreated every moment—by assigning units the responsibility for OJT in many occupations, combined with formal training outside of units, as appropriate and necessary. Thus, in flying units, senior pilots teach junior pilots how to become effective warfighters and train their own successors as flight leaders and instructor pilots. The same is true in aircraft maintenance: Senior maintenance personnel spend a great deal of time teaching young enlisted airmen to become qualified maintainers and—just like pilots—train their own successors as senior enlisted managers and trainers. This is the essence of relay training and is an element of all military units that is just as critical as the task of producing warfighting capabilities.

Life-cycle maintenance of equipment and facilities is also a key activity that helps to sustain a capable force over the long term. Outside observers often think of maintenance in terms of generating sorties and repairing broken parts to support only current operations. However, just as changing the oil in a car helps ensure its engine hums smoothly both now and over time, regular maintenance actions also enhance the long-term health of aircraft and tanks. In addition, a great deal of effort goes into the inspection and refurbishment of major items of equipment and their components. A large portion of these activities takes place in central depots, but operational units complete a significant amount. A number of technicians in Air Force squadrons are dedicated to life-cycle maintenance, sometimes drawing maintainers from the flight line or the backshops when necessary.

The tasks of teaching personnel how to perform all their individual assignments and of conducting life-cycle maintenance of equipment, then, must be accomplished *in addition* to maintaining current operational readiness. This view of readiness is more comprehensive than the traditional view because it better represents the day-to-day activities of military units. As noted, this is exactly the environment of military units, by design.[6]

Thus, a more complete assessment of the readiness of military units must focus on two separate tasks:

- the production of operational capabilities
- the continuous rejuvenation and upgrade of human and physical capital—the expertise and the tools required to deliver operational capabilities.

We have concluded that current readiness problems are primarily associated with the recreation of human capital. This is the key to

---

[6]There are exceptions, of course. For example, Air National Guard fighter units have a high complement of former active duty pilots with high levels of experience, and they tend to stay with their units for a very long time. Thus, these units are able to attain very high levels of operational capabilities and readiness status with far fewer flying hours than comparable active duty units. Much of their training is concentrated on continuation, i.e., the maintenance of high skills, as opposed to an active unit, which must pay much more attention to both continuation and upgrade training, i.e., the acquisition of higher-level skills, to create pilots capable of replacing skills lost when senior pilots rotate out of the unit and/or leave the service.

understanding why it is perfectly legitimate to have serious readiness problems despite being prepared to handle ongoing and contemplated warfighting contingencies. There is, of course, a fundamental relationship between the two elements of readiness. If the DoD does not pay sufficient attention to the need to recreate the human capital lost through rapid personnel turnover, there will be a potentially severe effect on operational capabilities of the future.

As previously noted, inadequate funding for some support items, such as maintenance and parts, can be compensated for by adjusting working hours. It is common to cover for such deficiencies through extra work, and in the absence of definitive standards for maintainers' working hours, the net result is considerable unpaid overtime. This interplay between management of material and human resources affects the readiness status of both—and affects the ability of a unit to meet all its taskings.

THE GREAT MISCONCEPTION ABOUT READINESS

JMRR is an elaborate system for making assessments regarding operational capabilities based on the monthly SORTS reports from units. However, senior service leadership lacks a comparable system for monitoring the continuous rejuvenation of human and physical capital that would allow them to intervene when necessary. Why is this? Why are so many expert observers convinced that the single most important question to ask about readiness is whether or not the services can execute current war plans?

We suggested above that this view of readiness is incomplete, but it is, in fact, misleading. The probable reason OSD, Congress, and the wider public tend to focus on operational readiness is that they believe this most demanding of all requirements will be the first to show the symptoms if readiness problems begin to emerge. If people cannot train enough, if their equipment is not of the highest standard, if too few personnel have been assigned, if supporting processes in maintenance and supply fall short, will not the immediate and most obvious effect be a reduction in the finely honed tools of war? This seems such a natural inference that one seldom stops to question it. *But it is false.* Indeed, readiness would be much simpler to monitor and manage if the highest warfighting skills were the first to indicate readiness problems. They are easier to track and measure, and the DoD would have an easier time avoiding unreadiness than it does in reality.

The reason one cannot expect to see operational readiness decline first lies in how the military services, by long tradition and deeply embedded culture, react to pervasive problems that gradually may affect their ability to go to war. If a unit commander has to choose between sacrificing the unit's operational status and slowing down and postponing the continuation and upgrade training of his personnel, he or she is conditioned by experience and collective military values to give the highest priority to the operational status of the unit. In fact, his superiors would probably redress him severely if he even tried to reverse those well-established priorities. The first priority of any military organization is to ensure that it is prepared to carry out any urgent operational tasks assigned. Almost everything will yield to this principle. Therefore, when taskings grow and budgets do not follow suit, units come under pressure. To protect their warfighting capabilities, they do not have much choice because they control very little in their environment. Equipment, personnel, facilities, support budgets, materiel, etc.—all the critical inputs they need to perform their duties—are handed to them and cannot easily be augmented. The only resource they typically can control is the working hours of unit members. So when asked to produce more with less, units turn to their personnel and ask for a more-intense work effort. They can do nothing else.

The result is that the entire chain of command, from unit leaders to the most senior decisionmakers in major commands and central staffs, shelters the production of current operational capabilities as far as possible. What must give, then, is the rejuvenation of capital, the other significant task assigned to units. Thus, when resources are not sufficient to allow the unit to perform all its tasks, the tasks that are not immediately required for operational performance are delayed or set aside. Over time, this will degrade operational capabilities, but it is possible to keep the traditional notion of readiness from showing signs of weakening at the expense of rejuvenating the force.

In sum, one of the great misconceptions about readiness is that the first place to look for readiness problems is the operational areas. This reflects a misunderstanding of how military organizations manage themselves, how they set their priorities when pressured, and the totality of tasks assigned to units. The implication is that readiness problems will manifest themselves first and foremost in hours of work inside units and in the pace and quality of upgrade training, especially of junior personnel. The next section summarizes recent RAND research on these issues as they relate to fighter wings in the Air Force.

ESTIMATING SOME CURRENT MAJOR READINESS PROBLEMS

To identify shortfalls in how the military preserves its knowledge base through unit training, the relevant activities must be measured against functional standards. Some of these activities lack such standards. And when standards do exist, they do not adequately capture the exigencies of day-to-day operations. The following discussion outlines observations and analyses that can help determine the right standards against which to measure the health of the force.

Operational units manage a pipeline that begins with junior, inexperienced trainees and ends with highly skilled, experienced pilots and maintainers. Personnel gain their skills through formal education and less-formal OJT. Senior pilots and maintainers serve as OJT trainers, imparting their knowledge and experience to junior personnel so the latter can eventually take the places of the former when they are reassigned or separate from military service. Therefore, maintaining a healthy personnel inventory requires having *personnel of all levels of experience at all times*. A unit having only top-notch, highly experienced pilots and maintainers may be a superb warfighting machine but it is not healthy. A unit must maintain enough senior pilots and maintainers to train junior personnel while meeting the workload (flying operational sorties, fixing the jets, and generating the sorties). Moreover, the unit must be able to absorb enough junior personnel to take the places of exiting senior personnel when the time comes.

To maintain this delicate balance between experienced and inexperienced personnel and in the mix of training and operational activities, the unit must have adequate levels of the following:

- a proper ratio of qualified instructor pilots to junior trainees

- flying-hour funding and aircraft availability rates to sustain sortie levels required for pilot training

- maintenance manning and materiel to support the requisite sortie rates

- a proper ratio of senior maintainers to provide adequate OJT for junior maintenance personnel.

Imbalances will arise when there are too few experienced pilots and maintainers training too many junior personnel while trying to meet stressing operational demands. Such imbalances are today common throughout the Air Force. Setting the right standards is the first step the Air Force must take to correct these imbalances.

Pilot Training and Flying Hours

The USAF bases the flying hours it assigns fighter units on the Ready Aircrew Program (RAP), which defines the minimum number of sorties per year per pilot based on his experience and classification. While RAP is a significant improvement over older methods for computing flying hours, it falls short in three areas.

First, it does not account for changes in individual and unit sortie requirements when the experience mix in the unit changes. In a single-seat jet, such as the F-16, every time an inexperienced pilot flies, an experienced instructor pilot or flight lead flies with him in a separate aircraft. Thus, the higher the proportion of inexperienced to experienced pilots in the unit, the more the experienced pilots must fly, thereby increasing total unit sorties required just to maintain the initial experience mix. Second, RAP is constrained by shortfalls in unit maintenance resources. Thus, rather than basing flying hours on what the existing inventory of pilots actually needs, RAP bases hours on what available maintenance resources dictate, thereby allowing planners to lose sight of what the actual requirement is. Finally, the attrition rate RAP assumes—the rate at which scheduled sorties must be aborted due to maintenance problems, inclement weather, and other factors—significantly underestimates observed rates at many wings.

RAND has developed a model for determining pilot training standards that are unconstrained by maintenance; use observed attrition rates; and, most importantly, *are adjusted according to pilot experi-*

ence mix in the squadron (see Taylor, Moore, and Roll, 2000). Using this model and adding an adjustment factor for ineffective sorties showed that a squadron with 18 aircraft requires an estimated 40-percent increase in required sorties over and above current allotments just to sustain its existing level of experience.[7] A significant portion of this increase is due to the need for senior pilots to fly more as trainers when the experience mix falls below a break point of around 52-percent experienced pilots. In sum, the current standard the USAF uses to determine flying hours required for fighter pilot training is inadequate. Under reasonable assumptions regarding the present experience mix and sortie attrition rates, the Air Force's current planning assumption that peacetime flying requires *fewer* sorties per available flying day than wartime operational missions is false.[8] Our analysis suggests that the standard for health *actually involves a higher peacetime tempo of daily operations than that in the current war plans.*

Unfortunately, fighter wings today are challenged even to execute the hours programmed, as the 388th Fighter Wing at Hill Air Force Base, Utah, exemplifies.[9] The wing fields three squadrons, each equipped with 18 F-16 Block-40 jets.[10] In fiscal year (FY) 1994 the wing's monthly sortie rate approached 21 per aircraft; by FY 1999, the rate had fallen over 25 percent, to about 15.3. This diminished sortie rate means that it takes a 100-hour pilot about 40-percent longer than it should to achieve the 500 hours of cockpit time necessary to be considered experienced. This means that, even though it should only take about 70 percent of a pilot's first assignment to become

---

[7]Official flying-hour planning assumes that 99 percent of all sorties are effective in meeting training objectives. While this is undoubtedly too high, there is no official estimate of the attrition rates of effective sorties. From pilot training records in one nondeploying F-16 wing, we found a 12-percent attrition rate—this will vary with local conditions. To this must be added the training backlog built up in units that deploy to peacetime contingencies.

[8]This analysis is reported in Dahlman, Thaler, and Kerchner (2002).

[9]For an in-depth discussion of these issues, see Dahlman and Thaler (2000). Our current research investigates the challenges facing airlifters and tankers at the 60th Air Mobility Wing at Travis Air Force Base, California.

[10]Block-40 jets are equipped with Low Altitude Navigation and Targeting Infrared for Night (LANTIRN) systems for ground attack; they also have air-to-air capability. At the time of writing, the 388th was in the process of increasing two of its three squadrons to 24 primary aircraft each.

experienced, it now takes well over 90 percent. This creates a follow-on training burden for the next unit to which the pilot is assigned. This means that the experience level of the fighter force will continue to fall precipitously over the next several years without substantial increases in flying hours. Thus, failing to rejuvenate the human capital as required will in turn yield a drop in operational capabilities, or readiness as commonly understood, over the coming years.

Maintainer Production and Training

As the 388th Fighter Wing's sortie rate has fallen over the past several years, Total Non-Mission Capable rates for the aircraft have soared from about 3 percent in FY 1994 to over 21 percent in FY 1999. The connection is clear: The sortie rate has dropped because jets have been increasingly unavailable while waiting for necessary maintenance and parts. Inadequate maintenance manning levels and experience mix are key factors in diminishing sorties available for pilot training.

The USAF bases maintenance manpower requirements on a computer simulation model that combines such inputs as wartime sortie rates and durations, break rates, and man-hours for repair tasks to estimate the maintenance manpower needed to support sorties in all kinds of aircraft.[11] Our research indicates that the manpower standards in maintenance are significantly below what would be required to support the sorties needed for adequate pilot training.

First, the computer simulation does not recognize the three factors that cause the Air Force to program too few flying hours for fighters to begin with: that training plans are already based on assumed constraints in maintenance, that a falling experience mix of pilots has raised sortie requirements, and that training plans assume too little attrition of training sorties both on deployments and at home station. Thus, the model is still based on the very suspect assumption that peacetime is less demanding than wartime.

Second, the computer simulation inadequately addresses the experience mix among maintainers, which exactly parallels the pilot

---

[11] This model is extensively examined and analyzed in Dahlman, Thaler, and Kerchner (2002).

seniority problem.[12] Skill mix can deteriorate quickly through unanticipated personnel losses, and in the last few years, highly skilled people have left and have been replaced by less-skilled junior personnel. This will reduce the productivity of the workforce and increase the amount of training senior maintainers must to.

These two shortfalls provide a maintenance manpower standard that underestimates what squadrons need on a day-to-day basis to generate sorties for pilot training *in conjunction with* properly absorbing junior maintainers and graduating them to higher skill levels. This problem is further compounded in that *personnel assignments* (the actual "faces" to fill the "spaces") do not suffice to fill the authorizations in many areas.

For example, since FY 1994, the number of maintenance personnel assigned to the 388th's flying squadrons has decreased by 12 percent.[13] The 5-level journeymen and 7-level craftsmen—those who do the bulk of the maintenance production *and* teach the 3-level apprentices—have decreased by 6 percent and 27 percent, respectively. The experienced maintainers contend that they are undermanned by 23 percent, while the 3-levels in their squadrons are overmanned by 42 percent. Undermanning and lack of experience have been particularly acute in avionics specialists, crew chiefs, and engine mechanics. The undermanning is even worse than it appears, because there are a number of tasks for which no positions have been authorized, so the persons who do the jobs must be taken away from other work.

These smaller numbers of experienced maintainers are spending 27 percent more time producing (maintaining aircraft, generating sorties) and 48 percent less time teaching than they did three to five years ago. The number of "trainer equivalents" (the number of 5- and 7-levels multiplied by the time they spend training others) has

---

[12]Junior maintainers enter operational units as 3-level apprentices and, through a combination of OJT, formal training, and promotion, become 5-level journeymen, 7-level craftsmen, and 9-level supervisors.

[13]Because of the difficulty of expressing the effects of maintainer shortfalls systematically, RAND asked the maintainers of the 388th a number of questions about manning, experience mix, absorption capacity, and how they spend their time under varying circumstances (contingencies, exercises, etc.). The results are documented in Dahlman and Thaler (2000). For further research into the effects of operational tempo on Air Force units, see Fossen et al. (1997).

dropped from an average of 27 to about 12 per flying squadron; trainees have doubled from three to six per trainer. The result is a 50-percent increase in the average time it takes for a 3-level to become a productive 5-level. Not surprisingly, the competing pressures seem to favor generating sorties to the extent possible—for pilot training as well as contingency and other operational demands. So, the 388th has too few experienced maintainers teaching and supervising too many 3-levels while trying to meet a highly demanding sortie generation schedule.

Under these circumstances, fighter wings face severe imbalances in pilot and maintainer training and production, and thus the USAF's ability to rejuvenate its human capital, and by extension its near- and far-term health, has degraded significantly. A key ingredient to reestablishing balance and a measure of health lies in remedying the maintainer shortfall. RAND has computed a rough but conservative estimate of its magnitude for the fighter units across the Air Force.

First, we estimated the additional maintenance manpower needed to sustain a higher peacetime sortie rate to support pilot training. Sustaining this higher sortie rate in peacetime would require a third maintenance shift, but it could be much smaller than the other two shifts.[14] A 4-percent increase in maintenance manpower across the fighter force would meet an estimate of peacetime flying that is similar to planned wartime flying.

Next, we computed the number of the additional authorizations needed to account for manpower engaged in OJT, which represents a significant workload. Under conservative assumptions, we estimated that this would require an additional 11 percent in authorizations.

These estimates do not account for the productivity shortfall created by inadequate experience or the need to meet the increasing requirement for teachers as the proportion of junior personnel in the unit increases. The results indicate that 5- and 7-level authorizations should increase by about 6 percent to compensate for the lower productivity of a less-experienced workforce and by an additional 7 percent to provide trainers for healthy absorption of the proportionally higher junior personnel.

---

[14]Fighter wings are authorized two shifts five days a week in peacetime.

Overall, authorizations for maintenance manpower in fighter units would need to be raised by about 28 percent to support pilot training requirements. Out of an FY 1999 maintenance force of about 22,000 in fighter and maintenance squadrons, this implies an increase of over 6,000 authorizations.

The magnitude of this shortfall in maintenance manpower suggests that the Air Force is having trouble training its pilots and sustaining their experience because of an inadequate flying-hour program and because of inadequate maintenance manning and diminishing experience. The maintainers have trouble gaining the needed experience because too few teachers are spending too little time teaching too many junior technicians. Thus, we found that current standards for pilot training and maintenance manning are not appropriate and cause a significant readiness shortfall. As a consequence, the USAF has been unable to define a readiness metric that can communicate its readiness problems to external decisionmakers.

Shortages of Parts

For aircraft, the status of the physical capital is determined by units' abilities to conduct sufficient maintenance to support the required sorties during peace and war, on home station and during deployments. Two interrelated areas, maintenance and supply, and two interrelated processes, setting functional requirements and funding them adequately, have to be properly established and monitored for units to meet their tasking to keep the physical capital healthy. Unfortunately, the Air Force has significant problems in both supply and maintenance, caused by difficulties both in the determination of functional requirements and in the supporting processes involving service organizations and financial management.

Maintenance requirements are, in principle, set by aircraft break rates associated with the most demanding sortie scenario, typically wartime. Given these break rates, demands for repairs and spare parts should be computable. Knowing the demand for spare parts should allow the Air Force to determine appropriate spares inventories for all items. And given the cost and lead-time for procurement, the annual budget for spares should be easy to estimate. Finally, internal financial management principles can be added so that incentives for economizing on parts can be introduced into maintenance.

Unfortunately, none of this works as described in practice. There are many intractable problems in the materiel processes that support the preservation of physical capital in the Air Force. These problems manifest themselves in chronic shortages of supplies. It is not possible to give these problems as much treatment as they deserve here, but we will describe a few of the most difficult material processes.

First, it has long been known that the stochastic variability of the break rates of various parts on aircraft is extraordinarily high. This makes it very hard to predict demand rates. One can add to this the effects of changing operating conditions, in particular the transition from peace to war. The effect of this is on different aircraft systems is different, but insufficient data are available for proper analysis. Automated Air Force data systems for tracking aircraft break rates are inadequate, providing incomplete and biased estimates of actual breaks and thus affecting estimates of both the mean and the variance of true break rates (see Dahlman, Thaler, and Kerchner, 2002).

Second, the Air Force uses a sophisticated set of inventory models to estimate the requirements for spare parts.[15] Since there is a significant but unknown gap—which varies by aircraft type—between aircraft break rates caused by sorties and demands for spares from the parts inventory, the connection with the operational demand for parts is not well established. Yet another important factor that is not well understood is the effect of aging aircraft on the demand for maintenance man-hours and spare parts. Thus, because of the intensive data requirements and inadequate analysis of variable demands, Air Force inventory models that determine spare parts procurements (the DO-41 models) only quantify a small subset of all required parts. For the remainder, budget considerations are determinative.

---

[15]Known as the DO-41 family of models, they include the Aircraft Availability Model and the Aircraft Sustainability Model, both of which are based on the Dynametric modeling approach. Spares budgets are further allocated through a model called DRIVE, designed to buy parts and allocate repair actions to the activity that maximizes the increase in the probability of making an aircraft mission-capable, per dollar spent. These are conceptually sophisticated models, which unfortunately cannot be fully implemented because of data problems.

The Air Force further lacks precise models to determine the size of War Reserve Spares Kits (an inventory of parts assigned to deploying flying squadrons), the requirement for spare engines, and the number of required backup aircraft. Hence, the funding in these critical areas is determined not by carefully analyzed requirements but by available budgets, which the Air Staff determines in the programming phase—which is the best anyone can do under the circumstances.

Finally, under the working capital fund concept, budgetary funds from congressional appropriations are allocated to the consuming units (wings). Wing maintenance organizations then purchase parts and maintenance services from provider organizations—supply and maintenance depots in the Air Force, and the Defense Logistics Agency for various consumable items. Customers have complained for a long time about prices being too high and too unstable. Providers have complained for an equally long time that customers of support maintenance and supply operations do not pay enough. And programmers have for years attempted, with limited success, to improve the efficiency and lower the budgets of support operations. The management and financial processes supporting such critical readiness activities and maintenance and supply are wanting.

In sum, requirements are not stated with sufficient reliability to ensure that adequate resources can be provided when and where they are needed. Again because of the budget cuts during the last decades, there is little doubt that the state of the capital assets is declining—and this is probably true for all the services, not just the Air Force.

REASONS FOR READINESS PROBLEMS: PLANNED AND UNPLANNED

This section diagnoses the broad manifestations of readiness problems across the service as a whole. The following is a list of the major readiness issues, representing a consensus view of managers, operators, and analysts inside and outside the Air Force:

- Programmed operations and maintenance savings have not materialized or cannot be sustained as envisioned.

- Anticipated Transportation Working Capital Fund revenues have not materialized because military customers have turned to commercial transportation.

- Aircraft throughout the fleet are aging, with numerous structural and corrosion problems, and require expensive upgrade programs for engines and avionics.

- There is a shortage of spare parts that is due both to shortfalls in funding and to technological problems.

- Ranges and airspace are increasingly constrained.

Some shortfalls have emerged as the Air Force has experienced force drawdowns while becoming more heavily tasked to various deployments around the world. For example,

- Home station units and units returning from deployments have suffered various materiel, personnel, and training shortfalls that cannot be made up when they return home.

- The net continuation and upgrade training of both pilots and maintainers has decreased.

- The effects of temporary duty assignments on deployments have been uneven and unacceptably long in some occupations and units.

- A strong economy has had negative effects on recruiting and retention of personnel.

While the lists above focus on major Air Force readiness problems, there is every reason to believe that they have counterparts in the other services, especially the Army. Budget pressures have affected all the services, and the readiness problems that relentless but largely unsuccessful attempts at infrastructure reductions have caused therefore manifest in all the services. However, the Navy and the Marine Corps have been less affected by the many deployments since Operation Desert Storm, for two major reasons: One is that the maritime services have, for a long time, structured themselves to cyclical deployments and so have a rotation base that makes them better prepared for the increased operational tempo of recent years. The other is that, because of the associated requirements, much of the burden for supporting the deployments has fallen on the Army and the Air Force. This is not to say that the Navy and the Marine Corps have been unaffected by deployments, only that they have been affected somewhat less than the other two services.

Planned Readiness Shortfalls

To understand why overall readiness has slipped in recent years, it is useful to make a distinction between planned and unplanned readiness problems. It may come as a surprise to some that certain readiness problems are actually *planned*, yet this is an important factor in all military planning and operations. Very few military units are at peak readiness, i.e., programmed and resourced to all known requirements. In the Army, units in the 82nd Airborne Division are authorized more than 100 percent of their required personnel to ensure that they can deploy at full strength, given that absences occur at all times. Special operations units in all the services are authorized to operate at full requirements. Submarines, by virtue of their harsh operating conditions, are also fully funded.

But this is not the norm for the rest of the military: *It is the exception for all known requirements to be funded.* Programmers and budget planners in the services, in consultation with operational and functional experts from the warfighting side of the house, deliberately plan funding to be less than the analytically known, accepted requirements. Deliberate risk is built into peacetime funding for the very simple reason that the defense budget is not—and never has been, not even in the heyday of the Reagan administration's deliberate support of a military buildup—capable of supporting all known requirements. The bill is simply too large. Military operations are so costly that it is not possible to allow all units to attain their highest level of proficiency.

In the Air Force, the practice is as follows, and there are parallels in the other services: First, a great deal of effort goes into determining operational and functional requirements (although, as indicated earlier, we believe that important requirements are not set correctly and, in some areas, seem to be nonexistent or forgotten). These analytically based operational and functional requirements are then presented to the resource planners and managers at the Air Staff. Through a process involving several teams of professional experts, the Air Staff arrives at something called "validated requirements." In many cases, these validated requirements are smaller than the analytically supported functional and operational requirements arrived at in the field. Sometimes, the reason is that the Air Staff may not accept the analytical models as the best representation of actual requirements; at other times, the reason may be that they deliberately find a workaround that saves money. For example, a unit may be assigned fewer backup aircraft and fewer maintenance personnel than strictly required because the Air Staff instructs war planners to move aircraft and personnel from certain training units if they are needed for a mobilization. Whatever the reasons, the planning process is designed to identify opportunities for building "acceptable" risks into resource determinations.

Then, decisionmakers on the Air Staff decide which validated requirements the USAF can afford to fund (authorize). At this point in the programming process, very difficult and sensitive judgments are necessary, *because the anticipated top line in the presidential or congressional budget cannot possibly fund all the validated requirements.* There are still too many requirements for the budget sack to hold, and decisionmakers have to decide what the bag can actually carry and what they have to leave out. Since all requirements, at this stage, have been validated—that is, they *should* be funded if the Air Force is to be as capable and ready as desired, given the best judgment of the organization as a whole—decisionmakers face only very difficult decisions. All accounts get shaved, including readiness.

In practice, this means that not even first-to-go warfighting units, and the units that support them, receive full funding. As noted above, SORTS designates units at C-1 status when resources managers have provided funding at 90 percent or better of all validated requirements. The Air Force makes it a practice to fund most units at *low* C-1 (close to 90 percent)—the only major exceptions being certain critical activities for which risks are simply not allowed, such as special operations and nuclear-capable units. Shortfalls are spread like peanut butter across most operational units.

This is the essence of *planned* readiness problems. Readiness cannot be funded to the extent desired. Top-line budget constraints force deliberate risk-taking, even in such a critical area. This is how and why the Air Force (and its sister services) actually plans for certain readiness shortfalls. There is nothing absolute about readiness when resource programmers and financial managers look through all the program elements; at that stage, everything is a "requirement," and almost everything will take a beating because almost every activity must be fair game when there is a top line that simply *cannot* be exceeded.

One very important implication for readiness management and readiness assessments immediately follows from the fundamental programming and financial management principle of balancing shortfalls by allocating resources as prudently as possible: *There will never be only one reason for readiness shortfalls.* There will be many. If programmers and financial managers have done their job correctly, it should, in principle, be impossible to determine *the* reason for planned readiness shortfalls. In the Air Force, for example, there should be equivalent shortfalls in pilot training, backup aircraft, flying hours, training munitions, range availability, aircraft availability, maintenance manpower, maintenance funding, depot repair times, part supply, upgrades and modernization of the aircraft, and base facilities. Real problems in resource management occur if one account has too much, while others have too little. Of what use are abundant ranges or beautiful facilities if they come at the cost of even more cutbacks in flying hours? Why have a supply system with shelves full of expensive items if the price is an even more severe shortage of maintenance personnel to put the parts in the aircraft?

Notice what this means: At any one time, there may be one absolutely pressing issue to fix—a binding constraint on readiness. Maintenance manpower may be such a binding constraint—this seems to be the most important readiness problem for the Air Force right now. But if one takes care of that single issue, one immediately runs into another. Perhaps it will be parts—not enough parts to fix the aircraft. Well, that too can be fixed, with more funding. This will not, however, fix the readiness problem because there is an engine problem right behind the parts problem.

Thus, there will never be just one readiness problem. The only way to address readiness problems is to look at a long series of interrelated functional and operational activities as an integrated whole. Each of the activities is complicated and subject to stochastic variability in performance. When many subactivities form an integrated chain from inputs through stages of production to a final output, the statistical variances at each stage compound. Ensuring readiness therefore requires deliberately building in buffers at every stage. This is not possible in the present budget climate. Instead every stage is squeezed to the minimum. The result is a sensitive and complex chain of interrelated activities that has no robustness designed into it. Even planned readiness problems will then crop up in many places, by the nature of the uncertainty in each stage.

Unplanned Readiness Problems

If planned readiness problems were the only issue, the military services would face a daunting task. But there is much more. The services have also been caught off guard by various unforeseen events that have caused a new set of readiness and resource issues. One area already mentioned is poorly understood, often unarticulated requirements (such as in pilot and maintainer training, as well as material processes), which tend therefore to be underfunded. It is an unfortunate but elementary fact that any activity that cannot clearly articulate the pain associated with too few resources will end up paying for it when the program and the budget are put together. This has, for too long now, been the case with unit-level human capital rejuvenation. Because of a poor understanding of the analytical requirement, the validated requirement is too low, and the funding is even lower. The result is an unintended and hitherto unappreciated level of pain in all unit activities that manifests itself through excessive overtime especially for senior personnel. This suggests that the USAF needs to pay much closer attention to the determination of functional and operational requirements. It is unfortunate that even senior decisionmakers have an incomplete understanding of some critical activities inside their own services. Without pretending to exhaust all other unforeseen causes of readiness problems, we will discuss four of the most obvious.

Wages and Working Conditions. The gap between wages in the military and those in the civilian economy continues to grow quickly. Until recently, a rip-roaring economy had raised all employment-related factors in the commercial sector—plentiful jobs, good wages, added benefits, excellent training opportunities, good promotion prospects, job stability, and high returns to higher education, to name a few. These factors would put enormous pressure on military recruiting and retention even under the best of circumstances. This means that even if the military environment had been able to offer the same quality of working conditions as ten years ago, there would still be recruiting and retention problems, because the commercial sector has raised its offers of employment and pay that much faster.[16] By running harder, that sector would have been ahead even if the military had been running at a steady pace.

But a steady pace has proven impossible. In fact, the burden on military personnel increased significantly during the 1990s.[17] Not only have many of them been required to deploy to various difficult places around the world for extended periods, but *all of them* have been asked to perform their jobs with less. The relentless budget pressures, combined with more taskings, have created a situation in which the working conditions in the military in reality have deteriorated relative to those of ten years ago. It is no longer the same place to work. This makes the effects of an improving workplace in the commercial sector that much more severe. The relative gap has increased both because the economy has been roaring ahead and because the military has become a less desirable working environment. Thus, recruiting is more difficult, and it is much harder to retain skilled personnel.

The Effects of Deployments. The more-frequent contingencies have created a variety of readiness problems. Perhaps it comes as a surprise to some that deployments hurt readiness. How can deployments, a version of going to war, hurt warfighting capabilities? In particular, when a unit participates in a shooting war, such as Operation Allied Force in Kosovo, surely that *improves* warfighting skills? Yet deployments can actually hurt both operational readiness and rejuvenation.

Let us first look at rejuvenation. Squadrons in the USAF are deployed to contingencies in full or in part; when only part of a squadron is

---

[16]In a working paper from 2000, Richard Fullerton of the Air Force Academy has demonstrated that the falling retention rate of pilots is primarily due to the wage gap, not to deployments, as often stated.

[17]During the drawdown in the early 1990s, the Air Force deliberately pursued a policy of retaining senior personnel, both pilots and aircraft maintainers. This caused an increased seniority mix for both categories. Many pilots have been coming to the ends of their service commitments, which has, quite predictably, aggravated the current problems (although the intense demand for pilots from airlines could not have been fully predicted). For maintainers, the exit rate among career technicians has been significantly higher than predicted. Now, in both communities, the Air Force has to deal with a shortfall in senior grades that is much greater than anticipated.

deployed, it engages in "split operations," whereby part of the unit remains at home station. The deploying part of the squadron will bring the squadron's best jets (and, if needed, it may borrow from another squadron), a full complement of pilots, and the more experienced maintainers. They leave behind the less-robust aircraft and an undermanned, less-experienced maintenance crew and pilots who still need sorties for training. To strengthen the deploying part of the squadron, the part at home station is "broken," a perennial problem with split operations across the Air Force. Under these circumstances, the ability of the squadron to produce training sorties at home station is substantially reduced. In addition, the deployed detachment, for natural and understandable reasons, is given clear priority for local parts, as well as in requisitions submitted through the supply system. The experience mix of maintenance personnel becomes more imbalanced at home station, and maintainers at home work harder to generate fewer sorties, but even this curtails maintainer training. This is in addition to the systemic problems facing the units on a day-to-day basis.

Deployments also affect the warfighting capabilities of deploying units, often degrading certain operational capabilities. Air Force units actually being employed against an enemy do have the opportunity to deliver missiles on specific targets in a hostile environment under challenging conditions, thus sharpening many important facets of their warfighting skills. The units return much more skilled at planning and executing missions involving strikes. But during an extended deployment, the units also miss various other required training opportunities. When a dual-capability aircraft deploys, one of its missions tends to suffer; for example, the air-to-air mission of the F-16 typically degrades because of the lack of a suitable threat and the lack of training ranges. This holds true also for air superiority aircraft, such as the F-15. Most Air Force weapon systems have elements that cannot be kept current during deployments, such as nuclear missions or simulator work on procedures relating to degraded systems. Thus, deployments may help some units acquire higher skills in some—often essential—capabilities but may simultaneously cause other essential skills to atrophy to some degree.

The way the Air Force has been structured plays an important role in the difficulties it faces with contingency deployments. The service has not been structured or designed for repeated deployments but for MTWs. This means that the Air Force expects to prepare for war in peacetime and to make an all-out effort during a (one hopes) short war. All continuation and upgrade training would virtually cease during the war, and USAF units would return home "broken" and in need of a long period of reconstitution. But when the USAF must deploy more or less continuously, there is little time for reconstitution, and whatever recovery can be made is in addition to the unit's ongoing peacetime training and is mostly drawn from resources organic to the unit (especially, working people harder). So, it becomes ever more challenging to make a dent in any backlog in training and maintenance.[18] To help ease this burden, the Air Force has developed a concept called the aerospace expeditionary force (AEF). Of the ten AEFs—packages of USAF platforms and capabilities—two would be on call every 90 days (forming a 15-month cycle). A two-week recovery period is built into the schedule.[19]

Budget Rules. The third unplanned source of readiness problems relates to the budget rules, as well as to the contentious game between Congress and the president over control of military operations abroad. This tends to create resource shortfalls that inevitably result in readiness problems. Congressional funding for contingency operations is usually insufficient and almost always too late. The president, using his powers as commander in chief, authorizes support of a peacetime contingency. The Secretary of Defense immediately authorizes transfer funding from internal sources in the defense

---

[18]There is an additional problem that is not addressed here: A rotating deployment schedule must also account for the possibility that a large percentage of people will not be available for deployment. This percentage is small in wartime, a few percentage points mostly caused by medical conditions, but can be very large in peacetime contingencies. In the Army, estimates range from 35 to 40 percent of the people assigned to deployable units (see Polich, Orvis, and Hix, 2000). Because of different deployment concepts, there are reasons to believe that the percentage would be smaller in the Air Force and even smaller in the Navy and the Marine Corps.

[19]The effects of the AEF concept on readiness are likely to be minimal. Ordering flying units into a structure of numbered AEFs will not do anything, per se, to reduce the demand for deployments and does not change total assets available for deployments. Thus, if there is a gap between taskings and available resources, the AEF construct will do nothing to close it. The effects on home-station parts of a split unit and the training cycle problems discussed in the text above will remain. Even the extent to which AEFs will improve predictability of deployments for individuals is in question, as it has always been Air Force practice to spread out deployment tasks as evenly as possible among available units. Furthermore, the Air Force has not yet decided how to report the readiness status of the ten expeditionary forces.

budget. This means that deploying and supporting activities get priority for funds and that all other accounts get to be short-term bill payers within a given defense budget. The administration then asks Congress for a supplemental budget authorization, to restore the funds just diverted to the recent contingency. Congress, which faces competing claims for scarce budget resources no less than the military services do, usually provides funds (if it supports the operation in principle), but usually less than requested. Depending on how popular the president's action is, the degree of support can be a high share of the total cost of the contingency deployments, but it may be significantly less. That is, the military services are typically asked to support a significant share of the cost of the contingency from their previously appropriated funds.

This means that many of the funds that were moved from various accounts, including those that support readiness and operations, will never be restored. The effect of the unplanned contingency is to force an unanticipated cutback in readiness, much beyond what was contemplated when the president's budget was presented to Congress. Service decisionmakers now have to recompute all their carefully balanced accounts to determine which activities will be bill payers—and some of that will inevitably come out of readiness again.

Add to this the institutional difficulty rooted in the budget process that any congressional supplemental appropriation typically cannot be enacted until at least half the budget year is already past. That means that the additional funds often reach the military services only in the last quarter of the fiscal year. Since the funds are one-year appropriations, they do not carry over but have to be committed before the end of the fiscal year. The result may be a spending frenzy within the military services, with no possibility of actually putting the funds to use where they would do the most good for readiness. It is simply not possible to recover all the lost training or maintenance in such a short time, and the discretionary funds Congress had provided then migrate to other worthy activities.

One solution to this dilemma would of course be to put into the annual defense budget a contingency account that could be drawn on only if the president authorizes an overseas deployment. That is not feasible for several reasons. First, Congress would not cede that degree of budgetary authority to the executive branch. Doing so would be tantamount to relinquishing Congress's constitutional role in funding all overseas engagements, a prerogative naturally guarded jealously by those who, over the years, have supported the War Powers Act. Secondly, Congress is not enamored of appropriated but unexpended funds. If it appropriated contingency funds, but no contingencies emerged, the result would be that total expenditures during the fiscal year would be less than they could have been, a situation absolutely abhorrent to most members of Congress. That money could have been used somewhere, by some committee with unfunded projects, and there would be much anguish if the money were not spent.

A better solution would be for Congress to appropriate all the costs of the contingencies early in the budget process. That is clearly what the military services would prefer and would benefit from the most. Again, such a simple proposal runs afoul of the constitutional reality that Congress may not wish to support every president's engagement in foreign contingencies. Congress does not view its institutional responsibility as simply opening the checkbook every time a president decides to do something that some may consider to be foreign adventurism. Therefore, the budget process will always cause unforeseen readiness problems when contingencies and deployments occur. This simply cannot be avoided.

Consequences of Budget Decisions. Finally, the DoD often is itself inadvertently the cause of many unforeseen readiness problems. This is not to suggest that there is either ill will or incompetence among defense managers; rather, the problems are institutional and perhaps even intractable.

One problem is in the way the DoD has approached budget cuts over the last ten years or so. As a public activity that produces an output that can be measured only qualitatively—deterrence of and victory in war through superiority in equipment, training, and concepts of operation—the DoD has no "bottom line" in terms of a "profit statement." All it has is a top line, a total budget that it must justify and spend as wisely as possible.

Budget pressures grew at the end of the Cold War to reduce all nonessential expenditures. In particular, a series of defense management initiatives focused on "infrastructure" activities. This led to cuts in the Future Years Defense Budget being taken in a number of accounts, such as base support, number of bases, real property maintenance, supply operations, maintenance activities, civilian personnel, support personnel in many areas, and staff cuts. These anticipated savings were thought to be achievable through a series of "efficiencies" that could be instituted through improved management at all levels throughout the department.

Few of these savings actually materialized, and the effects have been felt throughout all activities in the military services. One of the most important reasons for the relentless drive to take out anticipated savings in the top line was not to *reduce* the total budget; rather, the stated goal was always to protect the funds for needed modernization of weapon systems. Modernization is the heart of future capabilities, but unfortunately, the programmed but unrealized savings could only be financed by reducing the pace of modernization. All the military services have been forced to scale back their acquisition plans.

But the cuts have not worked out as planned. Readiness was never supposed to be a bill payer, yet when the forecast budget savings never showed up, readiness accounts were cut as well. In the Air Force, maintenance and supply in particular have been hurt because they were reduced to deliver savings, and the realized shortfalls have not been made up once it became clear that the savings would not materialize.

There is a great lesson here, one that is extremely difficult to assimilate because it is the source of disappointment and frustration: Planned management efficiencies deliver perhaps a third to a half of anticipated savings, if experience is to be a guide to the future. That is all that can be expected, no more. Thus, if decisionmakers in DoD plan to protect modernization and readiness by taking undiluted savings from better management of infrastructure, it might be better to rethink this plan. With the lessons of the 1990s as background, this is now foreseeable, predictable, and quantifiable and should become part of the budget planning process. Management initiatives that anticipate large future savings should be carefully reviewed outside the DoD, because the institutional incentives for accepting these difficult and hard-hitting implications are simply too deep.

PROGRAMMERS VERSUS OPERATORS: WHO SHOULD BE IN CHARGE?

This leaves some additional factors that actually are within the purview of internal service management prerogatives. One relates to mobilization practices, whereby deploying units improve their organic capabilities by borrowing resources from sister units who remain at home station in a state of reduced readiness. This practice—termed *robusting* in the Air Force, *cross-leveling* in the Army, and *cross-decking* in the Navy—is caused by the habit of providing peacetime resources at less than the full amount of resources needed in wartime—programming at low C-1 (or even less in certain cases) but deploying at the equivalent of high C-1.

One thing is eminently clear: The deployment standards will not be compromised. If that requires moving resources from nondeployers to protect deployers, that will be done. All the services abide by this overriding principle. Readiness problems are diverted so that operational standards can be met, and other force elements will have to suffer the consequences. There is an institutional reason that makes this practice endemic: Programmers determine resource levels during peacetime, but operators decide what will deploy to contingencies—and these two communities operate under different and sometimes incompatible principles. Programmers want to spread problems around the force, as noted above, but operators will not allow deploying units to go with anything less than a full complement of experienced people and reliable equipment. This creates a resource disconnect, both in total budgets and in the allocation of resources across units.

When these practices cause significant problems, the service leadership could contemplate two actions. First, the leadership can issue guidance that programming priorities will be reordered and that peacetime programming levels for operational units must be at higher levels—closer to high C-1 for personnel, equipment, and dollars in operations and maintenance. The guidance would naturally also have to identify the bill payers within the service budget—where are the extra resources for readiness going to come from within a given budget top line? Second, the leadership could order deployment planners to cease robusting deploying units and instead to act in accordance with the programming guidance already in place.

Units resourced at 90 percent of known requirements should deploy at 90 percent, no more, and operators should accept whatever risks are attendant to that level of contingency resourcing. Thus, when the top line is too low, the choices are either to tell the programmers to resource as operators habitually deploy or to tell the operators to deploy as programmers have resourced them.

The simple choice put so starkly—program as you go or go as you program—is actually an extraordinarily difficult one to make, given the institutional and cultural factors in the military. Peacetime budget pressures force economies in programming, which means telling operators to tighten their belts for the sake of the service's overall goals in all areas of the budget. At the same time, there is a very understandable attitude in wartime that whatever the operator states as a requirement, the service will do its utmost to provide. The contradiction between telling military operators to live with peacetime frugality and wartime largesse is perhaps not so painful if one only has to contemplate the odd major war. The signals get crossed only when troops are sent to a series of costly peacetime deployments, because the units then have to adjust to the contradictory signals every day rather than having a clear break between peacetime rules and wartime rules. It is time to face this as an ongoing reality that needs a better resolution, and only the senior service leadership can do this.

Failure to do so can skew assessments of the operational readiness of the Air Force. Using the current two-MTW construct, suppose that a major war breaks out in one theater. The theater's combatant commander will certainly demand as much as he can get to prosecute a quickly evolving shooting war, and the National Command Authorities will provide him what he needs in this dire situation. It is not unlikely that, as planned, the responding force would include roughly half the fighter force (ten fighter wing equivalents). As is its custom, the USAF will answer the call by deploying well-equipped squadrons with highly experienced personnel by robusting from nondeploying units—*units that are tagged to fight in a second MTW.*

Now suppose that, while U.S. forces are fully engaged in the first MTW, a second MTW starts brewing. On the face of it, the USAF should be well positioned to respond to the second MTW as well (is this not what it was sized for?). *But the half of the force that is dedicated to fighting the second MTW was drawn down by robusting units that deployed to the first MTW.* Units responding to the second MTW will then take longer to deploy and will likely be short of needed capabilities once they arrive. It is crucial to acknowledge that, despite the fact that the need to fight two MTWs is the sizing criterion of the fighter force, the ability of the USAF to respond to the second MTW may currently be at grave risk. The reason lies in a combination of inadequate resourcing of squadrons relative to current requirements and the deeply ingrained custom of unit and wing commanders to ensure that units they send to war are at peak capability. As we discussed at the beginning of this chapter, the readiness assessment the Joint Chiefs of Staff and OSD use—JMRR —*is not capable of identifying this problem because it is based on SORTS rather than on actual deployment practices.* JMRR fights the programmers', not the operators', war.[20]

Related to this is the requirement to build a more robust force that can sustain ongoing deployments. Even if peacetime operations were resourced to the same level at which operators prefer to deploy, deployments would still have unavoidable readiness implications. Training would still suffer; equipment would still wear out faster; and personnel would still require some recovery period after return to home station. The Navy has adapted to this, even if budget pressures have made it more and more difficult for it to protect its shore-based activities and keep them healthy as budgets have declined. But the Army and the Air Force (the AEF concept notwithstanding) are nowhere near having a healthy base for important elements of their forces. In the Air Force, this affects a number of weapon systems, especially but not exclusively low-density, high-demand units.[21] In the end, this is a force structure issue, and new force structures are very expensive.

---

[20]There are additional significant problems in JMRR. One is that it gives a potentially very inadequate view of how the warfighting CINC actually plans to fight the war. Another is that the services are quite likely to alter the TPFDL, perhaps significantly, by adding or moving units around when the war starts. Thus, the JMRR does not represent either the demand or the supply side of the force equation very well—it does not even do the old math well.

[21]These are, at present, intelligence and information gathering aircraft, airborne command and control, combat search and rescue, special operations, and some other systems.

MANAGING READINESS: REQUIREMENTS, RESOURCES, AND PROCESSES

If revised management priorities ultimately are the key to improved readiness, better information will be necessary in many areas than is currently available. Readiness problems arise when operators lack the tools they need to perform their tasks and assignments—another gap between requirements and resources. If requirements have a large stochastic element, budgets must be cushioned and/or proper resource reallocation processes must be developed.

Managing readiness requires

1. setting the standards of desired performance in particular areas

2. defining the metrics to measure actual performance

3. assessing whether actual performance is less than, equal to, or greater than the performance desired

4. defining and implementing remedies for matching actual performance with the standards, or, if that is impossible, at least systematically tracking and reporting shortfalls.

If a task is precise and given to evaluation, it may be easy to set the standards and derive simple and quantitative metrics. Then, the third step, the assessment, can be quick and automatic. However, if the task is complex and comprises many sequential steps, the number and complexity of standards and corresponding metrics could be daunting. Moreover, the standards and metrics could be qualitative, thereby requiring a larger degree of expert judgment. It may not be possible to set simple rules for how the assessment should be made, and it could be very difficult to make an integrated, "objective" assessment when *objective* may simply mean that different experts looking at the same data come to roughly the same conclusion.

RAND's readiness research has identified the need to set standards in many areas where they are painfully inadequate. For example, pilot training and maintenance manpower standards in the Air Force do not telegraph the appropriate requirements such that the personnel system that supplies the manning gets the right signals to act on, and the result is overwork and inefficiencies in maintenance units throughout the force. But once again, one may resolve the manpower problems in maintenance, only to find that supply problems immediately emerge as the next binding constraint or that maintenance equipment, facilities, and processes cannot adequately support operations even if manning is increased. The solution to this problem is to ensure that the standards are set in each area so that decisionmakers know precisely the amount of resources needed. At present, there is a series of severe disconnects in setting these standards.

There are great conceptual difficulties in setting the correct warfighting standards. In many cases, the CINC requirements are not, and possibly cannot be, stated in very precise operational terms. Targets and timetables must be flexible, and standards for various capabilities must be set at levels that afford operational flexibility to the combatant commanders. This places great responsibility on each military service to set its own internal standards of performance so that it can give the combatant commander enough to accomplish his assigned missions.

Yet, war plans are, in actuality, constrained by available resources. For longer-term planning, the Planning, Programming, and Budgeting System is supposed to provide the budget resources necessary to prepare for and execute the war plans. But in practice, this works the other way around: Available resources constrain execution. The CINC will take what he can get and allocate it to best effect. Thus, over the short term, there is very little guidance to be gleaned from war plans about the "correct" operational readiness standards. In effect, the plans require the service to provide what it can from what it already has, and hopefully this is enough to accomplish the missions.

The reality, then, is that operational standards are really determined by available budget resources. Instead of readiness standards setting the rules for what resources should be made available, available resources determine actual readiness standards. The services actually determine warfighting capabilities when they build their budgets each year, with the CINCs' inputs about their priorities.[22]

---

[22]There are many examples of this, such as the 1992 Air Force decision to mothball the F-111. This aircraft provided a unique capability, being able to fly at low level, in bad weather, for long distances, with a medium-weight bomb load. This aircraft would have been a great asset in the kinds of peacetime contingencies we have witnessed

This suggests a critical flaw in the way SORTS is used today. SORTS was designed to report on *authorized* levels of resources provided to a unit—people, aircraft, parts, etc. It was *not* designed as a readiness assessment system, even though it is used that way. SORTS provides insight into whether a unit is using the resources made available to it throughout the year. The system does not report on a unit's ability to meet true operational requirements. The baselines driving "C-status" may fluctuate with changes in authorized resources. When budgets decline, a unit may have fewer capabilities because it is authorized fewer resources, yet it can still report a "C-1" status, meaning that readiness can remain unchanged. In relation to operational requirements, however, readiness should have declined as well.

Actual budget practice is not as consistent as it should be. A logical chain of decisions should lead to certain capabilities being provided to the CINC. Once the size and composition of the force structure are determined, a series of decisions should be made that flow from numbers of aircraft to pilots and their training and to the monthly sortie rate of the aircraft for pilot training in peacetime. This should in turn drive the amount of maintenance needed, which then would determine the number of maintainers needed and the associated training. The technical characteristics of the aircraft, as well as its operations, should drive the requirement for supply and the supporting logistics center and contracting. Pilot training and all the supporting activities in maintenance and supply should then drive a requirement for infrastructure, i.e., bases and training areas. Such a logical sequence of computations from operational to functional standards and supporting activities could only be done by integrating the standards at one level with the implied ones at the next supporting level.[23]

---

over the last few years but was sent to the boneyard for one reason: It was too costly to maintain. The cost per flying hour was the highest of any weapon system at the time. This is but one example of how budgets determine operational capabilities and standards, popular belief to the contrary.

[23]The accounting principles required to support these steps exist in principle but have not been developed to a level of detail at which they can be applied to the budgeting of Air Force weapon systems. Called Activity Based Costing, these principles are at present used only at lower levels, typically at the initiative of some enterprising local individual. Efforts are under way in all the services to develop these accounting techniques further, but this is a slow and delicate process.

In practice, the Air Force does not compute resource requirements in this logical, analytical fashion. Rather, resource requirements are typically computed in one functional area separately from all others. Thus, within the major commands, operational divisions analyze the requirement for flying hours to support pilot training—across all weapon systems within the command. The logistics division computes resource requirements for maintenance manpower, equipment, facilities, and supplies—across all weapon systems within the command. The civil engineering division computes the requirements for ranges and facilities—across all activities within the command. The security forces division computes requirements for security guards and their supporting needs—across all weapon systems and facilities. And on and on in this fashion. Budgeting is, in reality, done by function across weapon systems, not by computing the total requirement for each weapon system separately.

This results in various anomalies. For example, some weapon systems have not been able to fly out all their allocated flying hours during the fiscal year because pilot training requirements had been funded without guaranteeing that the supporting activities in maintenance and supply were also adequately funded to ensure that the flying hours could actually be executed. Thus, in releasing the resources that controlled one binding constraint, no one ensured that the next binding constraint did not make it impossible to reach the goals of releasing the first constraint. Such stovepiping of resource management creates inefficiencies.

The answer is to set an integrated set of standards across functional stovepipes, within single weapon systems, and then to manage across the functional areas accordingly. The most obvious starting point is when setting integrated functional standards. Readiness will never be resolved without these. Once these have been defined in the correct areas, with considerable attention to the overlap and interrelations between functions, the metrics will suggest themselves. Standards for performance imply the appropriate metrics. The metrics will then allow evaluation of readiness and identification of bottlenecks, which will allow management to take corrective action. In practice, this may prove to be difficult and time-consuming, but there is no doubt that it can be done.

To implement an appropriate mechanism for readiness assessments, it is also necessary to revise the *processes* by which these assessments are made. At present, a unit's commanding officer assesses whether it is ready or not. This system has two significant shortcomings. First, as noted in the discussion of robusting and deployment practices, it is not typical that a unit goes to war just as it is. The typical modus operandi is to assess immediate warfighting requirements and then provide these resources by calling on other units to provide the missing elements of the package. In effect, the readiness of any one unit is determined by all the resources available to the entire weapon system across all units. It is therefore ultimately misleading to assess readiness on a unit basis—this does not reflect actual modes of deployment and operation. Second, it follows that it is also not appropriate to rely on the judgment of any one squadron commander to assess the readiness of the weapon system's capabilities. A squadron commander may be perfectly capable of assessing his or her unit's status, but since the readiness of any one unit ultimately depends on the resources that can be made available through robusting from other units, only a higher-level manager can make the correct assessment by collating information from all the units of that weapon system. In addition, the resource requirements must be determined and assessed from operations through support, and this implies informational processes and management prerogatives that cut across organizations and functions.

Clearly, the setting of operational and functional standards, the determination of appropriate metrics, the assessment of readiness (both operational and rejuvenation related), and the management of resources for deployments should be organized in a new and innovative way. The aim would be to reduce the effects of unplanned readiness problems and to leave the DoD only with as appropriately planned and executed a level of readiness as possible—in short, to minimize the "unreadiness" that it can never completely avoid.

CONCLUSION: THERE IS NO PARADOX

The introduction to this chapter posed a paradox of readiness: that the military services can be ready for war and still have ongoing peacetime readiness problems. One goal of this chapter was to show that this is not really a paradox at all. Readiness is the ability of units to perform all their tasks, of which the execution of wartime assignments is only one. Thus, unit personnel may face a daily struggle to keep human and physical assets at acceptable levels, seeing their capabilities slowly eroding, yet still be able to meet wartime taskings. This seems to be the state of much of the military today. Capturing these effects demands a different and expanded set of metrics than SORTS offers. Addressing the underlying causes of readiness problems requires a new set of management principles, focused on the need to define requirements better and to prepare budget submissions that cushion against the frequent occurrence of stochastic disturbances that affect the ability of units to carry out all their assigned functions.

In summary, four central tenets must be understood about the readiness complex. First, readiness is about matching resources to known operational and functional requirements, but severe disconnects can occur because resource programmers do not manage the execution of resources, and managers in charge of execution have so far proved unable to define minimum operational and functional standards adequately to resource programmers. Defining requirements is one of the greatest challenges even the most experienced functional managers face. There are extraordinarily difficult analytical and technical problems in almost every area—capabilities, equipment, personnel, training, and support functions, to name a few.

Second, readiness typically declines without immediately affecting warfighting capabilities. Recent research shows that what is termed "rejuvenation" will show the first signs of declining readiness; operational readiness will be the last to give. The widely held assumption that readiness will first affect warfighting capabilities is mistaken yet is the basis for existing readiness assessment systems. By implication, the services, the Joint Staff, and the Under Secretary of Defense for Readiness have much to accomplish.

Third, readiness problems *never* occur in just one single area but always in several interrelated areas in surprising and unforeseen ways. Resource programmers attempt—quite reasonably—to spread the pain and minimize excess resources at any one point in the chain of activities that produce readiness. This makes it very difficult, if not impossible, to predict where and how significant readiness problems will occur. This fact of defense management explains one of the most frustrating issues congressional and DoD resource managers face: If they put money against what seem to be the most important readiness problem today, why does readiness not improve significantly? The reason is that the only way to improve readiness is to add extra dollars to most or even all readiness-related accounts, as they will all be close to binding on readiness.

Finally, deployments diminish readiness—all kinds of deployments, including those that involve actual fighting. They use up resources, delay training, degrade equipment, and exhaust people. This will continue unless there is a proper rotation base and appropriate resourcing to avoid robusting and cross-leveling from home station units.

These central tenets point strongly to a question of resources and management. Resource managers own the programming, but operators own the execution—this is a significant management disconnect inside the military services. Couple that with an even more critical disconnect between the president's power to order deployments without congressional approval, on one hand, and Congress's disinclination to fund fully and promptly the budget costs of the president's decisions, on the other, and the result is an inevitable mismatch between resources and requirements.

Maintaining high levels of readiness is very expensive and must always compete for funding with both military (e.g., modernization) and nonmilitary expenditures.[24] Until recently, total defense budgets have been declining in real dollar terms. This has left a defense budget full of goals, some readiness related, that cannot be adequately funded within the overall spending targets set in legislation. As the heavy deployment demands of the 1990s have borne out, the military still plays a very active role in pursuing national security objectives. However, planning and future budget estimates in OSD and in Congress have not yet caught up with this reality.

---

[24]Exactly how expensive depends on one's definition of readiness. A simplistic number often suggested is "all of the operations and maintenance account." This would amount to over $100 billion, or about one-third of the defense budget. This does not count all the personnel costs and much of the acquisition budget that are driven by readiness considerations. Another overly simplistic estimate of the cost of readiness would be "the entire defense budget" because it represents the deterrent value of U.S. military capabilities. But above all, the budget really supports military activities, such as retiree benefits. The true cost depends on definitions.

REFERENCES

Betts, Richard K., *Military Readiness: Concepts, Choices, Consequences*, Washington, D.C.: The Brookings Institution, 1995.

Dahlman, Carl J., and David E. Thaler, *Assessing Unit Readiness: Case Study of an Air Force Fighter Wing*, Santa Monica, Calif.: RAND, DB-296-AF, 2000.

Dahlman, Carl J., David E. Thaler, and Robert Kerchner, *Setting Requirements for Maintenance Manpower in the U.S. Air Force*, Santa Monica, Calif.: RAND, MR-1436-AF, 2002.

DoD—*See* U.S. Department of Defense.

Fossen, Thomas, Lawrence M. Hanser, John Stillion, Mark N. Elliott, and S. Craig Moore, *What Helps and What Hurts: How 10 Activities Affect Readiness and Quality of Life at Three 8AF Wings*, Santa Monica, Calif.: RAND, DB-223-AF, 1997.

Fullerton, Richard, U.S. Air Force Academy, "Career Choice Under Uncertainty: The Case of Air Force Pilots," unpublished working paper, Colorado Springs, Colo., 2000.

Peters, Ralph, "Heavy Peace," *Parameters*, Spring 1999, pp. 71–79.

Polich, J. Michael, Bruce R. Orvis, and W. Michael Hix, "Small Deployments, Big Problems," Santa Monica, Calif.: RAND, IP-197, 2000.

Taylor, William W., S. Craig Moore, and C. Robert Roll, Jr., *The Air Force Pilot Shortage: A Crisis for Operational Units?* Santa Monica, Calif.: RAND, MR-1204-AF, 2000.

U.S. Department of Defense, *Dictionary of Military and Associated Terms*, Washington, D.C., Joint Publication 1-02, April 12, 2001 (as amended through October 15, 2001).

Made in the USA
Coppell, TX
29 January 2022